Perovskite Oxide for Solid Oxide Fuel Cells

FUEL CELLS AND HYDROGEN ENERGY

Series Editor:

Narottam P. Bansal
NASA Glenn Research Center
Cleveland, OH 44135
narottam.p.bansal@nasa.gov

Aims and Scope of the Series

During the plast couple of decades, notable developments have taken place in the science and technology of fuel cells and hydrogen energy. Most of the knowledge developed in this field is contained in individual journal articles, conference proceedings, research reports, etc. Our goal in developing this series is to organize this information and make it easily available to scientists, engineers, technologists, designers, technical managers, and graduate students. The book series is focused to ensure that those who are interested in this subject can find the information quickly and easily without having to search through the whole literature. The series includes all aspects of the materials, science, engineering, manufacturing, modeling, and applications. Fuel reforming and processing; sensors for hydrogen, hydrocarbons, and other gases will also be covered within the scope of this series. A number of volumes edited/authored by internationally respected researchers from various countries are planned for publication during the next few years.

Titles in this series

Perovskite Oxide for Solid Oxide Fuel Cells
T. Ishihara, ed.
ISBN 978-0-387-77707-8, 2009

Nanomaterials for Solid State Hydrogen Storage
R.A. Varin, T. Czujko, and Z. S. Wronski
ISBN 978-0-387-77711-5, 2009

Modeling Solid Oxide Fuel Cells: Methods, Procedures and Techniques
R. Bove and S. Ubertini, eds.
ISBN 978-1-4020-6994-9, 2008

Tatsumi Ishihara
Editor

Perovskite Oxide for Solid Oxide Fuel Cells

Springer

Editor
Tatsumi Ishihara
Faculty of Engineering
Department of Applied Chemistry
Kyushu University
744 Motooka
Nishi-ku, Fukuoka
819-0395 JAPAN
ishihara@cstf.kyushu-u.ac.jp

ISBN 978-0-387-77707-8 e-ISBN 978-0-387-77708-5
DOI 10.1007/978-0-387-77708-5
Springer Dordrecht Heidelberg London New York

Library of Congress Control Number: 2008936301

Printed on acid-free paper

Springer is part of Springer Science+Business Media (www.springer.com)

Preface

Fuel cell technology is quite promising for conversion of chemical energy of hydrocarbon fuels into electricity without forming air pollutants. There are several types of fuel cells: polymer electrolyte fuel cell (PEFC), phosphoric acid fuel cell (PAFC), molten carbonate fuel cell (MCFC), solid oxide fuel cell (SOFC), and alkaline fuel cell (AFC). Among these, SOFCs are the most efficient and have various advantages such as flexibility in fuel, high reliability, simple balance of plant (BOP), and a long history. Therefore, SOFC technology is attracting much attention as a power plant and is now close to marketing as a combined heat and power generation system. From the beginning of SOFC development, many perovskite oxides have been used for SOFC components; for example, $LaMnO_3$-based oxide for the cathode and $LaCrO_3$ for the interconnect are the most well known materials for SOFCs. The current SOFCs operate at temperatures higher than 1073 K. However, lowering the operating temperature of SOFCs is an important goal for further SOFC development. Reliability, durability, and stability of the SOFCs could be greatly improved by decreasing their operating temperature. In addition, a lower operating temperature is also beneficial for shortening the startup time and decreasing energy loss from heat radiation. For this purpose, faster oxide ion conductors are required to replace the conventional Y_2O_3-stabilized ZrO_2 electrolyte. A new class of electrolytes such as $LaGaO_3$ is considered to be highly useful for intermediate-temperature SOFCs.

Although a number of books on fuel cells have been published, a book focused on the materials aspects of SOFCs is not yet available. This book provides comprehensive and up-to-date information on the properties and performance of perovskite oxides for SOFCs. Individual chapters have been written by internationally renowned researchers in their respective fields. The book is primarily intended for use by researchers, engineers, managers, and other technical people working in the field of SOFCs. Also, the information contained in most of the chapters is fundamental enough for the book to be useful even as a text for a SOFC technology course at the graduate level. I hope that this book is able to contribute to the development of SOFCs from the material aspects. At present, global warming and the energy crisis are the most serious problems for sustained development of human society. I believe that SOFC technology would contribute in solving these issues.

I am grateful to Dr. Narottam Bansal, NASA Glenn Research Center, for the opportunity to edit this book and for proofreading the text. The support of Dr. Taner Akbay, Mitsubishi Materials Co. Ltd., in improving the quality of each chapter is also highly appreciated. Finally, I thank all the authors for their kind cooperation in spite of their busy schedules.

Fukuoka, Japan
August 2008

Tatsumi Ishihara

Contents

Contributors

Taner Akbay Mitsubishi Materials Corporation, Central Research Institute, 1002-14, Mukaiyama, Naka-shi, Ibaraki, 311-0102, Japan, akbay@mmc.co.jp

A. Berenov Department of Materials, Imperial College, London, London SW7 2AZ, UK, a.berenov@imperial.ac.jp

Teruhisa Horita National Institute of Advanced Industrial Science and Technology (AIST), AIST Central 5, 1-1-1 Higashi, Tsukuba, Ibaraki 305-8565, Japan, t.horita@aist.go.jp

J.T.S. Irvine School of Chemistry, University of St-Andrews, Fife, Scotland KY16 9ST, UK, jtsi@st-and.ac.uk

Tatsumi Ishihara Department of Applied Chemistry, Faculty of Engineering, Kyushu University, Motooka 744, Nishi-ku, Fukuoka 819-0395, Japan, ishihara@cstf.kyushu-u.ac.jp.

Naoki Ito Fuel Cell System Development Division, Toyota Motor Corporation, 1200 Mishuku, Susono, Shizuoka 410-1193, Japan, naoki@ito.tec.toyota.co.jp

Hiroyasu Iwahara Nagoya University, Furo-cho, Chigusaku, Nagoya, 464-8601, Japan, iwahara-h@m3.gyao.ne.jp

Tatsuya Kawada Graduate School of Environmental Studies, Tohoku University, 1-1 Aoba, Aramaki, Aoba-ku, Sendai 980-8579, Japan, kawada@ee.mech.tohoku.ac.jp

Akira Kawakami TOTO Ltd., Chigasaki, Kanagawa 253-8577, Japan, akira.kawakami@toto.co.jp

J.A. Kilner Department of Materials, Imperial College, London, London SW7 2AZ, UK, j.kilner@imperial.ac.uk

K.D. Kreuer Max-Planck-Institut für Festkörperforschung, Heisenbergstr. 1, D-70569 Stuttgart, Germany, Kreuer@fkf.mpg.de

Hiroshige Matsumoto INAMORI Frontier Research Center, Kyushu University, 744 Motooka, Nishiku, Fukuoka 819–0395, Japan, matsumoto@ifrc.kyushu-u.ac.jp

Truls Norby Department of Chemistry, Centre for Materials Science and Nanotechnology, University of Oslo, FERMiO, Gaustadalléen 21, NO-0349 Oslo, Norway, t.e.norby@kjemi.uio.no

J. Rossiny Department of Materials, Imperial College, London, London SW7 2AZ, UK, a.berenov@imperial.ac.jp

Masatomo Yashima Tokyo Institute of Technology, Yokohama 226–8502, Japan, yashima@materia.titech.ac.jp

Harumi Yokokawa Energy Technology Research Institute, National Institute of Advanced Industrial Science and Technology, Higashi 1-1-1, AIST Central No.5, Tsukuba, Ibaraki 305-8565, Japan, h-yokokawa@aist.go.jp

Chapter 1
Structure and Properties of Perovskite Oxides

Tatsumi Ishihara

1.1 Introduction

Oxide groups consisting of two or more different cations are called complex or mixed oxides, and many types of structures are known that are different from those of the simple oxides. In some special cases, oxides consisting of a single cation in different oxidation states are also classified as mixed oxides. For example, Eu_3O_4, a mixed oxide, consists of Eu(III) and Eu(II) in 6- or 8-coordination, respectively. However, the most typical structure of a mixed oxide consists simply of two or more different cations with different oxidation states, ionic radii, and coordination numbers. This diversity, which comes from the complexity of these structures, results in a larger number of different properties as compared to those of simple oxides. One of the most well known and important complex oxide structures is the spinel structure (AB_2O_4), which shows important magnetic properties. The structure of such oxides displays a most interesting complexity. Because the A and B ions in this structure are close in size, oxides of this type are typical examples of the versatility of mixed oxides. In the ideal case, one sixfold-coordinated ion occupies the A site and another sixfold-coordinated cation occupies the B site; however, in some cases, mixing of cations on A- and B-site ions occurs. In the most complex case of the spinel structure, the same cations occupy both sites with the structure in different environments. Therefore, a unique feature of mixed oxide compounds is the display of a variety of properties that are partially the result of the variety of the structures. In particular, among mixed oxides, the perovskite oxides are well known for displaying a multitude of structures and properties, which are briefly introduced in this chapter.

T. Ishihara (✉)
Department of Applied Chemistry, Faculty of Engineering, Kyushu University,
Motooka 744, Nishi-ku, Fukuoka, 819-0395, Japan
e-mail: ishihara@cstf.kyushu-u.ac.jp

T. Ishihara (ed.), *Perovskite Oxide for Solid Oxide Fuel Cells*,
Fuel Cells and Hydrogen Energy, DOI 10.1007/978-0-387-77708-5_1,

1.2 Structure of Perovskite Oxides

The typical chemical formula of the perovskite structure is ABO_3, where A and B denote two different cations. The ilmenite structure has the same composition as the perovskite one, i.e., ABO_3; however, A and B in this structure are cations of approximately the same size that occupy an octahedral site. Therefore, in spite of the fact that they share the same general chemical formula, structures classified as ilumenite or ilmenite-related structure (e.g., $LiSbO_3$) are different from perovskite.

Perovskite oxides comprise large families among the structures of oxide compounds, and several perovskite-related structures are currently recognized. Typical structures consist of large-sized 12-coordinated cations at the A site and small-sized 6-coordinated cations at the B site. Several complex halides and sulfides and many complex oxides have a perovskite structure. In particular, $(Mg,Fe)SiO_3$ or $CaSiO_3$ is thought to be the predominant compound in the geosphere [1, 2]. Perovskite compounds with different combinations of charged cations in the A and B sites, e.g., 1 + 5, 2 + 4, and 3 + 3, have been discovered. Even more complex combinations are observed, such as $Pb(B'_{1/2}B''_{1/2})O_3$, where B' = Sc, Fe and B" = Nb, Ta, or $La(B'_{1/2}B''_{1/2})O_3$, where B' = Ni, Mg, etc., and B" = Ru(IV) or Ir(IV). In addition, many ABO_3 compounds crystallize in polymorphic structures, which show only a small distortion from the most symmetrical form of the perovskite structure.

The ideal structure of perovskite, which is illustrated in Fig. 1.1, is a cubic lattice. Although few compounds have this ideal cubic structure, many oxides have slightly distorted variants with lower symmetry (e.g., hexagonal or orthorhombic). Furthermore, even though some compounds have ideal cubic structure, many oxides display slightly distorted variants with lower symmetry. Several examples of perovskite oxides are listed in Table 1.1, where it is clear that a large number of perovskite oxides have a rhombohedral lattice. Additionally, in many compounds a large extent of oxygen or cation deficiency has been observed. Because of the large lattice energy, many compounds are classified as perovskite oxides in spite of the large cation and/or oxygen deficiencies. There are various types of distortions in the perovskite structure that have strongly related to their properties, in particular, their ferromagnetic or ferroelectric properties.

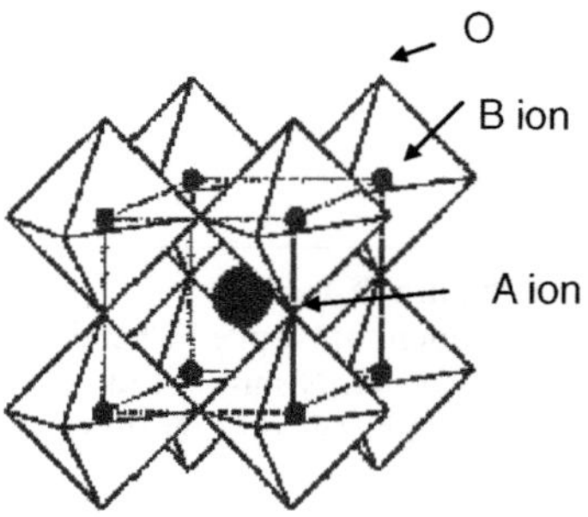

Fig. 1.1 Ideal perovskite structure

Table 1.1 Typical perovskite compounds

Compound	Lattice parameter/x10 nm a	b	c
		Cubic structure	
$KTaO_3$	3.989		
$NaTaO_3$	3.929		
$NaNbO_3$	3.949		
$BaMnO_3$	4.040		
$BaZrO_3$	4.193		
$SrTiO_3$	3.904		
$KMnF_3$	4.189		
$KFeF_3$	4.121		
		Tetragonal structure	
$BiAlO_3$	7.61		7.94
$PbSnO_3$	7.86		8.13
$BaTiO_3$	3.994		4.038
$PdTiO_3$	3.899		4.153
$TlMnCl_3$	5.02		5.04
		LaAlO3 type	
$LaAlO_3$	5.357	$\alpha = 60°\ 06'$	
$LaNiO_3$	5.461	$\alpha = 60°\ 05'$	
$BiFeO_3$	5.632	$\alpha = 60°\ 06'$	
$KNbO_3$	4.016	$\alpha = 60°\ 06'$	
		$GdFeO_3$ type	
$GdFeO_3$	5.346	5.616	7.668
$YFeO_3$	5.283	5.592	7.603
$NdGaO_3$	5.426	5.502	7.706
$CaTiO_3$	5.381	5.443	7.645
$NaMgF_3$	5.363	5.503	7.676

To understand the deviations from the ideal cubic structure, these ABO_3 oxides are first regarded as purely ionic crystals. In the case of the ideal structure, the following relationship between the radii of the A, B, and O^{2-} ions holds true:

$$r_A + r_O = \sqrt{2}(r_B + r_O)$$

Therefore, the deviation from the ideal structure in perovskite oxides can be expressed through the following so-called tolerance factor, t:

$$t = (r_A + r_O)/\sqrt{2}(r_B + r_O)$$

In perovskite-type compounds, the value of t lies between approximately 0.80 and 1.10. It is noted that the oxides with the lower t values crystallize in the ilmenite structure, which is a polymorph of the perovskite structure. It seems superfluous to say that for the ideal cubic structure the value of t is close to 1 or at least greater than 0.89. Figure 1.2 shows the crystal groups for $A^{2+}\ B^{4+}\ O_3$ and $A^{3+}\ B^{3+}\ O_3$ combinations, which are related to deviation from the ideal structure [3]. As the value of t decreases, the structure of the unit lattice is shifted

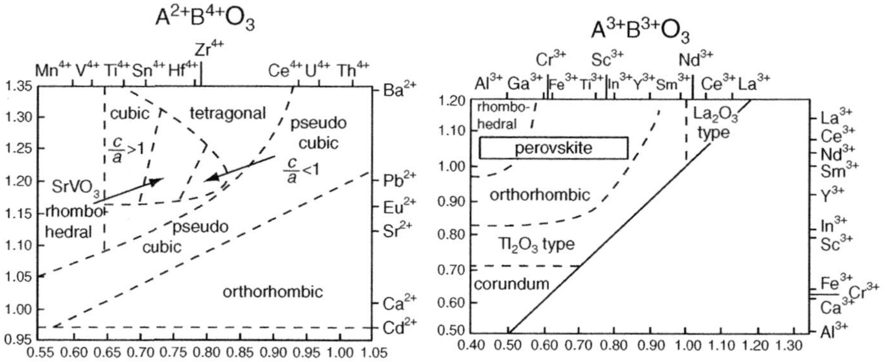

Fig. 1.2 The effect of ionic size of A- and B-site cations on the observed distortions of the perovskite structure

from cubic to triclinic as a result of the increased distortions. Figure 1.3 shows chemical elements that can be accommodated within the perovskite structure. It is evident that almost all elements except for noble gases can occupy either A or B lattice positions in the perovskite lattice, including dopants. The stability and the crystal group are mainly determined by the ratio of the ionic radii of the A and B cations. Indeed, the structure is dependent not only on the size but also on the nature of the A and B atoms. For example, $AMnO_3$ compounds crystallize in the perovskite structure when the A cation is La or Ce-Dy, whereas a new hexagonal structure with 5- and 7-coordination of Mn and A, respectively, is formed when A = Ho-Lu or Y if A = La or Ce-Dy [4]. Here, attention should be paid to the nature of the B atom, where the nature of the bond is highly

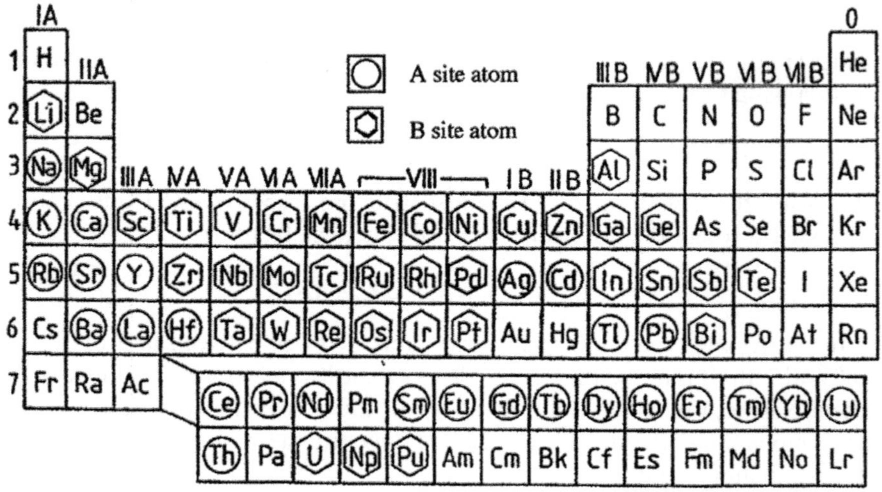

Fig. 1.3 Chemical elements that can occupy sites in the perovskite structure

covalent, and therefore the coordination numbers is lower than 6. The typical example of this type is $BaGeO_3$. In spite of a t value close to 1, i.e., ideal ionic size combination, $BaGeO_3$ crystallizes not in the perovskite structure but in the silicate-related one. This difference occurs because the preferred coordination number of Ge is 4. On the other hand, due to the progress in high-pressure technology, the synthesis of new Ge-based perovskite oxides has been reported [5]. As the coordination number of Ge increases with the pressure, perovskite structures with higher coordination numbers are preferred, and a typical example of this is $CaGeO_3$. Another group of interesting perovskite compounds is oxynitrides, i.e., $LaWO_{3-x}N_x$, $LaTiO_2N$, etc. Therefore, the value of t, which is determined by the ionic size, is an important index for the stability of perovskite structures; however, the contribution of the chemical nature, such as the coordinating number of the constituent elements, needs to be taken into account.

The formation of superstructures in the perovskites is discussed next. If a B-site cation is progressively replaced by a dopant, a large difference in ionic radii tends to lead to the formation of the superstructures rather than random arrangements of the two kinds of ions. The typical case of this is Ba_2CaWO_6, which is regarded as $Ba_2(CaW)O_6$. Similarly, in compounds with the general formula $Ba_3MTa_2O_9$, there is random distribution of M and Ta ion in the octahedral positions when M is Fe, Co, Ni, Zn, or Ca, whereas formation of a superstructure with hexagonal lattice is observed in $Ba_3SrTa_2O_9$. Another interesting type of superstructure observed in the perovskite is the ordering of cation vacancies located on A sites: e.g., MNb_3O_9 (M = La, Ce, Pr, Nb) and MTa_3O_9 (M = La, Ce, Pr, Nd, Sm, Gd, Dy, Ho, Y, Er). In these oxides, there is an octahedral framework of the ReO_3 type with incomplete occupancy of the 12-fold-coordinated A sites. Figure 1.4 shows the structure of $LaNb_3O_9$.

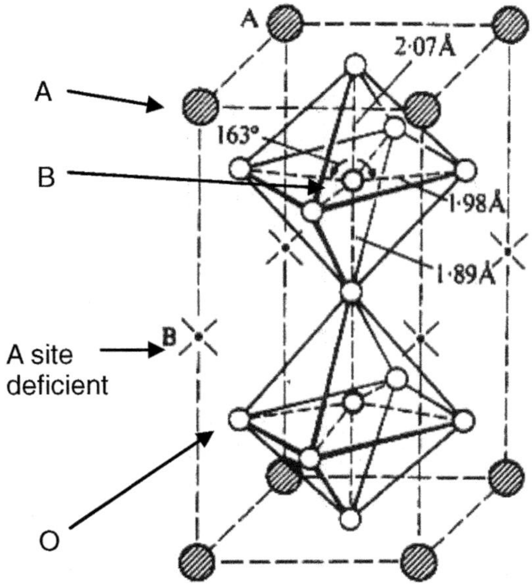

Fig. 1.4 Structure of $LaNb_3O_9$, A site-deficient perovskite oxide

The B sites of the perovskite structure are occupied by Nb ion, and two-thirds of the A sites remain vacant.

Other typical polymorphs of the perovskite structure are Brownmillerite ($A_2B_2O_5$) and K_2NiF_4 structures. Brownmillerite ($A_2B_2O_5$) is an oxygen-deficient type of perovskite in which the oxygen vacancy is ordered. The unit cell contains BO_6 and BO_4 units in an ordered arrangement. Because of the oxygen deficiency, the coordination number of A-site cations decreases to 8. The lattice parameter of the Brownmillerite structure relates to the cubic lattice parameter (a_p) of the ideal perovskite as $a = b = \sqrt{2}a_p$, $c = 4a_p$. Cu-based oxides or Ni-based oxides tend to adopt these oxygen-deficient structures because of the large amount of oxygen defects.

A combination of ordered B sites and oxygen defects is seen in K_2NiF_4 structures, which is well known as it shows superconducting properties. The K_2NiF_4 structures consist of two units, the $KNiF_3$ perovskite unit and the KF rock salt unit (Fig. 1.5), which are connected in series along the c-axis. As the rock salt structure is embedded into the c-axis direction, the K_2NiF_4 compound shows strong two-dimensional properties. Based on the intergrowth of the different numbers of $KNiF_3$ and KF units, there are many structures called Ruddelsden-Popper compounds with the general formula $(ABO_3)_nAO$ (Fig. 1.6); i.e., $Sr_3Ti_2O_7$ ($n = 2$), $Sr_4Ti_3O_{10}$ ($n = 3$). It is interesting to compare the isostructural Sr_2TiO_4 or Ca_2MnO_4 with $SrTiO_3$ or $CaMnO_3$, which crystallize in the perovskite structures. Two different A cations forming the perovskite and the rock salt units are also possible, and LaO•nSrFeO$_3$ is the typical example of this arrangement. Another interesting variant of these K_2NiF_4 structures occurs when two different anions occupy the two building blocks

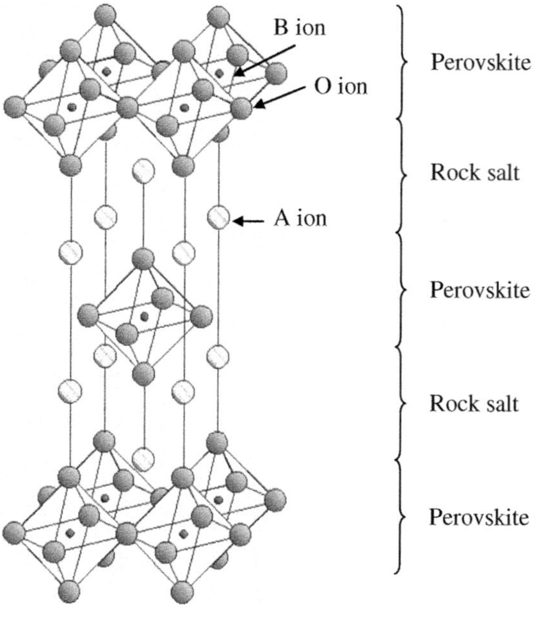

Fig. 1.5 K_2NiF_4 structure, a perovskite-related structure

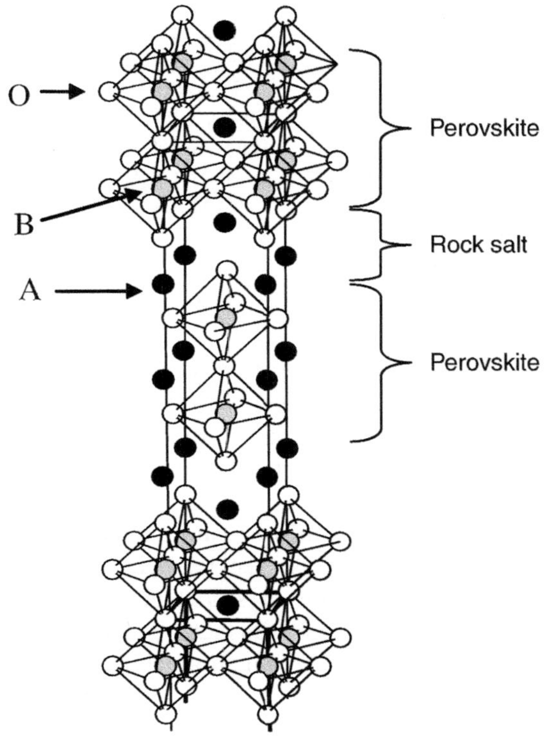

Fig. 1.6 Ruddelsden-Popper structure, another type of perovskite-related structure

exclusively, i.e., $SrFeO_3{\bullet}SrF$ or $KNbO_3{\bullet}KF$. In any case, it is evident that perovskite oxides comprise a large family of oxides. As a result, a variety of crystal structures and properties is expected in these compounds. For further detailed discussion on the perovskite-related oxides, the reader is referred to references [6–9]

1.3 Typical Properties of Perovskite Oxides

Because of the variety of structures and chemical compositions, perovskite oxides exhibit a large variety of properties. Well-known properties of the perovskite oxides are ferroelectricity in $BaTiO_3$-based oxides and superconductivity in $Ba_2YCu_3O_7$, etc. In addition to these well-known properties, several perovskite oxides exhibit good electrical conductivity, which is are close to that of metals, and ionic conductivity, as well as mixed ionic and electronic conductivity. Based on these variations in electrical conducting property, perovskite oxides are chosen as the components for SOFC. It is also well known that several perovskite oxides exhibit high catalytic activity with respect to various reactions, in particular, oxidation reactions [10]. Table 1.2 provides examples of the typical properties of perovskite oxides. In this section, several typical properties of the perovskite oxides, namely, ferroelectricity, magnetism, superconductivity, and catalytic activity, are briefly discussed.

Table 1.2 Typical properties of perovskite oxides

Typical property	Typical compound
Ferromagnetic property	$BaTiO_3$, $PdTiO_3$
Piezoelectricity	$Pb(Zr, Ti)O_3$, $(Bi, Na)TiO_3$
Electrical conductivity	ReO_3, $SrFeO_3$, $LaCoO_3$, $LaNiO_3$, $LaCrO_3$
Superconductivity	$La_{0.9}Sr_{0.1}CuO_3$, $YBa_2Cu_3O_7$, $HgBa_2Ca_2Cu_2O_8$
Ion conductivity	$La(Ca)AlO_3$, $CaTiO_3$, $La(Sr)Ga(Mg)O_3$, $BaZrO_3$, $SrZrO_3$, $BaCeO_3$
Magnetic property	$LaMnO_3$, $LaFeO_3$, La_2NiMnO_6
Catalytic property	$LaCoO_3$, $LaMnO_3$, $BaCuO_3$
Electrode	$La_{0.6}Sr_{0.4}CoO_3$, $La_{0.8}Ca_{0.2}MnO_3$

Dielectric properties: Ferroelectricity, piezoelectricity, electrostriction, and pyroelectricity are special properties inherent to dielectric materials and are important properties of electroceramics. The best known property of perovskite oxides is ferroelectric behavior, where $BaTiO_3$, $PdZrO_3$, and their doped compounds are representative examples. The study of ferroelectricity in $BaTiO_3$ has a long history, and many detailed reviews have been published. Furthermore, because the ferroelectric behavior of $BaTiO_3$ has a strong relationship with the crystal structure, detailed studies of crystal structure have been reported for $BaTiO_3$. $BaTiO_3$ undergoes mainly three-phase transformation, that is, from monoclinic, to tetragonal, and to cubic, as the temperature increases. Above 303 K, $BaTiO_3$ crystallizes in the cubic perovskite structure, which does not show ferroelectric behavior. The high dielectric constant observed in $BaTiO_3$ can be explained on the basis of the anisotropy of the crystal structure. Figure 1.7 shows the crystal structure of $BaTiO_3$ using

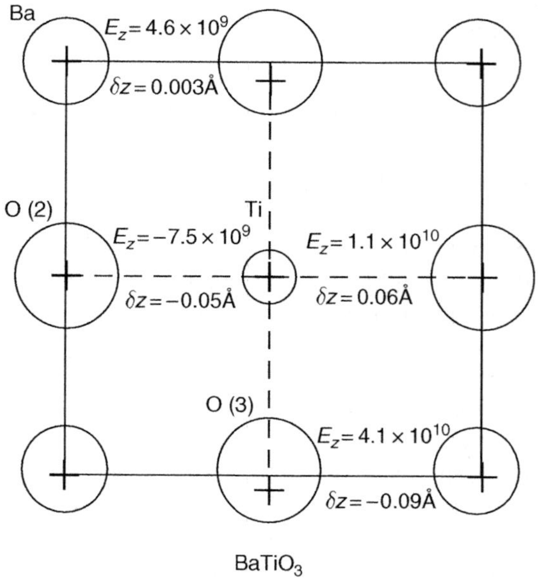

Fig. 1.7 Crystal structure of $BaTiO_3$ using Ewald method and local density of charge

the Ewald method as well as local charge density [11]. It is seen that the large negative potential is localized on the O_3 oxygen atom. When the electric field is applied, Ba^{2+} and Ti^{4+} cations move to the direction opposite to that of the oxygen atom. Thus, a net dipole moment is created in the unit cell. According to the Slater theory [11], the electrostatic field is strongly affected by the atoms located in O_3 sites; thus, a large dipole moment is generated in $BaTiO_3$.

Electrical conductivity and superconductivity: One of the most well known properties of perovskite oxides is superconductivity. In 1984, superconductivity was first reported by Bednorz and Müller in La-Ba-Cu-O perovskite oxide [12]. After their report, much attention was paid to new types of high-temperature oxide superconductors, mainly Cu-based oxides. As a result, several superconducting oxides with different A-site cations have been discovered. However, the presence of Cu on the B site was found to be essential for superconductivity to occur. High-temperature oxide superconductors of the $YBa_2Cu_3O_7$ system [13] and the $Bi_2Sr_2Ca_2Cu_3O_{10}$ system [14] were reported in 1987 and 1988, respectively, and currently the critical temperature of the superconducting transition (T_c) has been further increased to 130–155 K in the $HgBa_2Ca_2Cu_3O_{8+\delta}$ system [15]. As all high-temperature superconducting oxides are cuprites (Cu-based oxides), superconductivity is clearly related to the Cu-O layers. The critical temperature for superconductivity, T_c, is related to the number of Cu-O layers in the crystal structure:

One Cu-O layer: $T_c \sim 30$ K
Two Cu-O layers: $T_c \sim 90$ K
Three Cu-O layers: $T_c \sim 110$ K
Four Cu-O layers: $T_c \sim 120$ K

It is expected that further increase in the number of Cu-O layers may result in higher values of T_c. However, because of the low chemical stability, synthesis of five or more Cu-O layered compounds has not been successful so far. $YBa_2Cu_3O_7$ is one of the most important superconductor systems with high T_c, and detailed studies of its crystal structure have been performed. Also, the content of oxygen nonstoichiometry is an important factor for high T_c. When the value of d is smaller than 0.5, $YBa_2Cu_3O_{7-d}$ crystallizes in an orthorhombic structure, which is superconductive, whereas for $d > 0.5$, $YBa_2Cu_3O_{7-d}$ has a tetragonal structure, which does not exhibit superconductivity. Figure 1.8 shows the crystal structures of both oxygen-deficient phases in $YBa_2Cu_3O_{7-d}$. The main difference between the two structures is that the incorporation of oxygen in the lattice expands the b lattice parameter to a greater extend than the a lattice parameter. Those changes in crystal structure are related to the oxygen content, which is determined by the annealing temperature and oxygen partial pressure during postannealing treatment. As discussed, superconductivity in high T_c oxides is also dependent on the crystal structure; thus, the high chemical stability of the perovskite crystal structure could be effective in achieving high values of T_c.

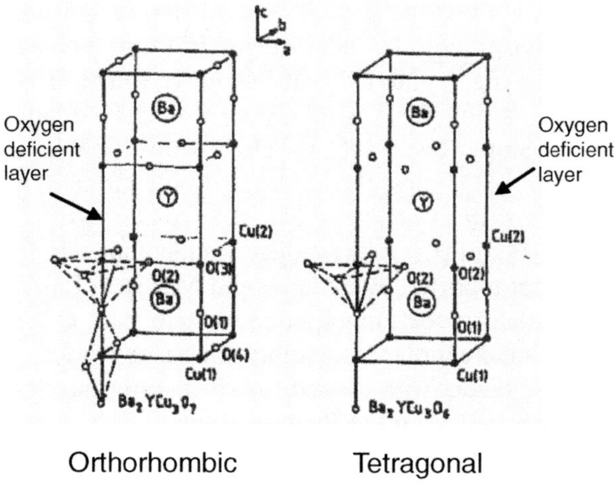

Fig. 1.8 Orthorhombic and tetragonal crystal structures of $BaY_2Cu_3O_7$, an oxygen-deficient perovskite

In addition to superconductivity, there are many perovskite oxides showing high electronic conductivity, which is close to those of metals such as Cu. The typical examples of such perovskite oxides are $LaCoO_3$ and $LaMnO_3$, which is now commonly used as a cathode in SOFC. These perovskite oxides shows superior hole conductivity, which is as high as $\sigma = 100/S/cm$. Doping of aliovalent cation on the A site is also highly effective in enhancing the electrical conductivity because of the increased number of mobile charge carriers generated by the charge compensation.

Catalytic activity: Because of the variety of component elements and their high chemical stability, perovskite oxides have been also extensively studied as catalysts for various reactions. Two types of research trends clearly emerged from these characteristics. The objective of the first trend is the development of oxidation catalysts or oxygen-activated catalysts as an alternative to catalyst containing precious metals, whereas the second trend regards perovskite as a model for active sites. The stability of the perovskite structure allows preparation of compounds with an unusual valence state of elements or a high extent of oxygen deficiency. Table 1.3 summarizes the reactions studied by using perovskite oxides as catalysts. Evidently, the high catalytic activity of perovskite oxides is based partially on the high surface activity to oxygen reduction ratio or oxygen activation resulting from the large number of oxygen vacancies present.

Among the various catalytic reactions studied, those applicable to environmental catalysis (e.g., automobile exhaust gas cleaning catalyst) attract particular attention. Initially, it was reported that perovskite oxide consisting of Cu, Co, Mn, or Fe exhibited superior activity to NO direct decomposition at higher temperatures [16–18]. The direct NO decomposition reaction ($2NO = N_2 + O_2$) is one of the "dream reactions" in the catalysis field. In this reaction, the ease of removal of surface oxygen as a product of

Table 1.3 Main catalytic reactions studied by using perovskite oxides

Catalytic reaction		Example
Oxidation	CO, lower hydrocarbon, Methanol Catalytic combustion	$LaCoO_3$, $LaMnO_3$
deNOx	Selective reduction	$LaAlO_3$, $SrTiO_3$
	NO decomposition	$BaMnO_3$, $SrFeO_3$, $YBa_2Cu_3O_7$
	NO absorption	$LaAlO_3$, $BaCeO_3$, $BaFeO_3$
Hydrogenation	C_2H_4 hydrogenation	$LaCoO_3$
CH_4 coupling	Oxidative CH_4 coupling	$BaTiO_3$, $Ba_{0.5}Sr_{0.5}Fe_{0.2}Co_{0.8}O_3$
Oxygen electrode	Oxygen reduction (alkaline solution)	$LaCoO_3$, $LaMnO_3$
	Oxygen generation (alkaline solution)	$LaCoO_3$, $LaFeO_3$
	Cathode for Solid Oxide Fuel Cell	$LaCoO_3$, $LaMnO_3$
	Oxygen sensor	$LaCoO_3$, $LaMnO_3$
Gas sensor	Oxygen sensor, Humidity sensor, Alcohol Sensor	$SrTiO_3$, $BaSnO_3$, $LaCr(Ti)O_3$, $GdCoO_3$

the reaction plays an important role, and due to the facility of oxygen deficiency present, perovskite oxides are active with respect to this reaction at high temperatures. It is pointed out that doping is highly effective in enhancing NO decomposition activity. Under an oxygen-enriched atmosphere (up to 5%), a relatively high NO decomposition activity was reported for $Ba(La)Mn(Mg)O_3$ perovskite [19].

Recently, another interesting application of perovskite oxides as automobile catalysts has been reported, namely, the so-called intelligent catalysts [20]. Up to now, three-way Pd-Rh-Pt catalysts have been widely used for the removal of NO, CO, and uncombusted hydrocarbons. To decrease the amount of precious metals, a catalyst consisting of fine particles with high surface-to-volume ratio is required. However, these fine particles are not stable under operating conditions and easily sinter, resulting in deactivation of the catalyst. To maintain a high dispersion state, the redox property of perovskite oxides has been proposed; i.e., under oxidation conditions, palladium is oxidized and exists as $LaFe_{0.57}Co_{0.38}Pd_{0.05}O_3$, and under reducing conditions, palladium is deposited as fine metallic particles with a radius of 1–3 nm. This cycling of the catalyst through oxidizing and reducing conditions results in the partial substitution of Pd into and deposition from the perovskite framework, thus maintaining a high dispersion state of Pd. This method was found to be highly effective in improving the long-term stability of Pd during removal of pollutants from exhaust gas (Fig. 1.9). The high dispersion state of Pd can be recovered by exposing the catalyst to an oxidation and reduction environment. As a result, this catalyst is called an intelligent catalyst. This unique property also originates from the high stability of the perovskite crystal structure in complex oxides.

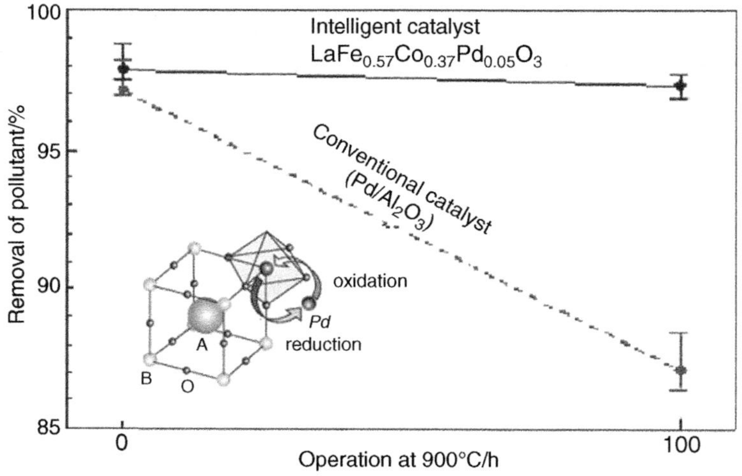

Fig. 1.9 Structure of "intelligent catalyst" and comparison of the catalytical activity of the intelligent catalyst and the conventional Al_2O_3-supported one for the removal of pollutants in exhaust gas

1.4 Preparation of Perovskite Oxide

Because the perovskite structure is stable at high temperatures and also stable in terms of thermodynamic equilibrium, the perovskite oxides form only at a temperature typically higher than 1273 K. The most simple and popular method for preparation of perovskite oxides is the so-called solid-state reaction method, when the starting compounds (often simple oxides and carbonates) are calcined at temperatures higher than 1273 K. However, because of the high temperature of the calcination, the Burumauer-Emmott-Teller (BET) surface area of the resulting perovskite powders is generally small, usually less than 10 m^2/g. The preparation of perovskite oxide powders with a large surface area, namely, fine particles, is strongly demanded in various fields, in particular, for catalyst and electrode application not only for solid oxide fuel cells (SOFC) but also for batteries and/or electrolysis. To obtain fine particles of perovskite oxides, some advanced synthetic methods that generally involve the use of organic compounds have been developed. However, the preparation of perovskite oxide powders with a large surface area is quite a difficult subject, and the BET surface area is generally smaller than 50 m^2/g. This restriction is easily understood by considering a simple relationship between the specific surface area (S) and the diameter of a spherical particle (D) [21]:

$$S = 6/(\rho D) \tag{1.1}$$

where ρ is the density of the sample. Figure 1.10 shows the relationship between the geometrical surface area (S) of a spherical body and radius (D): the density

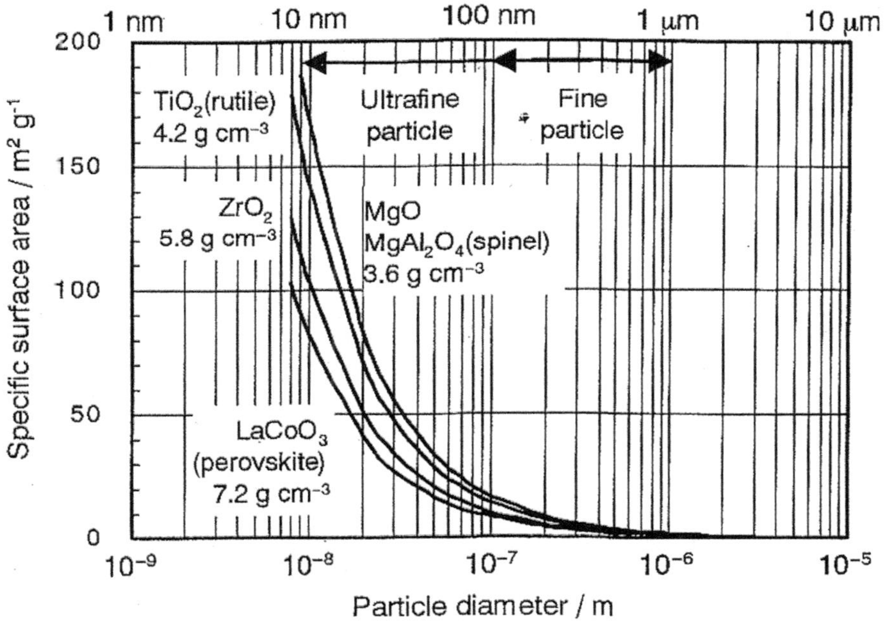

Fig. 1.10 Relationship between geometrical surface area (S) of a spherical body and radii (D)

of $LaCoO_3$ perovskite oxide is much lower than that of a general single oxide such as MgO or Al_2O_3. Therefore, for the purpose of obtaining a high surface area, such as 100 m^2/g, the required particle size of the perovskite oxide must be smaller than 10 nm, which is quite difficult to achieve.

Figure 1.11 summarizes the general procedure of the liquid-phase synthesis method used in the preparation of perovskite oxides with a large surface

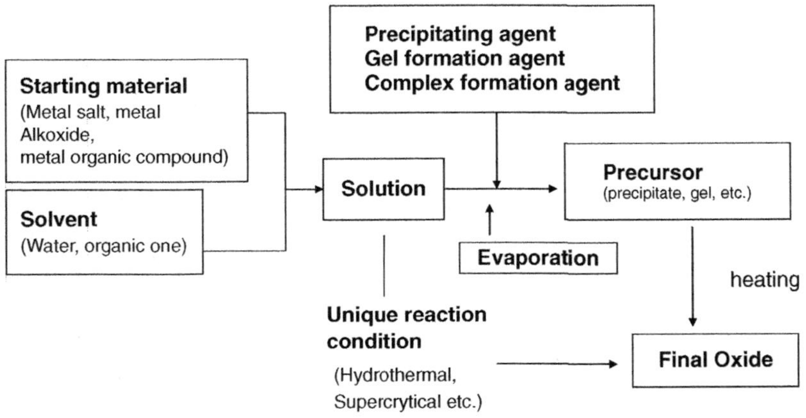

Fig. 1.11 General procedure of the liquid-phase synthesis method

Table 1.4 Proposed liquid phase synthesis method for perovskite oxides

Category	Method
Group I	
(Controlled evaporation or reactant decomposition rate)	Spray pyrosis, Spray (mist, aerosol) thermal decomposition, Freeze dry, Combustion synthesis, Microwave assisted method, Supercritical water
Group II	
(Usage of designed micro pore)	Antimicelle
Group III	
(Designed precursor)	Hydroxide precursor; Uniform precipitation, Sol gel method another precursor; Cyanide decomposition, Oxalic Acid method, EDTA-citrate complexing method, Pechini method

area. In this method, atomic-level dispersion of the component elements in the precursor solution is essential. Based on the dispersion method, the proposed liquid-phase preparation method could be classified into three groups (Table 1.4). The techniques classified into group I use energy such as ultrasonic vibration or supercritical conditions to achieve a high dispersion state. The application of microwave heating to a precursor containing $BaCl_2$, Ti isopropoxide, and KOH has been employed during the synthesis of $BaTiO_3$ fine particles. It has been reported that $BaTiO_3$ perovskite powder with a particle size of 20–30 nm was successfully prepared [22]. On the other hand, group II focuses on the usage of micelles, which limit the space for the perovskite precursor. $LaMnO_3$ prepared by using reverse micelles has been reported to possess high electrode activity when used at the anode of a metal-air battery. Finally, techniques in group III involve the usage of organic compounds for achieving atomic-level dispersion in the precursor solutions. In the most popular cases, the addition of ammonia is used to obtain uniform precipitates of perovskite precursors. However, because of the difference in the precipitation rates, it is difficult to obtain a precursor with uniform distribution of constituent elements at the atomic level.

Teraoka et al. reported the use of organic coordination compounds for the preparation of perovskites [23]. They found that addition of acetic acid or maleic acid is useful for obtaining finely powdered perovskite oxides by decreasing the crystallization temperature. Figure 1.12 shows the C_3H_8 oxidation rate of $LaMnO_3$ perovskite oxide prepared by various methods and compositions plotted against the BET surface area. It is evident that the C_3H_8 oxidation rate increases monotonically with increasing the BET surface area of $LaMnO_3$, and it can be easily understood how the preparation method is important for improving the surface activity of perovskites.

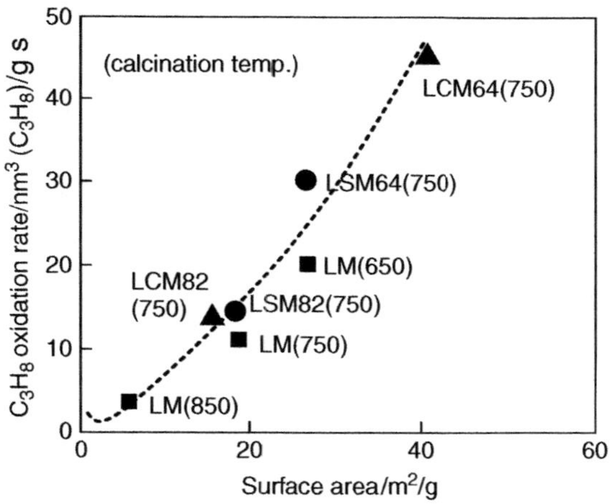

Fig. 1.12 C_3H_8 oxidation rate on $LaMnO_3$ perovskite oxide prepared by various methods plotted against the BET surface area LM, $LaMnO_3$; LSM82, $La_{0.8}Sr_{0.2}MnO_3$; LSM64, $La_{0.6}Sr_{0.4}MnO_3$; LCM82, $La_{0.8}Ca_{0.2}MnO_3$; LCM64, $La_{0.6}Ca_{0.4}MnO_3$.

1.5 Perovskite Oxides for Solid Oxide Fuel Cells (SOFCs)

As briefly discussed, because of their diversity in structures, chemical composition, and high chemical stability, perovskite oxides are widely used for preparing SOFC components. Particularly, the application of Co- and Mn-containing perovskites as cathodes has been extensively studied for reasons of their high electrical conductivity and catalytic activity for oxygen dissociation. In addition, $LaCrO_3$ is also regarded as a promising interconnector material for the tubular-type SOFC operating at higher temperatures.

Table 1.5 summarizes the important applications of perovskite oxides for SOFC technology. $LaCoO_3$ or $LaMnO_3$ is shown as a promising candidate for SOFC cathodes, and $LaGaO_3$-based oxides are suggested for the electrolyte. In addition, recently there were several reports on the application of Cr-based perovskites as the anode. Therefore, the concept of SOFCs based entirely on a perovskite component, an "all-perovskite SOFC," is also being considered. In contrast to the SOFCs using oxide ion-conducting electrolytes, the development of SOFCs using high-temperature proton-conducting electrolytes is slightly delayed, particularly as compared with development of polymer electrolyte-type fuel cells. However, the Toyota group has been quite successful

Table 1.5 Important materials for perovskite oxide for solid oxide fuel cell applications

Component	Typical materials
Cathode	La(Sr)MnO_3, La(Sr)CoO_3, $Sm_{0.5}Sr_{0.5}CoO_3$, La(Sr)Fe(Co)O_3
Electrolyte	La(Sr)Ga(Mg)O_3 (O^{2-}), $BaCeO_3$ (H^+), $BaZrO_3$(H^+), $SrZrO_3$(H^+) $Ba_2In_2O_5$(O^{2-})
Anode	$La_{1-x}Sr_xCr_{1-y}M_yO_3$ (M = Mn, Fe, Co, Ni), $SrTiO_3$
Interconnector	La(Ca)CrO_3

in demonstrating a high-power SOFC using a $BaCeO_3$-based electrolyte (see Chapter 14). Their data suggest that the proton-conducting perovskite oxides may be also an essential component in real SOFCs in the near future.

In this book, various aspects of perovskite oxides used for solid oxide fuel cells are reviewed from the point of view of materials. It is evident that perovskite oxides will be essential key materials in SOFC technology.

References

1. R.M. Hazen, Sci. Am. 258, 74 (1988)
2. T. Yagi, H.K. Mao, P.M. Bell, Phys. Chem. Miner. 3, 97 (1978)
3. F. Kanamura, Kikan Kagaku Sosetsu, No. 32, "Perovskite Related Compound", p. 9, ed. Japanese Society of Chemistry (1997)
4. S. Geller, J.B. Jeffries, P.J. Curlander, Acta Crystallogr. B31, 2770 (1975)
5. R.C. Liebermann, L.E.A. Jones, A.E. Ringwood, Phys. Earth Planet. Inter. 14, 165 (1977)
6. A.F. Well, "Structural Inorganic Chemistry", pp. 575 (5th ed.), Oxford University Press (1984)
7. A.F. Cotton, G. Wilkinson, "Advanced Inorganic Chemistry", John Wiley & Sons (1988)
8. F.S. Galasso, "Perovskites and High Tc Superconductors", Gordon and Breach, New York (1990)
9. R.H. Mitchell, T. Bay, "Perovskites Modern and Ancient", Ontario Almaz Press (2002)
10. H. Arai, T. Yamada, K. Eguchi, T. Seiyama, Appl. Catal. 26, 265 (1986)
11. J.C. Slater, Phys. Rev. 78, 748 (1950)
12. J.B. Bednorz, K.A. Muller, Z. Phys. B 64, 189 (1986)
13. P.H. Hor, R.L. Meng, Y.Q. Wang, L. Gao, Z.J. Huang, J. Bechtold, K. Forster, C.W. Chu, Phys. Rev. Lett. 58, 1891 (1987)
14. H. Maeda, Y. Tanaka, M. Fukutomi, T. Asano, Jpn. J. Appl. Phys. 27, L209 (1988)
15. L. Gao, Y.Y. Xue, F. Chen, Q. Xiong, R.L. Meng, D. Ramirez, C.W. Chu, J.H. Eggert, H.K. Mao, Phys. Rev. B50, 4260 (1994)
16. S. Shin, H. Arakawa, Y. Hatakeyama, K. Ogawa, K. Shimomura, Mater. Res. Bull. 14, 633 (1979)
17. Y. Teraoka, T. Harada, S. Kagawa, J. Chem. Soc., Faraday Trans. 1998, 94 (1887)
18. H. Yasuda, T. Nitadori, N. Mizuno, M. Misono, Bull. Chem. Soc. Jpn. 66, 3492 (1993)
19. H. Iwakuni, Y. Shinmyou, H. Yano, H. Matsumoto, T. Ishihara, Appl. Catal. B 74, 299 (2007)
20. Y. Nishihata, J. Mizuki, T. Akao, H. Tanaka, M. Uenishi, M. Kimura, T. Okamoto, N. Hamada, Nature 418, 164 (2002)
21. Y. Teraoka "Syokubai Gijyutsu no Doko to Tembo", Jpn. Catal. Soc. 2002, 23 (2002)
22. O. Palchik, J. Zhu, A. Gedanken, J. Mat. Chem. 10, 1251 (2000)
23. H. Kusaba, T. Asada, T. Kayama, K. Sasaki, Y. Teraoka, Syokubai 47(2), 171 (2005)

Chapter 2
Overview of Intermediate-Temperature Solid Oxide Fuel Cells

Harumi Yokokawa

2.1 Introduction

The first breakthrough in solid oxide fuel cell (SOFC) technology was achieved by Westinghouse Power Corporation (WHPC; currently Siemens Power Generation Corporation) [1] in the late 1980s in their efforts in establishing tubular SOFCs with the following technologically important points:

1. Optimizing the materials [yttrium-stabilized zirconia (YSZ) for the electrolyte, lanthanum strontium manganite for the cathode, nickel for the anode, and lanthanum magnesium chromite for the interconnect].
2. Adopting an excellent processing technology of electrochemical vapor deposition (EVD) [2] that has extraordinary advantages in fabricating dense films on porous materials or in anchoring nickel on YSZ.
3. Adopting a sealless tubular stack design to avoid usage of sealant materials.
4. Aiming for stationary applications.

This breakthrough leveraged up the development of the SOFC stacks/systems from the R&D stage to a more realistic stage with specifically targeted market sectors. The long operation life was successfully demonstrated, and also the high conversion efficiency from natural gas to electricity was demonstrated as 47% Lower Heating Value (LHV) for stationary 100-kW SOFC systems and as 52% for combined SOFC-gas turbine systems.

Immediately after the first breakthrough with sealless tubular cells, detailed analyses were made by Ackerman at the Argonne National Laboratory (ANL) to identify the merits and demerits of the sealless tubular cells [3]. The main disadvantages were pointed out as follows:

H. Yokokawa (✉)
Energy Technology Research Institute, National Institute of Advanced Industrial Science and Technology, Higashi 1-1-1, AIST Central No.5, Tsukuba, Ibaraki 305-8565, Japan
e-mail: h-yokokawa@aist.go.jp

T. Ishihara (ed.), *Perovskite Oxide for Solid Oxide Fuel Cells*,
Fuel Cells and Hydrogen Energy, DOI 10.1007/978-0-387-77708-5_2,

1. High fabrication costs because the EVD process utilizes metal chloride vapors in vacuum.
2. Low volumetric power densities because the electrical paths lie transversely along the cathode layer in tubular cells.

Since then, various attempts [4–6] have been made to investigate the following main points:

1. Planar cells to improve the power density.
2. Tubular cells to lower the fabrication cost or to increase the power density.

The next new wave in developing solid oxide fuel cells arose around the mid-1990s. One of the biggest achievements in this period was the discovery of a new oxide ion conductor, namely, lanthanum strontium gallium magnesium oxides (LSGM), by Ishihara in 1994 [7, 8]. Another important impact on SOFC technology was the proposal of using SOFCs as auxiliary power units for automotive applications by BMW and Delphi [9]. A similar proposal was made by ANL for monolithic SOFCs in the late 1980s [10] in their efforts to overcome the demerits of sealless tubular cells. Even so, the proposal by BMW/Delphi was based on the important trends in recent SOFC technology development; that is, lowering the operational temperature for using metal interconnects. In view of this, the discovery of a new electrolyte has further facilitated this trend.

Quite recently, a small SOFC system for residential application has been constructed by Kyocera and tested by Osaka Gas [11]. These test results indicate surprisingly high stack efficiencies, more than 50% Higher Heating Value (HHV) during the steady-state operation and 42%–48% LHV as the averaged system net efficiency over a 24-h service period in a residential house. Similarly, Mitsubishi Materials Corporation and The Kansai Electric Power Co., Inc. also achieved high conversion efficiencies by using a Co-doped LSGM (LSGMC) electrolyte. These achievements indicate that the development stage of the SOFC technology is apparently being stepped up and that a new era of SOFC developments has already started. The important key word for this new era is the "intermediate-temperature SOFCs." In this chapter, these recent developments associated with the intermediate-temperature SOFCs are reviewed with an emphasis on the stack/system development.

2.2 Characteristic Features of Solid Oxide Fuel Cells

2.2.1 Merits and Demerits of SOFCs

Solid oxide fuel cells make use of the high-temperature oxides as electrolyte. Figure 2.1 compares the conductivities of common electrolytes that are utilized in various fuel cells; namely, phosphoric acid fuel cells (PAFC), polymer electrolyte fuel cells (PEFC), molten carbonate fuel cells (MCFC), and solid

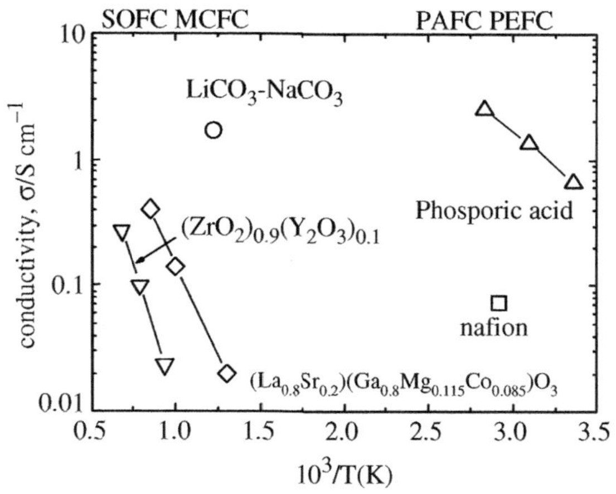

Fig. 2.1 The conductivity of typical electrolytes for fuel cells: PAFC, PEFC, MCFC, and SOFC [12]

oxide fuel cells (SOFC) [12]. When compared with liquid electrolytes such as molten alkali carbonates or phosphoric acid, the conductivity of solid electrolytes is not high; this implies that the solid electrolyte should be fabricated into a thin film with an appropriate technique. Another important feature appearing in Fig. 2.1 is that the activation energy of conductivity is quite large for solid oxide electrolytes; thus, the lower limit of operating temperature range for oxide electrolytes is dictated by the Joule loss. This limitation inevitably leads to a high-temperature operation for solid oxide fuel cells.

However, this condition provides some merits for solid oxide fuel cells. The most important one is that the operation temperature can be higher than the reforming temperature so that the heat required for the reforming process may be supplied from the SOFC exhaust heat, and this is one of the reasons why the efficiency of SOFCs can be high. For a similar reason, SOFCs are appropriate for hybrid systems with gas turbines in which further increase in efficiency can be expected by postcombustion of remaining fuels.

The demerits of high-temperature operation appear as higher thermal stresses and longer start-up times. Cells are made up of all solid components so that thermal stresses may result from thermal expansion coefficient mismatch among cell components or due to volume changes during redox cycles or chemical reactions. When the operation temperature is high, large temperature differences tend to develop, causing more severe conditions for thermal stresses. In addition, higher operating temperatures inevitably require longer start-up times.

All solid fuel cells have the important merit of long life expectation. In other words, fuel cells with liquid electrolytes suffer from degradation due to severe corrosion. In solid oxide fuel cells, the lifetime is not determined by such a degradation mechanism. On the other hand, however, another demerit may arise from the fact that solid oxide fuel cells are made up of all solids, implying difficulty in constructing SOFC stacks [4]. To ensure the gas tightness of SOFC stacks, it is essential to fabricate stacks by high-temperature sintering processes or by physically activated processes such as the EVD process. The latter is convenient for fabricating dense films on a porous substrate but expensive for competition with inexpensive gas engines. The former is economical but needs high-temperature exposure of materials, leading to deterioration of materials due to interdiffusion across interfaces. Even when stacks are well fabricated as gas-tight stacks, thermal stresses caused by temperature variations lead to mechanical weakness. The merits and demerits of SOFCs are shown briefly in Table 2.1.

Table 2.1 Merits and demerits of solid oxide fuel cells

	Merits	Demerits	Solutions
High-temperature operation	High conversion efficiency Hybrid system with gas turbines Cogeneration system	Thermal stress Starting up time	Sealless tubular
All solid	Long life No need for electrolyte management or water management	Difficulty in stacking cells Volume changes cause degradation	Monolithic cells Anode support cells Micro tubes
Electrochemical cells	Little NO_x/SO_x emission No use of precious metals	Needs for fuel treatment Little scale merits High fabrication cost	1 MW class Several times 10 kW Several kilowatts (kW)
Membrane reactor	CO_2 removal	Difficulty of 100% fuel utilization	Hybrid system with gas turbine (GT)

These features of SOFCs affect selection of appropriate materials that must meet a number of physicochemical requirements. An additional but important requirement is materials compatibility to achieve chemical and mechanical stability. For example, even when excellent performances are measured for electrode materials, they cannot be used if their compatibility with the electrolyte is not good. A typical example is $LaCoO_3$, which exhibits excellent electrochemical activity; however, the reactivity of this material with YSZ is significant and the thermal expansion mismatch with YSZ is large. In view of this, the selection of the electrolyte material mainly dictates the additional requirements for other materials.

2.2.2 *Issues for Intermediate-Temperature SOFCs*

The first breakthrough in solid oxide fuel cells by WHPC was made by using yttria-stabilized zirconia (YSZ) so that the operation temperature could be around 900°–1000°C.

Technological motivation for lowering the operation temperature can be summarized as follows:

1. In the beginning, strong interest arose in utilization of metal interconnects instead of $LaCrO_3$-based oxide interconnects [13]. Because of the severe corrosion of metals at high temperatures, operation temperature needs to be lowered.
2. Thermodynamic conversion efficiency increases with decreasing temperature for reformed gas (a mixture of CO and hydrogen).
3. Sealing technique becomes less difficult with lowered temperature.
4. For a small system, radiation heat loss becomes less severe by decreasing temperature. Hence, heat management becomes easier at lower temperatures [11].

On the other hand, decreasing operation temperature gives rise to additional materials issues as follows:

1. The oxide ionic conductivity decreases rapidly with decreasing temperature. As indicated in Fig. 2.1, the activation energy for the ionic conductivity is high so that the ionic conductivity drop is rather significant. To establish intermediate-temperature SOFCs, it is essential to have faster oxide ion conductors or to have a good method of fabricating a thinner electrolyte film. In view of these concerns, anode-supported cells are one of the possible technological solutions.
2. Usually, the electrode activity also decreases drastically with decreasing temperature, which makes it necessary to utilize more active electrode materials.
3. For the anode, nickel is still the best choice for operation in the intermediate-temperature region. Most frequently observed effects on nickel anodes are sulfur poisoning. It is well known that degradation caused by hydrogen sulfide becomes more severe with decreasing temperature.
4. For the cathode, Cr poisoning is severe on the lanthanum strontium manganites and becomes worse with decreasing temperature, against expectations.

In what follows, the materials aspects of intermediate-temperature SOFCs are described.

2.2.2.1 Electrolytes and Conversion Efficiency

The conduction properties of electrolytes are the most important factors in determining the operational temperature [11]. Here, the conversion efficiency is described in terms of conduction properties. The oxide ion conductivity determines the area-specific resistance contributed by the electrolyte. The contribution increases with increasing thickness of the electrolyte plate (film).

The oxide ion conductivity exhibits an Arrhenius-type behavior (Fig. 2.2). For YSZ, no oxygen potential dependence is observed over a wide oxygen partial pressure range applicable to solid oxide fuel cells. For electron and hole conductivities, the oxygen potential dependence is given by the following equation:

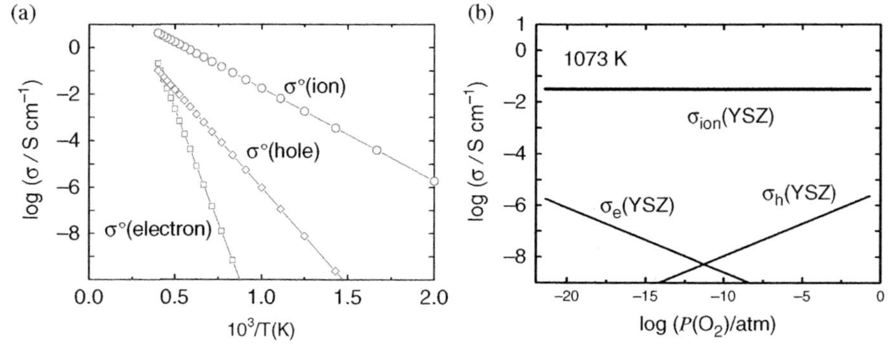

Fig. 2.2 Characteristic features of conductivities of YSZ as functions of (**a**) temperature and (**b**) Oxygen potential [14]

$$\sigma(\mathrm{el}) = \sigma^{\mathrm{o}}{}_{\mathrm{electron}}\, p(\mathrm{O}_2)^{-1/4} + \sigma^{\mathrm{o}}{}_{\mathrm{hole}}\, p(\mathrm{O}_2)^{1/4} \tag{2.1}$$

Here, $\sigma^{\circ}{}_{\mathrm{electron}}$ and $\sigma^{\circ}{}_{\mathrm{hole}}$ are the normalized contribution of electrons and holes at 1 atm oxygen partial pressure. Their temperature dependencies are compared in Fig. 2.2(a) together with the oxide ion conductivity. Since the activation energies for electron and hole conductions are larger than that of the oxide ion conduction, the contribution of electron conduction becomes large when temperature increases. At 1073 K, as illustrated in Fig. 2.2(b), the oxide ion conductivity is several orders of magnitude higher than those of electrons and holes, indicating that YSZ is an excellent electrolyte for fuel cells. At temperatures as high as 2000 K, however, YSZ may no longer be utilized as an electrolyte but can be characterized as a mixed conductor.

The electrical properties determine the energy conversion losses that occur inside the electrolyte plate. The conversion losses can be expressed as a deviation from the theoretical conversion based on the Gibbs energy. In Fig. 2.3(a), the conversion losses are plotted as a function of products of current density (J^{ex}) and electrolyte thickness (L). For large values of $J^{\mathrm{ex}}L$, the conversion efficiency, 1-η (electrolyte), decreases with increasing $J^{\mathrm{ex}}L$; this is known as the Joule effect. On the other hand, even for small values of $J^{\mathrm{ex}}L$, 1-η (electrolyte) decreases with decreasing $J^{\mathrm{ex}}L$; this is called the shorting effect caused by electronic conduction. With this, oxide ions are transported and take part in electrochemical reactions without generating electricity, which can be regarded as an ordinary chemical reaction of fuel with permeated oxygen gas (oxide ion and holes). In view of this, the shorting effect can also be called the oxygen permeation effect. By combining the Joule effect and the shorting effect, the deviation from the Gibbs energy conversion efficiency can be characterized by a curve with a maximum point.

A similar maximum behavior is observed even for the temperature dependence when the thickness of electrolyte and the current density are fixed as shown in Fig. 2.3(b). In this figure, three different electrolyte materials are

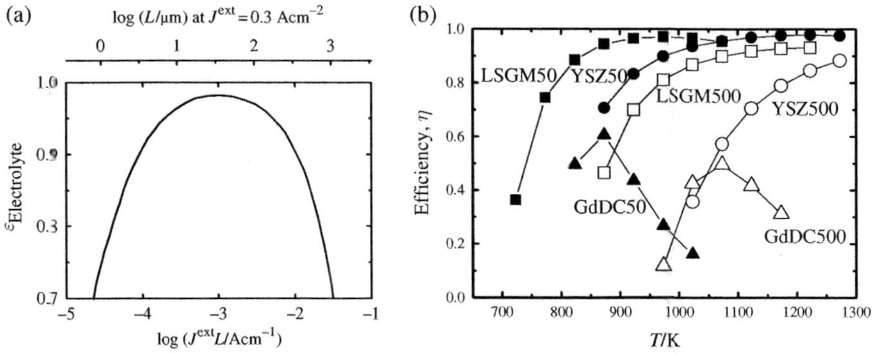

Fig. 2.3 Gibbs energy-based conversion loss occurring in electrolytes due to Joule effects and shorting effects: (**a**) as a function of products of current density and thickness at a given temperature (1273 K) [14] and (**b**) as a function of temperature for a given current density (0.3 A/cm^2) with parameter of thickness (μm) [12, 14]

compared with each other. As shown in Fig. 2.2(b), the electronic contributions in YSZ are small so that the efficiency lowering is very small over a wide temperature range. For $(La_{0.8}Sr_{0.2})(Ga_{0.8}Mg_{0.2})O_{2.8}$ (LSGM), the region of high efficiency extends to a lower temperature than for YSZ, because the oxide ionic conductivity of LSGM is higher so that the Joule effect is smaller. For Gd-doped ceria (GDC), the efficiency at high temperatures is quite low due to the large contribution of the electronic conduction.

The energy conversion efficiency is usually discussed in terms of the enthalpy-based conversion rate. For example, theoretical efficiency is defined as the ratio of the Gibbs energy change to the enthalpy change for fuel cell reaction. Therefore, the values just discussed should be transferred to the enthalpy-based ones. For this purpose, the thermodynamically theoretical conversion efficiency should be defined in a manner that enables comparison with other energy convertors such as heat engines. Here, we start with methane as the common fuel. In Fig. 2.4(a), we compare several cases as a function of temperature:

1. Carnot efficiency is usually defined as w/q, where w is work to be done during one cycle, whereas q is high-temperature heat to be used. In the present case, we start with methane chemical energy (enthalpy), which has a higher quality than high-temperature heat. To create high-temperature heat from the chemical energy of methane, we lose some part of it during the combustion process. Line (1) in Fig. 2.4(a) shows the Carnot efficiency after subtracting this effect.
2. Line (2) is the ratio of Gibbs energy to enthalpy for the complete direct oxidation of methane:

$$CH_4(g) + 2O_2(g) = CO_2(g) + 2H_2O(g) \tag{2.2}$$

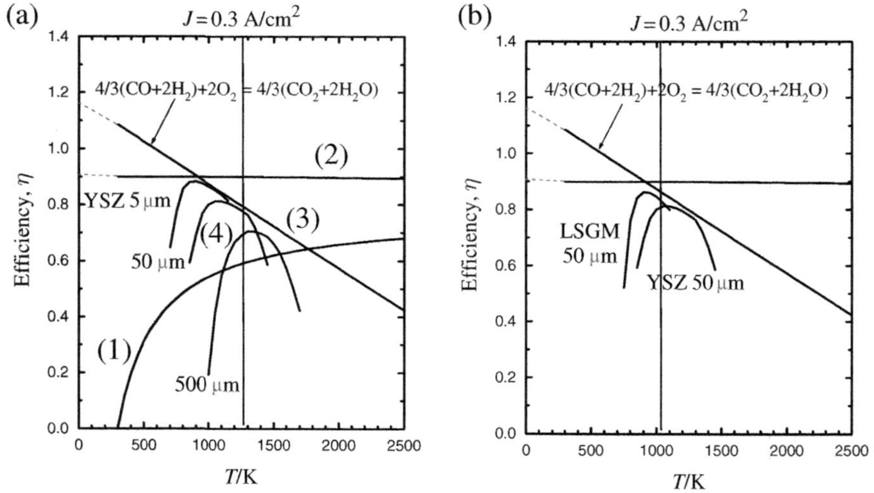

Fig. 2.4 Comparison in conversion efficiency: (**a**) (1) Carnot efficiency after correction of self-heating of methane combustion, (2) methane direct oxidation, (3) oxidation of reformed gas, (4) oxidation of reformed gas after correction for conductive properties of YSZ in given thickness; (**b**) comparison between YSZ and LSGM in the same thickness of 50 μm

Because these two quantities, ΔH and ΔG, have no temperature dependence, the derived efficiency is high and independent of temperature. In actual fuel cells, this reaction could not proceed, because nickel anodes are not active to the direct oxidation of methane.

3. Line (3) is for electrochemical oxidation of hydrogen and CO after reformation process. Methane reforming can be achieved by using water vapor or anode-circulated gases. Here, we assume that the anode gas is circulated, simply because we would like to utilize the methane alone as a starting material for comparison purposes. The anode gas circulation makes it easy to compare the case with Carnot cycles in which external water is not used. Assumption is also made for the point that the heat required for the reforming process is supplied from the heat emitted from the fuel cells. This is the essentially different point from the reforming process connected to the PEFC system, in which additional fuel should be burned to supply the required heat. This benefit of utilizing internal heat appears as the feature that the Gibbs energy conversion rate for the reformed gas shifts to quite high values in the vicinity of the reforming temperatures around 900 K in Fig. 2.4(a). Even so, the Gibbs energy-based conversion rate decreases with increasing temperature above the reforming temperature; this implies that the solid oxide fuel cells to be operated around the reforming temperature are expected to have the highest conversion efficiencies.
4. Lines (4) are combined effects of the Gibbs energy-based rate [line (3)] and the lowering effects resulting from the electrolyte conductivity properties

from the Gibbs energy-based rate. For YSZ electrolyte, three values corresponding to different thickness values are taken from those in Fig. 2.3(b). Figure 2.4(a) shows, in a clearer manner, the effect of electrolyte thickness. For YSZ, the thickness of 50 μm adopted first by WHPC in the EVD process provides rather good efficiency even at a temperature lower than 1273 K. The thinner YSZ can provide more efficient SOFC systems. In view of this, the anode-supported cells are of strong interest and importance for developing the intermediate-temperature SOFCs.

In Fig. 2.4(b), comparison is made between YSZ and LSGM for identical thickness values of 50 μm, which indicates the superiority of higher oxide ion conductors in the intermediate-temperature SOFCs. Particularly, technological conditions are quite different. For YSZ, anode-supported cells are inevitably required, whereas the self-supported cells can be operated around 1073 K for LSGM. Actually, Mitsubishi Materials Corporation successfully designed and manufactured SOFC systems based on the self-supporting LSGMC cells, confirming that the obtained efficiency is high, as is described in the following sections.

2.2.2.2 Cathode

Relationship with YSZ and Cr Poisoning

In the first generation of SOFC to be operated around 1273 K, the lanthanum strontium manganites [$(La_{1-x}Sr_x)MnO_3$, LSM] have been well investigated because of their higher cathode activity and compatibility with the YSZ electrolyte [15]. Since the chemical stability was much more important in the first generation, LSM has been utilized widely in actual stacks. When LSM is used for intermediate-temperature SOFCs, it has been found that the performance of LSM on the anode support cells is degraded rapidly with decreasing temperature. In addition, LSM is poor against the Cr poisoning [16], which is caused by the chromium-containing vapors emitted from Cr_2O_3 oxide scale on the metal interconnects.

Lanthanum strontium cobaltite [$(La_{1-x}Sr_x)CoO_3$, LSC] was the first perovskite-type oxide investigated as a SOFC cathode in 1969 [17]. Even in this attempt, it was found that LSC degraded rapidly because of a chemical reaction with YSZ. Since then, major investigations on cathodes moved to the lanthanum strontium manganites. The recent trend of lowering operation temperature, however, leads again to the investigation of LSC, $(La_{1-x}Sr_x)FeO_3$ (LSF), and $(La_{1-x}Sr_x)(Co_{1-y}Fe_y)O_3$ (LSCF) by using the interlayer made up of doped ceria between YSZ and those perovskite cathodes.

It is interesting to see the work by Matsuzaki and Yasuda [18], who investigated Cr poisoning using different combinations of electrolyte and cathodes; as electrolyte, they selected YSZ and samarium-doped ceria (SDC), and LSM and LSCF were selected as cathodes. As shown in Fig. 2.5, a potential drop from Cr poisoning is largest and most rapid for LSM/YSZ, whereas LSCF/

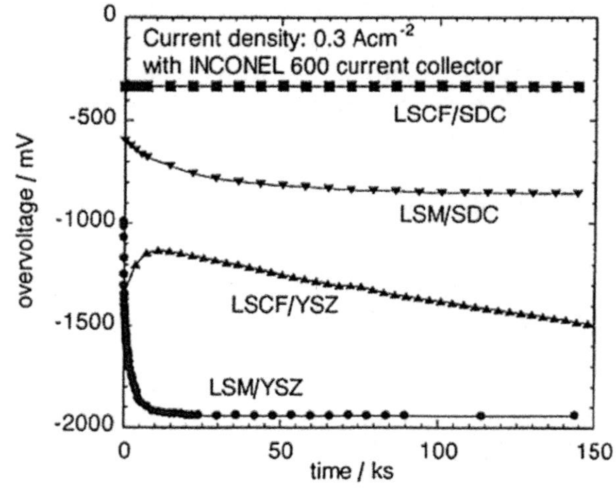

Fig. 2.5 Cr poisoning for different combinations of electrolyte and electrode, measured with a cathode half-cell in contact with a plate of INCONEL 600 by Matsuzaki and Yasuda [18]

SDC showed no decrease from Cr poisoning. These results indicate that the identification of electrochemical reaction mechanism is crucial in understanding Cr poisoning.

Table 2.2 summarizes and compares various features of perovskite cathodes from the aspect of valence stability. Valence stability is directly related with chemical stability and also indirectly with electrochemical activity through oxide ion conductivity.

Table 2.2 Comparison among perovskite cathodes (LSM, LSF, LSC) in their trade-off relationship between chemical stability and performance with emphasis on the reaction with YSZ and Cr vapors [21]

Items	LSM	LSF	LSC
Valence stability	Mn^{4+} stable	Fe^{4+} unstable	Co^{4+}/Co^{3+} unstable
O^2 conductive	Quite slow	Fast	Fast
Cathode mechanism	Three-phase boundary	Surfaces	
Reactivity with YSZ	Stable (A-site deficient)	$SrZrO_3$ formation	$La_2Zr_2O_7$ $SrZrO_3$ formation
Reactivity with Cr	Cr^{3+} substitute	$SrCrO_4$/ Cr^{3+} substitute	$SrCrO_4/Cr^{3+}$ Cr^{4+} substitute
Cr poisoning	Significant	Not seen in early stage, but degradation due to $SrCrO_4$	

The valence stability itself is defined as the thermodynamic properties, so that it is natural to expect that the chemical stability of perovskite oxides is related to the valence stability in addition to the stabilization energy of double oxides from the constituent oxides. Particularly, the reaction of perovskite oxides with YSZ has been well examined experimentally as well as thermodynamically.

The formation of $La_2Zr_2O_7/SrZrO_3$ at the interfaces is accompanied with the reduction of transition metal oxides and precipitation of compounds with reduced valence ions [19]. Recently, chemical reactions of perovskite oxides with chromium-containing vapors have been analyzed in a similar manner, and it has been found that reactivity with chromium vapor is of the same order among the LSM, LSF, and LSC as reactions with YSZ [20, 21]. That is, the Sr component in the perovskite oxides can react with Cr vapors to form $SrCrO_4$ for LSF and LSC but not for LSM, which exhibits the most severe Cr poisoning effect. This result implies that chemical reactivity alone cannot explain the Cr poisoning effect, because LSM exhibits most severe Cr poisoning effect, although LSM is most stable against reactions with Cr vapors.

The oxide ion vacancies in the perovskite ABO_3 oxides are formed as a result of reduction of the B-site ions on substitution of Sr^{2+} ions to the La^{3+} (A) sites; this will lead to mixed conductivity of lanthanum strontium transition metal oxides. Even so, the oxide ion vacancy formation is competing with the oxidation of transition metal ions in the B sites. The latter depends on the valence stability of the transition metal ions. For the case of $(La,Sr)MnO_3$, the tetra valence of manganese ions is stable in the perovskite lattice so that the Sr substitution gives rise to the oxidation of manganese ions from 3+ to 4+ and to essentially no oxide ion vacancy formation. As a result, the oxide ion conductivity in LSM is not high. This fact affects the reaction mechanism; that is, only the three-phase boundaries (TPB) are electrochemical active sites for the LSM cathode, whereas the high oxide ion conductivity in LSF and LSC provides wider distribution of electrochemically active sites. This difference on the distribution of active sites in relationship to the oxide ion conductivity leads to different features in the oxygen flow and associated with the oxygen potential distribution.

This difference in the oxygen flow and distributions of active sites and of oxygen potential provides a good basis of explaining the difference in the Cr poisoning. For LSM, the oxygen flow has to be concentrated in the TPB where Cr tends to be deposited, whereas the chemical reaction with Cr vapors can take place at any point of the LSF or LSC surfaces, but the oxygen flow can be changed to avoid such reaction sites. For long-term stability, however, $SrCrO_4$ formation should be avoided to maintain mechanical stability as well as chemical stability.

It is generally considered that the cathodes for intermediate-temperature SOFCs should be electrochemically more active than those cathodes in the first generation, namely, LSM. When LSF, LSC, or LSCF is used as cathode, an interlayer made of doped ceria becomes inevitable to avoid chemical reactions between cathode and YSZ. In addition, such cathodes should be also protected against Cr vapors. This approach gives rise to a complicated layer structure across the electrolyte to current collector and makes it difficult to fabricate such complicated layers. For example, the following points are important:

1. Doped ceria–YSZ interface. Solid solutions between doped ceria and YSZ provide interesting systems for investigating the transport and related properties [22, 23]. That is, the ionic conductivity has a minimum in the middle of the Ce concentration range, whereas the electronic conductivity has a maximum. Similarly, the surface exchange reaction rate exhibit strong concentration dependence. This property suggests that when the doped ceria–YSZ interface was prepared in a well-bonded state at interfaces by sintering at high temperatures, interdiffusion takes place across the interface, forming a layer with high electrical resistance. Furthermore, interdiffusion sometimes gives rise to Kirkendall pores in the ceria side because of the differences in diffusivities of zirconia and ceria [24].
2. Strictly speaking, interfaces between perovskite cathode and doped ceria are not thermodynamically stable, and some chemical reactions can take place [21, 25]. In addition, cation diffusion can occur. In particular, Sr diffusion through doped ceria is important. There are some differences between LSF and LSC as far as reactivity and interdiffusion are concerned; that is, no products are formed for the diffusion couple between Gd-doped ceria (GDC) and LSC, because $GdCoO_3$ exhibits no thermodynamic stability. In other interfaces, there arises a driving force of forming another perovskite phase from the dopant in ceria and the B-site ions (Fe or Co ions) in the perovskite; this is accompanied with Sr diffusion.
3. To obtain a stable interface between cathode and metal interconnect, it is essential to adopt a coating layer on the interconnect to prevent the migration of Cr from the metal alloys.

Compatibility with LSGM

Immediately after the discovery of LSGM by Ishihara [7,8], it became clear that interdiffusion associated with LSGM is significant between LSGM and cathode electrode [26]; this makes it difficult to prepare cathode-supported cells in which the cathode–electrolyte interfaces are exposed to high temperatures. Currently, $(Sm,Sr)CoO_3$ is widely utilized as a cathode material on the basis of Ishihara's results [27].

Because $(Sm,Sr)CoO_3$ exhibits similar features to those of $(La,Sr)CoO_3$, reactions with Cr vapors are also technologically important issues. That is, immediate degradation of SSC cathodes is not expected, but the formation of $SrCrO_4$ leads to changes in microstructure and other properties. Particularly, the thermal expansion coefficients of the Sr-depleted cobaltites and of the formed $SrCrO_4$ are both large.

2.2.2.3 Anode

Even for the intermediate-temperature SOFCs, Ni is the best anode as far as the current technological status is concerned. Although a number of investigations have been made on oxide anodes, nickel cermet (*cer*amic-*met*al) anodes exhibit excellent ability of dissociating hydrogen bonds. As the oxide component of cermet anodes, YSZ is frequently used. In recent years, ScSZ or doped ceria have attracted attention because characteristic features against carbon deposition or sulfur poisoning can be improved by the use of these oxides instead of YSZ.

Nickel Anode

Technological issues associated with nickel anodes can be summarized as follows.

1. Sintering: In the first-generation SOFCs, sintering of nickel anodes and the associated degradation are one of the major issues because the high operation temperature promotes sintering during long operation times. Furthermore, nickel microstructure can be heavily damaged to form metastable Ni-C liquids in the presence of carbon. In the intermediate-temperature SOFCs, however, these mechanisms of sintering or change in microstructure caused by the Ni-C liquids are expected to diminish.
2. Carbon deposition: Nickel is weak against carbon deposition even in the intermediate-temperature region. There is an apparent effect of the oxide mixing on the carbon deposition behavior among various cermet anodes. Figure 2.6 depicts the different features of patterned Ni on YSZ or SDC without any electrochemical reactions under an atmosphere that is thermodynamically favorable to carbon deposition [28]. Surface species on nickel were detected by secondary ion mass spectrometry (SIMS), indicating that these are

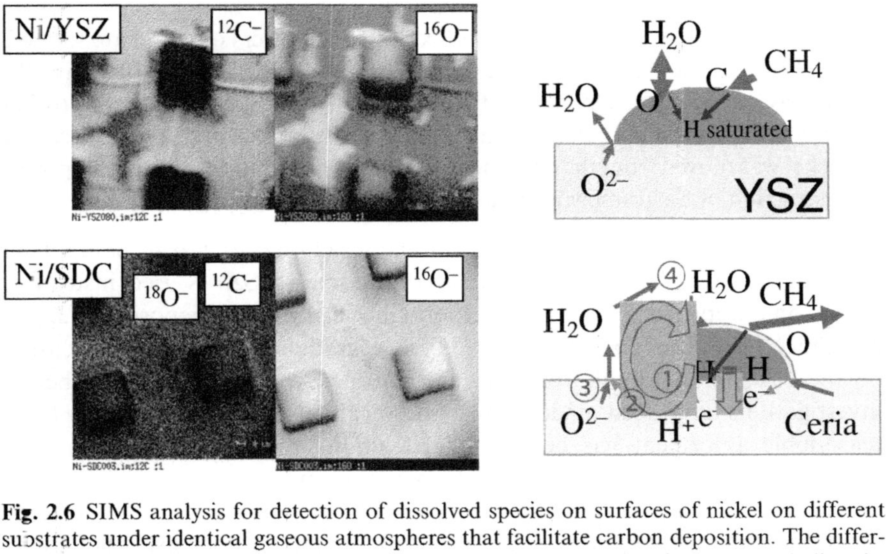

Fig. 2.6 SIMS analysis for detection of dissolved species on surfaces of nickel on different substrates under identical gaseous atmospheres that facilitate carbon deposition. The difference between YSZ and SDC can be explained by using a mass transfer model including the dissolution of water into SDC together with enhanced surface exchange reaction rates [28]

not adsorbed species but dissolved atoms. For Ni/YSZ, carbon covers almost the entire surface of nickel and only a small amount of oxygen is present on the surface. Under the same condition, Ni/SDC exhibits quite different features of nickel surface. That is, the nickel surface is covered by oxygen instead of carbon. This observation can be reasonably explained by considering the mass transfer mechanism in which nonnegligible water solubility in ceria, and enhanced surface reaction at the ceria surface, can be accounted for as different features. Under a polarization of the Ni/YSZ combination, a similar coverage of oxygen was observed on nickel, indicating that the above mechanism is closely related with the anode reaction mechanism.

3. Sulfur poisoning: From the earlier stages of the development of SOFCs, it has been well known that, in the presence of a small amount of hydrogen sulfide, the anode activity is lowered but will recover after switching back to non-hydrogen sulfide fuels [29]. In addition to this reversible lowering activity, nickel anodes show irreversible degradation at higher concentration of H_2S or at lower temperatures.
4. Redox cycle tolerance [30]: As anode-supported cells have been investigated extensively, redox cycles are recognized as quite important. One reason is that the anode-supported cells inevitably have a sealing problem on their edges. Because the anode is used as the supporting body, its mechanical stability becomes crucial. Another reason originates from the purge gas. When nitrogen is used as a purge gas, nickel anodes are always protected against reoxidation. However, for cases where nitrogen cannot be used due to system requirements, etc., stability during redox cycles becomes also a crucial technological matter. This phenomenon is closely related with diffusion of Ni and reconstruction of microstructure on reduction from NiO to Ni; this is because diffusion of Ni in the metal phase is faster than Ni^{2+} ions in the oxide. On the reduction of NiO in a mixture of NiO and YSZ (or other oxides), fine powers of nickel are formed, and then the electrical path will be established using powders by diffusion in the framework of YSZ. On reoxidation of nickel, NiO does not move so that volume expansion on oxidation takes place in the framework of YSZ. Because nickel was moved from the original position, the reoxidation gives rise to partial destruction of the framework as a result of a single redox cycle.

These features are closely related with the selection of the oxide component in cermet anodes. When Sc_2O_3-stabilized zirconia (ScSZ) is used instead of YSZ, some improvements have been obtained for carbon deposition [31] or resistance for sulfur poisoning [32]. These degradations should be discussed on the basis of the anode reaction mechanism. Even so, a large number of investigations have been made on reaction mechanisms, but unfortunately no reasonable agreement has been obtained among researchers. Here, a brief discussion is made about the role of the oxide component.

The surface reaction rate and the water solubility in ScSZ are found to be about the same as those of YSZ [33]; this implies that merits of using ScSZ may originate from properties such as the oxide ion conductivity or the cation diffusivity affecting the microstructure of cermet anodes. In particular, higher oxide ion conductivity values positively affect the anode activities against carbon deposition or sulfur poisoning. As to carbon deposition, the water vapor emitted from active sites may have strong effects of avoiding carbon deposition by transferring oxygen atoms from the electrolyte to the nickel surface. When the current density is the same, the same amount of water vapor should be emitted. So, effects of higher oxide ion conductivity appear only in the distribution of electrochemically active sites. When the oxide ion conductivity is low, only the TPB located at the bottom of the anode layer becomes active, whereas the TPB even far from the bottom can be active when the oxide ion conductivity is high in the oxide component of cermet anodes. For the case of sulfur poisoning, the equilibrium shift should be considered as a function of anode overpotential as well as fuel utilization. Here, the overpotential should be related to the oxide ion conductivity.

Nickel Anode with LSGM Electrolyte

For the LSGM electrolyte, the doped ceria is used as the oxide component in cermet anodes. The interface between doped ceria and LSGM is rather stable, although some interdiffusion occurs. One of the biggest issues associated with the nickel anode used together with LSGM electrolyte is that the dissolution of NiO into perovskite phase takes place significantly during the high-temperature sintering process of cells; this occurs because in air the $LaNiO_3$ perovskite phase is rather stable so that NiO can be easily dissolved into the LSGM/LSGMC phases. In a worst case, NiO can penetrate completely to the cathode side. This phenomenon should be avoided, because NiO in LSGM can be reduced again to Ni metal by hydrogen so that the reduced Ni can cause electronic shorting paths inside the LSGM electrolyte. Figure 2.7 shows the distribution of

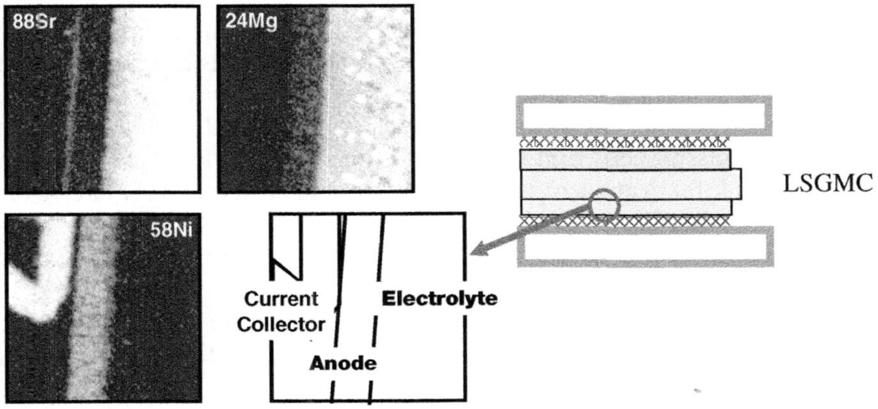

Fig. 2.7 The elemental distribution detected with SIMS technique after 24-h operation of sealless disk-type cells made by Mitsubishi Materials Corp. [34]

elements in an actual LSGMC-based cell operated for 24 h obtained by secondary ion mass spectroscopy (SIMS) technique. The cell was fabricated and tested by Mitsubishi Materials Corporation [34]. It is clearly seen that the Ni dissolution into the LSGMC layer was successfully prevented during the fabrication process.

Oxide Anodes

Recently efforts have been made on oxide anodes. The main reason for such investigations is to overcome the demerits of Ni cermet anodes as just described. Although the oxide anodes should be in service under a reducing atmosphere, the fabrication is usually performed in air so that oxide anodes should be stable at both oxidative and reductive atmospheres. This requirement is similar to those for oxide interconnects, implicitly indicating that material selection becomes severe to meet the chemical stability requirement.

Doped ceria and doped lanthanum chromites were investigated a long time ago because ceria is a mixed conductor in a reducing atmosphere, whereas lanthanum chromites are typical candidates for oxide interconnects. Neither of the materials shows good performance as an anode. In recent years, other types of perovskite oxides have attracted attention, as is described in other chapters of this book. The basic trade-off relationship associated with oxide anodes is stability versus performance.

2.2.2.4 Metal Interconnects

The reasons to utilize metal interconnects [13] instead of oxide interconnects [35, 36] may be listed as follows:

1. Material cost: La in the oxide interconnect is expensive, whereas ferritic alloys can be regarded as inexpensive.
2. Difficulty in fabricating $LaCrO_3$-based interconnects: Particularly, sintering in air is the most challenging. Although no SOFC stacks can be fabricated without establishing an appropriate technology for fabricating dense oxide interconnects, only a few manufacturers have succeeded in sintering oxide interconnects properly and constructing them into SOFC stacks. On the other hand, fabrication of metals is usually much easier than that of the $LaCrO_3$-based oxides. For oxide dispersed alloys such as $Cr_5Fe_1Y_2O_3$, however, special technology is required to fabricate these into a shape for SOFC stacks.
3. High thermal conductivity: Management of temperature distribution inside stacks is essential in solid oxide fuel cells to protect the fragile ceramic components.
4. High mechanical stability: To moderate the thermal stresses in ceramic systems, it is essential to shorten the relaxation times for thermal fluctuations by using materials with low thermal expansion coefficients and high thermal

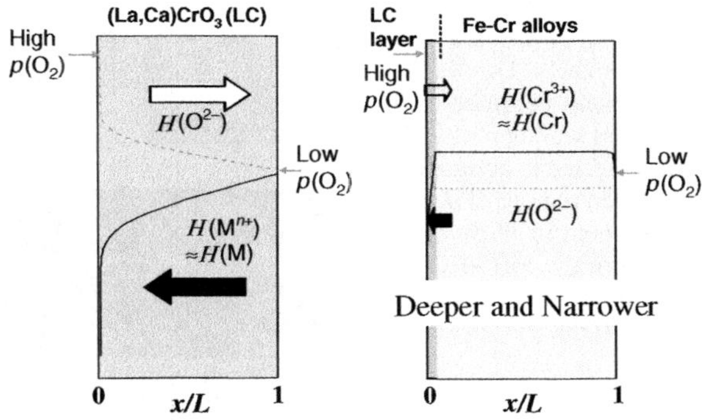

Fig. 2.8 Oxygen potential distributions in the oxide interconnects and the metal interconnects. Inside the oxide, the oxygen potential distribution is determined by the oxide ion and electron conductivity, whereas the surface oxide scale determines the main features of the metal interconnects

conductivities. Because YSZ electrolyte cannot meet such requirements by itself, it becomes essential to use metal components in SOFC stacks.

From the physicochemical point of view, the aforementioned interconnects can be compared in terms of their oxygen potential distribution (Fig. 2.8). In the $LaCrO_3$-based interconnect, a steep oxygen potential gradient appears in a thin layer on the air side, mainly because of extremely small oxide ion vacancy concentration and, hence, low oxide ion conductivity in the oxidative side. In the metal interconnect, the major steep oxygen potential drops appear in the oxide scale region in both fuel and air sides, and correspondingly the oxygen potential values inside metals are maintained at quite a low level. This finding implies that in the metal interconnect, the control of the mass transfer in the oxide scale vicinity is essential in judging the appropriateness of the materials.

Technological issues associated with metal interconnects can be summarized as follows:

1. Thermal expansion coefficients: For high-temperature utilization, Ni-Cr-based alloys are excellent from the anticorrosion point of view. Even so, such Ni-Cr alloys have high thermal expansion coefficients dictating a larger match with YSZ. Cr-based alloys developed by Siemens/Plansee have an essentially similar thermal expansion coefficient as YSZ. This alloy was utilized by Sulzer Hexis. Alternatively, Fe-Cr ferritic alloys are frequently utilized. Although complete matching in thermal expansion coefficient with YSZ is not obtained, Ni-Cr alloys provide considerable improvement.
2. Stable oxide scale: Corrosion affects SOFC stacks in two distinct ways. First, it increases the electrical resistivity. In normal configuration of cells, the electrical path usually penetrates across the oxide scale, which implies that growth of oxide scale makes a contribution to increase the area-specific resistivity. Another aspect of oxidation is related to mechanical stability. Growth of an oxide scale is inevitably accompanied with a volume change,

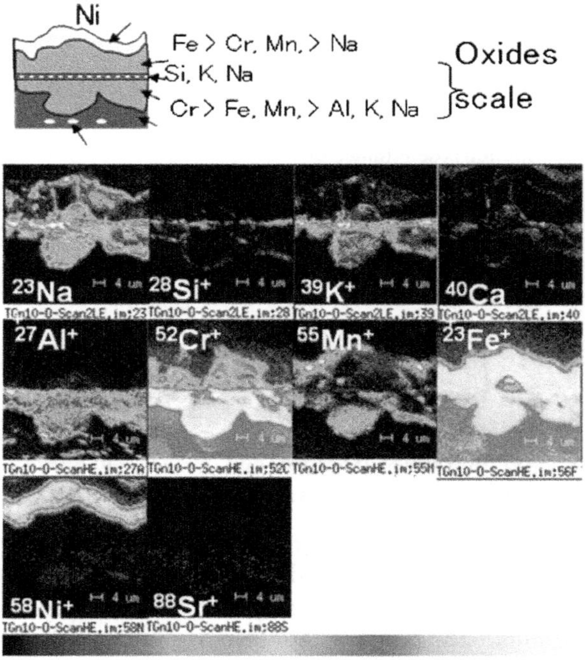

Fig. 2.9 Anomalous oxidation of metal interconnects. The Na component migrated from the glass sealing materials was detected in anomalously corroded regions. The iron component in alloys moved out from the corroded area to anomalously expanded regions [37]

causing mechanical instability. From these, the oxide scale of a metal interconnect should be thin and electrically conductive.

3. Anomalous oxidation: In some cases, oxide scale made up of Cr_2O_3 gets broken and can no longer serve as a protective layer; as a result, the iron component in alloys may become anomalously oxidized away from the scale. Typical features of such a phenomenon is shown in Fig. 2.9, in which anomalous oxidation of ferritic alloys in the presence of glass sealing materials is analyzed by using the SIMS technique detecting several times 10 ppm of the Na component [37]. In this particular experiment, the Na component is thought to have migrated from the glass sealing materials. Even so, Na contamination can commonly take place. It is a phenomenon similar to hot corrosion caused mainly by NaCl and/or Na_2SO_4.
4. Chromium poisoning: In the air side of the interconnect, Cr volatilization becomes an issue, because the perovskite cathodes tend to exhibit the Cr poisoning effect. In particular, the manganite cathodes show severe Cr poisoning. To avoid this, several attempts have been made on the metal interconnect side. Cr volatilization depends on the Cr_2O_3 activity of the oxide scale. One way is to form a spinel phase on the inner Cr_2O_3 scale. In typical ferritic alloys, $MnCr_2O_4$ is formed as the outer oxide scale. Cr poisoning does not cease even for such a case. To stop Cr volatilization completely, a spinel phase containing no chromium should be coated on the surface of metal interconnects.

2.2.3 Stack Design

The stack design and its fabrication process are quite important in developing the SOFC systems. For the case of the first-generation SOFCs, selection of materials was accompanied by adoption of the electrochemical vapor deposition technique and the sealless tubular design. For the intermediate-temperature SOFCs, utilization of metals is a key in selecting the fabrication technique and the design. Typical designs are as follows:

1. Sealless planar: Electrochemical cells made up of electrolyte, cathode, and anode are stacked with metal interconnects without using sealing materials. For example, disk-type planar sealless stacks have been constructed by Mitsubishi Materials Corporation. For this purpose, fuel and air are introduced through the central part of the respective cells. Outside the cells, the remaining fuel becomes combusted with air. Self-supporting cells are usually used for this design.
2. Planar cells with sealant: Fuel and air are separately controlled using sealing materials. Anode-supported cells are used for this design, which makes it necessary to seal properly the edge part of anode-supported cells.
3. Flattened tubes: Anode-supported or cathode-supported tubes are flattened so that there is no need to seal the side of the tubes. When both ends are open, at least the entrances are sealed. When one end is closed, there is a need for an additional pipe to introduce air or fuel. Normally, interconnect materials are fabricated simultaneously. For this purpose, oxide interconnects are more appropriate. Figure 2.10 shows the flattened tube cells fabricated by Kyocera to be operated at 750°C.
4. Micro tubes: Micro tubes without oxide interconnects are fabricated with cathodes and anodes. Current collection becomes a key issue in this design.
5. Metal-supported cell: Since metal-supported cells are still in the early stages of development, there is no defined stack design associated with metal-supported cells. There is a possibility of achieving gas tightness without using sealing materials.
6. Segmented in series: Fabrication process is complicated in this design. In addition, interconnects are also key materials. When the oxide interconnect

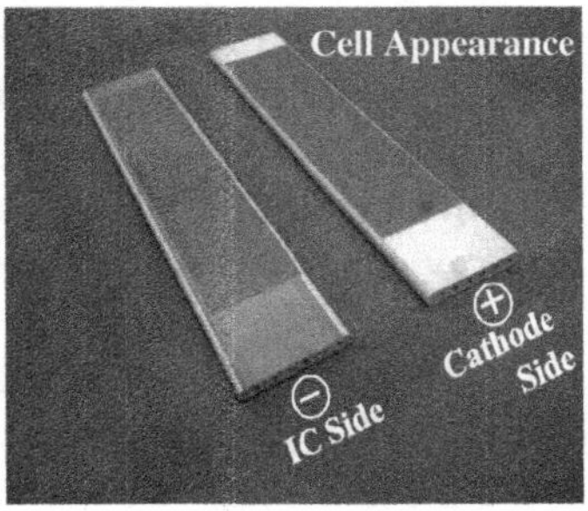

Fig. 2.10 Flattened tube-type cells for a small SOFC cogeneration system by Kyocera (courtesy of Kyocera)

is adopted, the same ceramic processing can be applied. In other words, proper manufacturing techniques addressing the problems in air sintering are essential. When metal interconnects are used, it becomes crucial to find appropriate methods for simultaneous fabrication of metals and ceramics.

2.3 Development of Intermediate Temperature SOFC Stacks/Systems

2.3.1 Kyocera/Osaka Gas

After fundamental investigations on SOFC materials for a long period of time, Kyocera started the development of a small SOFC cogeneration system for residential houses in 2001. Although similar developments have been made by Sulzer Hexis for the last decade in cooperation with utility companies/local governments in Germany, there are some important differences between the two SOFC systems; that is, the Sulzer system is based on supplying 1 kW electricity and 2 kW heat for domestic utilization, whereas the Kyocera system has focused on those small systems with high conversion efficiency of generating electricity. This difference is mainly the result of strong requirements for electricity rather than heat in the Japanese market. To achieve this requirement technologically, lowering the operation temperature is effective to reduce the heat losses of the SOFC stacks and therefore to maintain high efficiency even in smaller systems.

Kyocera adopted the flattened tube design shown in Fig. 2.10. The YSZ electrolyte is fabricated on a cermet anode substrate having gas channels for fuel flow. One side of the anode flattened tube is coated with the $LaCrO_3$-based interconnect. In Fig. 2.10, both sides of the SOFC are shown. These flattened tubes are unique in the sense that they use metals as cell-to-cell connection. This design should be compared with sealless tubular cells by Westinghouse Power Corp., in which the cell-to-cell connection is made with nickel felt. Nickel is thermodynamically stable in a fuel atmosphere so that the adoption of nickel makes it possible to establish stable and effective connection among tubes. In other words, WHPC adopted the cathode-supported tube and made the fuel side as the outer side of the tubes to realize a thermodynamically stable cell-to-cell connection. On the other hand, Kyocera adopted the reverse; flattened tubes were made as anode supported, and, as a result, the cell-to-cell connection should be made on the air side. For this design, therefore, utilization of (non-precious) metal connection becomes the technological key point. In this sense, the lowering of operation temperature is essential.

They built the 1-kW SOFC cogeneration system for residential houses and tested one system in an actual house together with Osaka Gas in 2005 [11]. The start-up time is typically 2 h. A typical result of 24 h operation is shown in Fig. 2.11. During the night, electricity demand is essentially zero except for the refrigerator. Power demand during daytime fluctuates between 500 and

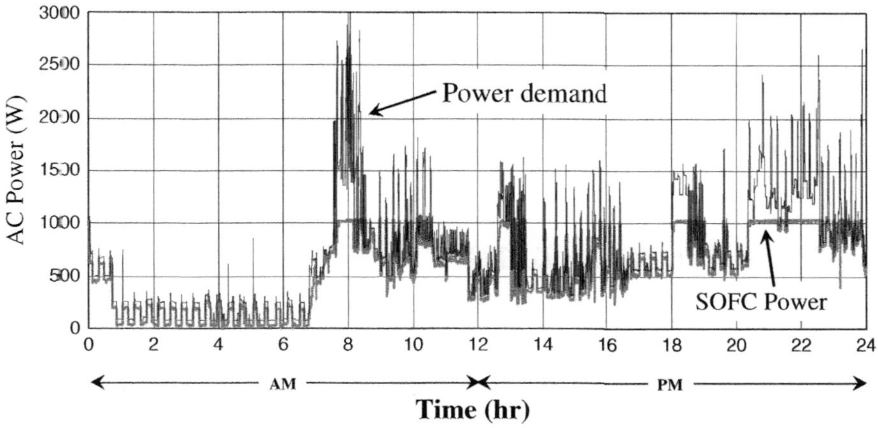

Fig. 2.11 One day's AC power trend, "Family power demand vs. SOFC power," in the automatic operation mode following the family power demand (under 1 kW) [11]

2500 W. The SOFC system supplies 1 kW, and the rest of the power demand is supplied from grid electricity. The characteristic load-following feature of the cogeneration system is shown in Fig. 2.11. These tests confirm that the system efficiency for a residential house is 43%–48% LHV as an average value for the 24-h test period.

2.3.2 Mitsubishi Materials Corporation

Mitsubishi Materials Corporation has developed the SOFC stack/system by using the LSGMC electrolyte. As described earlier, one of the merits of using LSGMC is its high oxide ion conductivity. Although the activation energy of the oxide ionic conductivity of LSGM becomes large with decreasing temperature, and hence the ionic conductivity drops rapidly at lower temperatures, this behavior is improved by Co doping of LSGM. Although the electronic conductivity is also increased by Co doping, the total benefit can be expected from analyses for the deviation from Gibbs energy-based efficiency given in Figs. 2.3 and 2.4 [14].

The high conductivity of LSGMC makes it possible to fabricate self-supported cells and operate them at intermediate temperatures; this in turn makes it easy to fabricate cathode and anode on the LSGMC electrolyte. A sealless stacking method is adopted; electrochemical cells are stacked with metal interconnects using current collectors for cathode and anode sides. Because of this simple configuration and materials, their stack performance is similar to the sum of respective cell performance. In other words, loss in stacking is very small.

As described in a separate chapter, the performance of their 1-, 3-, and 10-kW systems operated around 800°C is quite remarkable, and the systems exhibit high conversion efficiencies. Typical stack efficiency is more than 50% HHV [38]. This result is consistent with the arguments made in Section 2.1 that a strong impact can be expected from the increase in ionic conductivity.

2.3.3 Micro SOFCs by TOTO

TOTO attempted to fabricate micro tubes in various types. Eventually, they adopted the anode-supported tubular cells with the LSGM electrolyte. The adoption of anode-supported cells much thinner LSGM electrolyte can be used, leading to higher benefits in performance. On the other hand, technological difficulty in the fabrication process becomes more visible to avoid inter-diffusion between cell components. Details are also described in a separate chapter of this book.

2.4 Perspective

2.4.1 Applications

The development of SOFC systems is governed first by materials selection. However, breakthrough by WHPC on the sealless tubular cells indicated that not only materials selection, but also materials processing techniques together with stack design, are inevitably not separable and should be considered together from the early stages of development. In this sense, the main application of WHCP stacks was a stationary power generator for a few hundred kilowatts.

Recent achievement by Kyocera reminds us another aspect of applications: "For what purpose are the SOFC systems applied?" This question is the most important point of the SOFC development. They started with the concept of applications to the residential houses and then selected the operation temperature, plausible cost, and lifetime. On the basis of such system requirements, they started to construct their stacks.

One of their most important achievements is that they demonstrated that a small system can be fabricated and operated as an efficient energy conversion device. There was an argument about the self-thermal sustainability of small SOFC stacks, and it was thought that 1 kW is not sufficient to achieve self-thermal sustainability as well as high conversion efficiency simultaneously. Kyocera has demonstrated that this argument should be made by considering temperature as a parameter, and lowering the operating temperature makes it possible to construct a small but efficient SOFC system. Another impact of the Kyocera system is that they apply the SOFC system to residential houses by adopting appropriate load-following mode. In gas turbine systems connected to grid electricity, the daily start-and-stop (DSS) operation mode is common, so that whether this DSS operation mode can be applied was thought to be a basic criterion for adaptability of SOFC systems to the electricity market, particularly in Japan. The Kyocera system demonstrates that the DSS operation mode is not necessarily required. Instead, an effective load-following feature can be sufficient for providing power in low-demand periods. In view of this feature, the Kyocera system expands the applicability of SOFC systems not only for nearly steady-state but also for highly transient applications.

In Figs. 2.12 and 2.13, some characteristic features of SOFC stacks and SOFC systems are compared. In Fig. 2.12, the volumetric power density is plotted as a function of stack power. In this evaluation, the gas manifold parts are excluded. It is apparent that high volumetric power density can be achieved in a rather small stack. For tubular stacks aiming at larger systems,

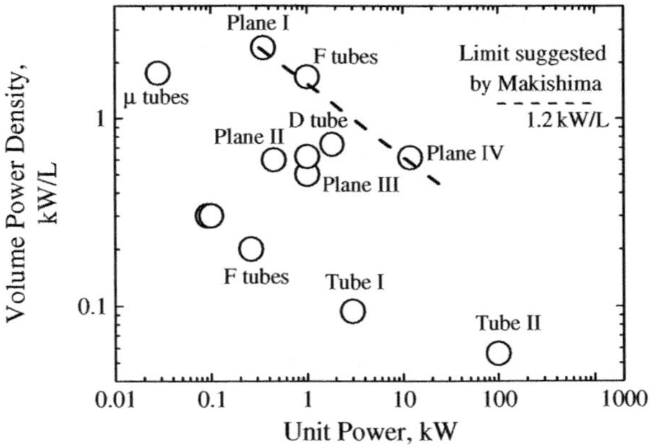

Fig. 2.12 Comparison of volumetric power density of core stack portion among recently developed SOFC stacks as a function of stack power. A limit of 1.2 kW/l is suggested by Makishima for large continuous chemical reactors such as iron blast furnaces

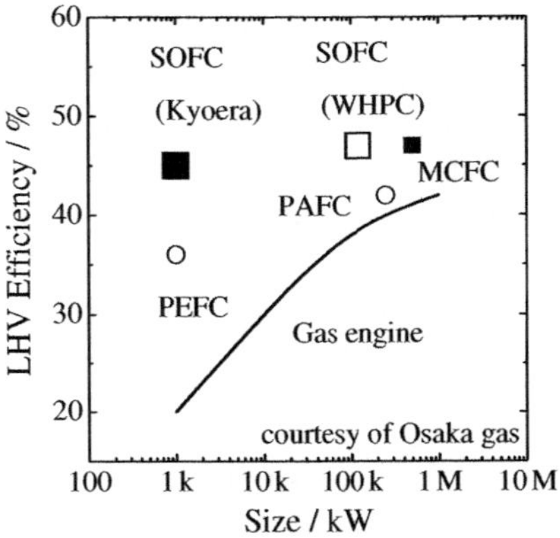

Fig. 2.13 Comparison in efficiency among solid oxide fuel cells, polymer electrolyte fuel cells and gas engines as a function of size of generators [Courtesy of Osaka Gas]

the volumetric power density is rather low. However, it should be noted here that the inner portions of tubes can be regarded as paths for introducing air or fuel. For planar stacks, with increasing stack power size, the cell-to-cell distance should be widened to keep the transfer of fuel and air. In Fig. 2.12, a limiting value of 1.2 kW/l is provided for comparison. This limit was proposed by Makishima through considerations on the pattern dynamics on chemical reactors to be operated continuously. For example, the energy density of an iron blast furnace is given in this magnitude. This limiting value tells us that not only the reaction itself but also transport of chemical species to such reaction sites or from those sites is quite important to maintain the continuous chemical reactors. When reactions and mass transfer in fuel cells are compared with the value, the following features should be taken into account:

1. All electrochemical reaction sites are distributed in a two-dimensional manner.
2. All electrochemical reactions and mass transfer phenomena are combined by the electrical chemical path having a three-dimensional network over the entire volume through atomistic reactions. Since electron mobility in metals is much faster than that in ionic species, electron movement in the fuel cell systems leads to a situation where any given reaction site can be connected with the next through electron transport. As a result, the current density distribution is governed by this kind of connection. This is an important difference between fuel cell reactors and normal chemical reactors in which the atomic transport is governed by corresponding reaction rates and mobilities.

These features make it more difficult to construct and operate larger SOFC stacks.

In Fig. 2.13, comparison is made in the system efficiency among various electricity generating systems as a function of system output power. Gas engine systems currently have some market share so that these systems can be regarded as plausible competitors of the fuel cell systems. The most important feature of gas engines is that the efficiency exhibits a strong size dependence. That is, the efficiency of a nearly 1-kW system is as small as 20%, whereas that of a 100-kW system is nearly 40%. Note that the earlier 100-kW SOFC system provided 47% LHV efficiency during a steady-state operation. Because gas engines have the merit of low cost, the cost competition will be quite severe. On the other hand, in a kilowatt size range, difference in efficiency is quite large so that the SOFC systems may have certain merits compared to the inexpensive gas engine systems and also against the PEFC system, which exhibits an efficiency of 32%. In this sense, a 1-kW SOFC cogeneration system can open a new era of development and demonstration of stationary SOFCs.

In 1998, BMW and Delphi proposed to use SOFCs as auxiliary power units (APU). For this purpose, severe requirements such as low volumetric and/or mass-specific power density, rapid start-up times, and stability for frequent thermal cycles should be met. For this purpose, SOFC stacks with novel designs, processing techniques, and materials will be necessary.

2.4.2 Fuel Flexibility and Reliability in Relationship to Intermediate-Temperature SOFCs

Fuels are important in fuel cell systems. Particularly, hydrocarbon fuels such as natural gas are the most important fuels. Technological issues associated with hydrocarbon fuels are carbon deposition and sulfur poisoning on nickel anodes. Both factors exhibit temperature dependence:

1. Carbon deposition is a result of competition among thermal decomposition reactions of hydrocarbons, reforming reaction with water vapor and electrochemical reactions. By lowering temperature below the decomposition temperature, carbon deposition due to the decomposition ceases, whereas carbon deposition takes place when CO becomes thermodynamically unstable at low temperatures. In this sense, carbon deposition region depends on temperature, composition, and pressure.
2. Sulfur poisoning is particularly important when lowering temperature. The interaction of nickel with sulfur is strong at lower temperatures; that is, solubility of sulfur in Ni increases with lowering temperature and then nickel sulfides can be formed. Actually, Ni cermet anode performance decreases with lowering temperature in the presence of hydrogen sulfide.

Similarly, for certain degradation mechanisms, temperature becomes a dominant factor. Particularly, cathode materials have the tendency of being more reactive to CO_2, $CrO_3(g)$, etc. In view of this, in the intermediate-temperature SOFCs, it becomes important to know degradation mechanisms associated with materials utilized in such SOFCs.

2.4.3 Hybrid Systems

Recent achievements in the development of intermediate-temperature SOFCs will open a new era even for the hybrid systems combined with gas turbines or other engines. When hybrid systems are designed on the basis of gas turbines, systems will be large and operation temperature will be higher to promote higher efficiencies in the gas turbine side. On the other hand, when a hybrid system is built mainly on fuel cells, smaller size and lower operation temperature will become attractive. Although gas turbine technology has matured, fuel cell technology has just started to evolve so that there is a room for small and low-temperature fuel cells with increased efficiency. Thus, appropriate size and operation temperature for a plausible hybrid systems is still an open issue and will depend on successful development of fuel cells in the near future. For the time being, a MW-size hybrid system is considered to be a typical target.

2.5 Summary

Solid oxide fuel cell technology has been reviewed from the point of view of lowering the operation temperatures. This process is closely related with the development of new materials, processing techniques, and stack designs. The most striking fact is that the recent achievement by Kyocera may indicate the start of a new era in the development and demonstration of SOFC systems for stationary applications.

References

1. S.C. Singhal, "Recent Progress in Tubular Solid Oxide Fuel Cell Technology," in Solid Oxide Fuel Cells V, U. Stimming et al. Ed., The Electrochemical Society, Inc., PV 97–40, pp. 37–50 (1997)
2. A.O. Isenberg, Solid State Ionics 3/4, 431 (1981)
3. D.C. Fee, S.A. Zwick, J.P. Ackerman, "Solid Oxide Fuel Cell Performance," in Proc. Conf. High Temperature Solid Oxide Electrolytes, "Anion Conductors," F.J. Salzano Ed., Brookhaven National Laboratory, Vol. 1, pp. 29–38 (1983)
4. H. Yokokawa, N. Sakai, "History of High Temperature Fuel Cell Development," Fuel Cell Handbook, Vol. 1, pp. 217–266 (2003)
5. S.C. Singhal, K. Kendall Ed. "High Temperature Solid Oxide Fuel Cells, Fundamentals, Design and Applications," Elsevier (2003)
6. N.Q. Minh, T. Takahashi, "Science and Technology of Ceramic Fuel Cells," Elsevier, Amsterdam (1995)
7. T. Ishihara, H. Matsuda, Y. Takita, J. Am. Chem. Soc. 116, 3801 (1994)
8. T. Ishihara, "Chapter 79. Novel electrolytes operating at 400–600 C," Fuel Cell Handbook Vol. 4, pp. 1109–1122 (2003)
9. S. Mukerjee et al. "Solid Oxide Fuel Cell Auxilliary Power Unit – A New Paradigm in Electric Supply for Transportation," Solid Oxide Fuel Cells VII(SOFC VII), H. Yokokawa and S.C. Singhal Eds., The Electrochemical Society, PV 2001-16, pp. 173–179 (2001)
10. D.C. Fee et al., "Monolithic Fuel Cell Development," Fuel Cell Seminar, Tucson, Arizona, pp. 40–43 (1986)
11. M. Suzuki, T. Sogi, K. Higaki, T. Ono, N. Takahashi, K. Shimazu, T. Shigehisa, "Development of SOFC Residential Cogeneration System at Osaka Gas and Kyocera," SOFC X, ECS Trans. 7(1) 27–30 (2007)
12. H. Yokokawa, "Recent Developments in Solid Oxide Fuel Cell Materials," Fuel Cells Fundam. Syst. 1(2), 1–15 (2001)
13. K. Hilpert, W.J. Quadakkers, L. Singheiser, "Chapter 74. Interconnects," Fuel Cell Handbook Vol. 4, pp. 1037–1054 (2003)
14. H. Yokokawa, N. Sakai, T. Horita, K. Yamaji, M.E. Brito, "Solid Oxide Electrolytes for High Temperature Fuel Cells," Electrochemistry 73, 20–30 (2005)
15. H. Yokokawa, T. Horita, "Chapter 5. Cathode", in High Temperature Solid Oxide Fuel Cells Fundamentals, Design and Application," S.C. Singhal and K. Kendall Eds., Elsevier, pp. 119–147 (2003)
16. S. Taniguchi, M. Kadowaki, H. Kawamura, T. Yasuo, Y. Akiyama, Y. Miyake, T. Saitoh, J. Power Sources 55, 73–79 (1995)
17. C.S. Tedmon, Jr., H.S. Spacil, S.P. Mitoff, J. Electrochem. Soc. 116, 1170 (1969)
18. Y. Matsuzaki, I. Yasuda, J. Electrochem. Soc. 148, A126 (2001)
19. H. Yokokawa, "Understanding Materials Compatibility," Annu. Rev. Mater. Res. 33, 581–610 (2003)

20. H. Yokokawa, T. Horita, N. Sakai, J. Yamaji, M.E. Brito, Y.P. Xiong, H. Kishimoto, "Thermodynamic Considerations on Cr Poisoning in SOFC Cathodes," Solid State Ionics 177, 3193–3198 (2006)
21. H. Yokokawa, H. Sakai, T. Horita, K. Yamaji, M.E. Brito, H. Kishimoto, "Thermodynamic and Kinetic Considerations on Degradations in Solid Oxide Fuel Cell Cathodes," J. Alloy Comp. 452, 41–47 (2008)
22. H Yokokawa, N. Sakai, T. Horita, K. Yamaji, Y.-P. Xiong, "Thermodynamic Correlation Among Defects in Ceria-Zirconia Solid Solutions," High Temperature Materials: A Symposium in Honor of the 65th Birthday of Professor Wayne L. Worell, The Electrochemical Soc. Inc., PV 2002-5,p.26–37, (2002)
23. H. Yokokawa, T. Horita, N. Sakai, K. Yamaji, M.E. Brito, Y.P. Xiong, H. Kishimoto, "Protons in Ceria and Their Roles in SOFC Electrode Reactions from Thermodynamic and SIMS Analyses," Solid State Ionics 174, 205–221 (2004)
24. H. Mitsuyasu, Y. Nonaka, K. Eguchi, "Analysis on Solid State Reaction at the Interface of Yttria-Doped Ceria/Yttria-Stabilized Zirconia," Solid State Ionics 113–115, 279–284 (1998)
25. N. Sakai, H. Kishimoto, K. Yamaji, T. Horita, M.E. Brito, H. Yokokawa, "Degradation Behavior at Interface of LSCF Cathodes and Rare Earth Doped Ceria," SOFC X, ECS Transactions, 7(1) 389–398 (2007)
26. K. Huang, M. Feng, J.B. Goodenough, C. Milliken, J. Electrochem. Soc. 144, 3620 (1997)
27. T. Ishihara, M. Honda, T. Shibayama, H. Minami, H. Nishiguchi, Y. Takita, "Intermediate Temperature Solid Oxide Fuel Cells Using a New $LaGaO_3$ Based Oxide Ion Conductor," J. Electrochem. Soc. 145(9), 3177–3183 (1998)
28. T. Horita, H. Kishimoto, K. Yamaji, N. Sakai, Y.P. Xiong, M.E. Brito, H. Yokokawa, "Active parts for CH_4 decomposition and electrochemical oxidation at metla/oxide interface by isotope labeling-secondary ion mass spectrometry," Solid State Ionics 177, 3179–3185 (2006)
29. Y. Matsuzaki, I. Yasuda, Solid State Ionics 132, 261–269 (2000)
30. G. Robert, A. Kaiser, E. Batawi, "Anode Substrate Design for Redox-Stable ASE Cells," Proc. 6th European Solid Oxide Fuel Cell Forum, European Fuel Cell Forum, Vol. 1, pp. 193–200 (2004)
31. H. Kishimoto, K. Yamaji, T. Horita, Y.-P. Xiong, N. Sakai, M. E. Brito, H. Yokokawa, "Reaction Process in the Ni-ScSZ Anode for Hydrocarbon Fueled SOFCs," J. Electrochem. Soc. 153(6), A982–A988 (2006)
32. K. Sasaki, K. Suzuki, A. Iyoshi, M. Uchimura, N. Imamura, H. Kusaba, Y. Teraoka, H. Fuchino, K. Tsujimoto, Y. Uchida, N. Jingo, "H_2S poisoning of Solid Oxide Fuel Cells," J. Electrochem. Soc. 153(11), A2023–A2029 (2006)
33. N. Sakai, K. Yamaji, T. Horita, H. Yokokawa, unpublished data
34. H. Yokokawa, T. Watanabe, A. Ueno, K. Hoshino, "Investigation on Degradation in Long-Term Operations of Four Different Stacks/Modules," in Solid Oxide Fuel Cells 10 (SOFC-X) K. Eguchi, S.C. Singhal, H. Yokokawa, J. Mizusaki Eds., ECS Transactions, 7(1), 133–140 (2007)
35. N. Sakai, H. Yokokawa, T. Horita, K. Yamaji, "Lanthanum Chromite-Based Interconnects as Key Materials for SOFC Stack Development," Int. J. Appl. Ceram. Technol. 1, 23–30 (2004)
36. N. Sakai, K. Yamaji, T. Horita, Y.P. Xiong, H. Yokokawa, "Rare-Earth Materials for Solid Oxide Fuel Cells (SOFC)," Handbook on the Physics and Chemistry of Rare Earths Vol. 35, K.A. Gschneidner, Jr., J.-C.G. Bünzli, V.K. Pecharsky Eds., pp. 1–43 (2005)
37. K. Ogasawara, H. Kameda, Y. Matsuzaki, T. Sakurai, T. Uehara, A. Toji, N. Sakai, K. Yamaji, T. Horita, H. Yokokawa, "Chemical Stability of Ferritic Alloy Interconnect for SOFCs," J. Electrochem. Soc. 154(7), B657–B663 (2007)
38. M. Shibata, N. Murakami, T. Akbay, H. Eto, K. Hosoi, H. Nakajima, J. Kano, F. Nishiwaki, T. Inagaki, S. Yamasaki, "Development of Intermediate-Temperature SOFC Modules and Systems," Solid Oxide Fuel Cells 10 (SOFC-X), ECS Transactions, 7(1), 77–83 (2007)

Chapter 3
Ionic Conduction in Perovskite-Type Compounds

Hiroyasu Iwahara

3.1 Introduction

A perovskite-type oxide typically expressed by ABO_3 is structurally stable because of its well-balanced geometrical array of constituent atoms and their valences, as described in Chapter 1, which means that the deviation from its strict stoichiometric composition is allowed to a considerable extent, keeping the original perovskite-type structure. Thus, a nonstoichiometric perovskite such as oxygen-deficient $ABO_{3-\delta}$, A-deficient $A_{1-\delta}BO_3$, or B-deficient $AB_{1-\delta}O_3$ often appears, where δ expresses the number of deficient atoms per unit formula. In the first case, an oxygen vacancy would be formed, and in the second and third cases deviation from stoichiometric composition (A:B = 1:1) would result in the formation of some lattice imperfections. Also, it is possible to partially substitute a foreign atom M for A or B in ABO_3 forming $A_{1-x}M_xBO_{3-\delta}$ or $AB1_{-x}M_xO_{3-\delta}$. If the valence of M is different from A or B, lattice defects would be formed to maintain the electrical neutrality of the crystal.

A perovskite structure is tolerant of a certain difference in size or valence of foreign atoms, and it forms various kinds of defect structure according to the kind of inserted atoms and the formation environment such as atmosphere and temperature. Not only the single perovskite type but also various kinds of derivatives, so-called perovskite-related compounds, form different kinds of lattice defects. They have also tolerance to accept foreign atoms as an impurity or to deviate from their stoichiometric composition, giving rise to some kinds of lattice defects.

Such an adaptable structure of the perovskite-type oxide suggests that some constituent ions in the crystal will be mobile from one site to another if the energy needed to overcome the barrier to jump from one site to the other is small. In fact, several kinds of ions have been found to be mobile in perovskite and perovskite-related compounds during the past four decades. Oxide ions

H. Iwahara (✉)
Nagoya University, Furo-cho, Chigusaku, Nagoya, 464-8601, Japan
e-mail: iwahara-h@m3.gyao.ne.jp

T. Ishihara (ed.), *Perovskite Oxide for Solid Oxide Fuel Cells*,
Fuel Cells and Hydrogen Energy, DOI 10.1007/978-0-387-77708-5_3,

and protons are the representative conducting ions in perovskite oxide. Also, the lithium ion is mobile in some perovskite-type oxides. In addition, several non-oxide perovskite compounds such as halides are known to be ionic conductors. Furthermore, some kinds of antiperovskite non-oxide compounds are good silver ion conductors. Regarding oxide ion conduction and proton conduction, detailed descriptions are given in Chapters 4 and 11, respectively. In this chapter, an outline of ionic conduction in perovskite compounds and a brief history of the research in this field are described, introducing some examples of the studies.

3.2 Conduction Behavior of Perovskite-Type Compounds

Ionic species that contribute to high conductivity of perovskite-type compounds are rather limited. They are listed in Table 3.1 with representative compounds, their conductivities, and distinctive features. The mobile ionic species, except hydrogen, are host components of the compounds, whereas protons are unique in that they are incorporated from water vapor or hydrogen gas in the ambient atmosphere at elevated temperature. Of these, oxide ion conductors are best known, and oxide ionic conduction in various kinds of perovskite and perovskite-related oxides has been studied.

Confirmation of ionic conduction in the electrically conductive oxides can be made in different ways. One of the most convenient methods is to examine the electromotive force (*emf*) of an electrochemical oxygen concentration cell:

$$\text{Pt, } O_2(P_{O2}(1))/\text{oxide specimen}/O_2(P_{O2}(2)), \text{ Pt cell} \quad [1]$$

using the oxide sample as an electrolyte diaphragm at elevated temperature. The concept of the oxygen concentration cell is schematically shown in Fig. 3.1.

Table 3.1 Examples of ionic conduction in perovskite-type compounds

Mobile ions	Examples of ionic conductor	Conductivity / Scm^{-1} (at°C)	Remarks
O^{2-}	$La_{0.9}Sr_{0.1}Ga_{0.8}Mg_{0.2}O_{2.85}$	$1.5x10^{-1}$ (800°C)	Doped single perovskite oxide
H^+	$SrCe_{0.95}Yb_{0.05}O_{3-\alpha}$	$1x10^{-2}$ (900°C)	Doped Single perovskite oxide under hydrogen-containing atmosphere
Li^+	$La_{0.51}Li_{0.34}TiO_{2.94}$	$1.4x10^{-3}$ (27°C)	Host oxide: $La_{2/3}TiO_3$ A-site deficient perovskite
Cl^-	$CsPbCl_3$	$1.2x10^{-3}$ (500°C)	Non-oxide perovskite
Br^-	$CsPbBr_3$	$8 x 10^{-4}$ (500°C)	Non-oxide perovskite
Ag^+	Ag_3SI	$1 x 10^{-2}$ (25°C)	Anti-perovskite-type structure. Non-oxide anti-perovskite Averaged structure for Ag

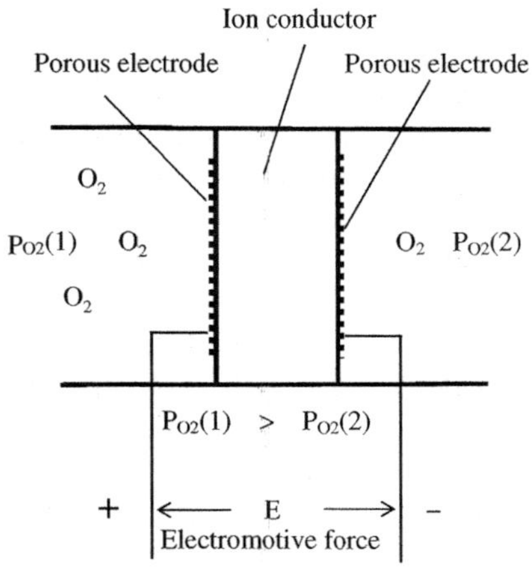

Fig. 3.1 Concept of an electrochemical oxygen concentration cell

If the observed emf is close to the theoretical value E_o calculated from Nernst's equation given by Eq. (3.1), the oxide can be regarded as an ionic conductor.

$$E_0 = \frac{RT}{4F} \ln \frac{P_{O2}(1)}{P_{O2}(2)} \tag{3.1}$$

If no emf is observed, the charge carriers in the oxide would be electrons or electron holes. When emf is not zero but smaller than E_o in Eq. (3.1), conduction in the oxide would be partially ionic and partially electronic. However, readers should note that this method does not give any information about which ions in the oxide are mobile and that the ratio of observed to theoretical emf, E/E_o, does not always give the correct value of ionic transport number. To determine which ions are mobile, additional experiments are necessary, such as electrochemical mass transport measurement or tracer technique.

Transport behavior of mobile ions in perovskite-type compounds has been investigated by many researchers using various experimental methods and computer modeling. It is acceptable that the transport mechanism of mobile host ions in the perovskite is based on hopping via vacant sites of the ionic species. In that case, tolerance factor $\boldsymbol{t}$ and free volume $\boldsymbol{v}$ of the crystal are thought to be important factors affecting the mobility of the ions. The tolerance factor is a measure of symmetry of the crystal structure of the perovskite and is written as follows:

$$t = \frac{r_A + r_X}{\sqrt{2}(r_B + r_X)} \tag{3.2}$$

where r_A, r_B, and r_X are ionic radii of A, B, and X ions, respectively, in ABX3. Empirically and qualitatively, good ionic conductors of perovskite compounds have a relatively large value of t. The free volume is a measure of degree of packing in a unit cell, which is given by the difference between measured unit cell volume and the sum of ionic volume in the unit cell calculated from known ionic radii. In many cases, the free volume of the good ionic conductors is large, although quantitative studies have not yet been completed. Of course, other factors such as the concentration of vacancy, ionic radius, and polarizability are also important for good ionic conduction. Detailed discussions are presented in the following chapters.

As described in the Introduction section, many perovskite-type compounds can deviate from their stoichiometric composition to a considerable extent because of the strong stability of the perovskite-type structure. This deviation results in the formation of electronic defects such as excess electrons or electron holes, which cause n-type or p-type electronic conduction. Thus, it should be noted that, in general, perovskite-type compounds are have facile electronic conduction, and that the ionic conduction is often accompanied by electronic conduction. As the concentration of the electronic defects depends on the deviation of A/B and/or (A + B)/X from their stoichiometric ratio, the content of impurity, atmosphere and temperature, and contribution of electronic conduction varies with those conditions.

In general, electronic conductivity of an oxide electrolyte at elevated temperature is influenced by partial pressure of oxygen, P_{O2}, in the atmosphere; n-type electronic conductivity σ_n increases with decreasing P_{O2}, whereas p-type electronic conductivity σ_p increases with increasing P_{O2}. It is known that P_{O2} dependence of σ_n or σ_p is given by the following:

$$\sigma_n = \sigma_n^o \exp P_{O2}^{-1/n} \tag{3.3}$$

and

$$\sigma_p = \sigma_p^o \exp P_{O2}^{1/n} \tag{3.4}$$

where n is some natural number, and σ_n° and σ_p° are the constants that are independent of the partial pressure of oxygen [1]. It is accepted that the ionic conductivity σ_i itself is independent of P_{O2} for many oxide electrolytes. Accordingly, the total conductivity σ is expressed as follows:

$$\sigma = \sigma_i + \sigma_n^o \exp P_{O2}^{-1/n} + \sigma_p^o \exp P_{O2}^{1/n} \tag{3.5}$$

The relationship between each logarithm of conductivity and logarithm of P_{O2} is illustrated in Fig. 3.2. The hatched regions indicate the mixed conduction domains, and between them there exists an ionic conduction domain where electronic conductivity is negligibly low. The outer sides of the mixed conduction domains are electronic conduction regions. An example of the experimental

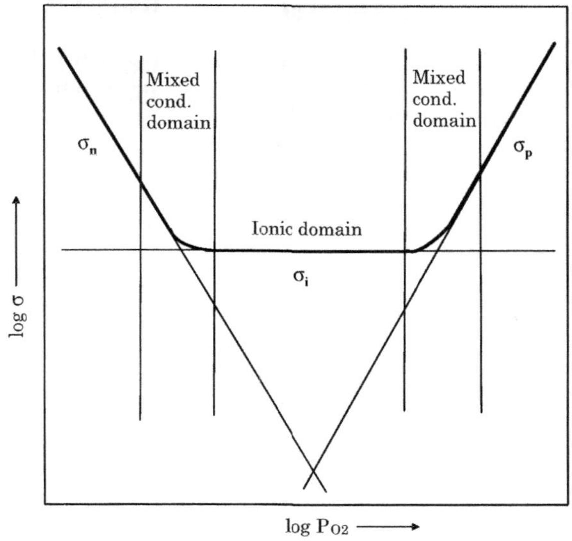

Fig. 3.2 Dependence of conductivity on partial pressure of oxygen

results on $\sigma - P_{O2}$ plots is shown in the following section for oxide ion–electron mixed conductors.

3.3 Early Studies on Ionic Conduction in Perovskite-Type Oxides

Most of the possible combinations of large A cations and smaller B ions, which is needed to form perovskite-type oxides ABO_3, had been tried by 1955, as described by F.S. Galasso in his famous book [2] entitled *Structure, Properties and Preparation of Perovskite-Type Compounds*, published in 1969. This book compiled almost all available data at that time concerning structure, properties, and preparation of perovskite-type compounds. In this book, although lattice defects in the perovskite-type crystal were described, the author did not touch on ionic conduction in the perovskite except for a very brief description of $BaTiO_3$. However, in the 1960s, several pioneering studies on ionic conduction in perovskite-type oxides were performed.

Ionic conduction in perovskite-type oxides was first a source of interest in ferroelectric materials. S. Swanson showed that DC conductivity of $BaTiO_3$ ceramics was significantly influenced by their fabrication history, which suggests that there would be an intimate relationship between the solid-state reactions of raw materials and ionic conduction [3]. In the 1960s, when research and development of perovskite-type oxides as a dielectric or ferroelectric material such as $BaTiO_3$ and $PbTi_{1-x}Zr_xO_3$ had become active, some of the researchers paid attention to the conduction behavior of these perovskite-type oxides. They considered that ionic conduction in the ferroelectric materials would be affected by their manufacturing and the characteristics of pyroelectric properties [4, 5].

In 1961, Stephenson and Flanagan thought that the anomalous pyroelectric behavior revealed in lead zirconate titanate (PZT) would probably be caused by ionic conduction in the oxides [5]. To test the existence of ionic conduction in this oxide, they adopted the electrochemical oxygen concentration cell method that had been reported by Kuikkola and Wagner [6] 4 years previously for stabilized zirconias and which subsequently became a very familiar method to solid-state electrochemists.

Stephenson and Flanagan constructed the concentration cell using PZT as a diaphragm of two oxygen electrode chambers.

$$Pt,\ Pb,\ PbO/PbZ_{r0.53}Ti_{0.47}O_3/Cu,\ Cu_2O/Pt\ \text{Cell} \qquad [2]$$

The electromotive force (emf) of this cell in purified nitrogen gas has been measured. The observed emf was 0.205 ± 0.005 V at 250°C, which is close to the theoretical value calculated from Nernst's equation given in Eq. (3.1), where $P_{O2}(1)$ and $P_{O2}(2)$ were thermodynamic oxygen partial pressures in each gas chamber. This result means that conduction was purely ionic, as described in the previous section. The authors reported that there was no conclusive evidence as to whether the ionic conduction is caused by a cation or an anion migration. However, from the observed cell behavior, the authors considered that at least some of the conduction is contributed by oxygen ion migration.

Heckman et al. [7] studied the conduction properties of polycrystalline lead zirconate titanates $Pb(Zr_xTi_{1-x})O_3$ + 1 w% Nb_2O_5, constructing electrochemical oxygen concentration cells

$$Pt\ (1.0\ \text{atm}\ O_2)\ /\ PTZ\ /\ Pt\ (0.01\ \text{atm}\ O_2)\text{Cell} \qquad [3]$$

in the temperature range from room temperature (r.t.) to 600°C, and they published the results 2 years after Stephenson's report. The cells showed a stable emf, suggesting that the specimens exhibit ionic conduction. However, in contrast to Stephenson's report that the conduction was entirely ionic, the results indicated that the conduction was partly assigned to an ionic state and partly an electronic one, which is dependent on temperature and the composition of specimens. The difference in the ionic contribution in the conduction between the two experiments would be caused by the differences in the sample composition and the condition of oxygen concentration cells. Stephenson's cell [2] had a lower average partial pressure of oxygen than that of Heckman's cell [3]. In any case, the absolute values of ionic conductivity are low; the specific resistivity of 1 w% Nb_2O_5-contained $Pb(Zr_xTi_{1-x})O_3$ is reported to be 5×10^8 to 1×10^{10} ohm cm at 300°C.

Subsequently, Ezis et al. studied the dependence of the transport number of ions on composition x in $Pb(Zr_xTi_{1-x})O_3$ in detail using an electrochemical oxygen concentration cell [8]. Their experiment showed that the transport number decreased with increasing value of x over the range $0.05 < x < 0.35$ and that electronic conduction became dominant above $x = 0.40$. Ezis et al. considered that the reduction of T_i to the trivalent state might be a mechanism for defect formation, which tended to counter the effects of Nb^{5+} included in all specimens investigated.

In addition to the studies on PZT, Glower et al. examined the ionic conduction in a ferroelectric material ,$Ca_{0.1}Ba_{0.9}TiO_3$, by means of the same method as that used for PTZ [9]. They showed that the conduction in this oxide is ionic below 300°C, but electronic conduction appears sharply near 300°C and becomes predominant above 500°C. They identified the mobile ions as calcium ions by using so-called activation analysis, which was a kind of irradiation analysis of the surface of the solid after passing DC current across the solid conductor.

An important study of charge carrier in a typical ferroelectric perovskite-type oxide $BaTiO_3$ was published in 1964 by Glower and Heckman [10]. The title of the paper "Conduction – Ionic or Electronic in $BaTiO_3$" was attractive for materials researchers in those days. In their paper, they reported about conduction species in barium titanate as follows: "Usual practice of authors has been to make the tacit assumption that conduction is exclusively electronic. There is no priori assurance that this is true." Also, they wrote " In fact, to our knowledge, no one has to date addressed himself to the primary problem of transport, i.e., does conduction occur via the motion of electrons or of ions?" The purpose of their study was to answer this essential question from experimental results. They applied Cell [1] to $BaTiO_3$ specimens for different oxygen partial pressures p_1 and p_2:

$$\text{Pt, } O_2(p_1) \text{ / } BaTiO_3 \text{ specimen / } O_2(P_2), \text{ Pt Cell} \qquad [4]$$

The cell using a pure single crystal showed no significant emf in any cell examined, indicating that its conduction was not ionic but electronic. On the other hand, the single crystal containing 0.1 at% of iron showed strong emf dependency on temperature and oxygen partial pressures at two electrodes. The values were less than the theoretical one calculated from Nersnt's equation, suggesting that the conduction was partially ionic. $BaTiO_3$ showed theoretical emf below 250°C, indicating that the conduction was purely ionic but at 540°C, emf of the concentration cell was far lower than the theoretical one, suggesting that contribution of ionic conduction to total one is rather small. Thus, it was clarified that the contribution of ionic conduction in $BaTiO_3$ depends on its purity and morphology, single crystal or polycrystal ceramic. The conduction in a pure single crystal is electronic and that in impure single crystal and polycrystalline ceramic is partially ionic, but the ionic contribution decreases with increasing temperature.

However, gas concentration cell experiments give essentially no information about which kind of ion among the constituent ions is mobile in the solids. To investigate this, Glower and Heckman have also applied the activation analysis method to the surface of an iron-containing single crystal of $BaTiO_3$ before and after applying DC voltage of 22 V/mil for 52 h. The result showed that the iron ions were mobile and contributed to ionic conduction but that titanium was not mobile.

These studies were not intended to seek a good ionic conductor but to confirm the conduction species to clarify the phenomena characteristic to the ferroelectric or pyroelectric materials. It should be noted that the conductivity of ferroelectric materials mentioned above is very low and most of the researchers in those days took no notice of the value of conductivity itself. Studies on highly conductive ionic conductors of perovskite-type compounds were started in the second half of the 1960s to search for a good oxide ion-conducting electrolyte for fuel cells and oxygen sensors. These are described in the following sections.

3.4 Oxide Ion Conduction

When a cation A or B in a perovskite-type oxide ABO_3 is partially replaced by a cation M of lower valence, it sometimes gives rise to a relatively large number of oxygen ion vacancies in the lattice so as to maintain the electrical neutrality of the crystal. Chemical composition of the oxide can be expressed as $A_{1-x}M_xBO_{3-\alpha}$ or $AB_{1-x}M_xO_{3-\alpha}$, where α is an average number of oxygen deficiencies per unit formula. In such crystals, appreciable oxide ion conduction may be expected at elevated temperatures if the energy needed for the oxygen ions to jump from their original sites to adjacent vacant sites is not so high. In this case, similarly to the case of stabilized zirconia, oxide ions can migrate through the crystal lattice with the assistance of oxide ion vacancies.

There are several methods to confirm oxide ion conduction in the oxide specimens. One of the most convenient methods is to examine the discharge performance of the oxygen concentration cell. If the conduction in the sample is oxygen ionic, a steady and stable current with a reasonable value can be taken from the oxygen concentration cell, which is composed of the specimen oxide as a solid electrolyte. Therefore, for example, one can regard the conduction as the oxide ion conduction, if the oxygen concentration cell shown in Cell [1] gives rise to stable emf and a steady and reasonable current can be taken from the cell.

Historically, oxygen ion conduction in the perovskite-type oxide was reported for the first time by Stephenson and Flanagan [4], as described in the previous section (3.2). They tried to construct a fuel cell using modification of oxygen concentration cell shown in Cell [3]. In the experiment, PZT was used as an electrolyte, and hydrogen gas and oxygen gas were supplied to each side of the electrolyte separately. They observed a steady terminal voltage of 0.81 V at 325°C and 0.41 V at 700°C. The current output was estimated to be a few mA/cm^2. The authors wrote as follows: "These results would indicate that at least some of the conduction is by oxygen ion migration." This study would historically be the first experiments to confirm oxygen ion conduction in perovskite-type oxide and be the first experiment of SOFC using a perovskite-type oxide as a solid electrolyte, although their intention was not to develop a fuel cell and its solid electrolyte.

The research for developing the good ionic conductors with perovskite-type structure was first considered by van Gool, who was known as the first to propose a one-chamber solid oxide fuel cell. In 1965, he published a paper about one-chamber fuel cells entitled "The possible use of surface migration in fuel cells and heterogeneous catalysis" [11]. In this paper, he touched on the oxygen-deficient perovskite-type oxides as a candidate for an oxide ion conductor applicable to a fuel cell electrolyte. However, he thought that the perovskite structure seemed to be less favorable because A and O in ABO_3 would make a closed packing structure in which the ion migration might be difficult.

Studies on highly conductive ionic conductors of perovskite-type compounds were started in the second half of the 1960s to find a superior electrolyte for fuel cells and sensors. From the analogy of oxygen ion conduction in fluorite-type oxides such as stabilized zirconias, it was thought that considerable concentration of oxygen vacancy would be essential for high oxygen ion conductivity. The present author and coworker have paid attention to the solid solution based on $LaAlO_3$ that is composed of large-sized trivalent cation La and a small-sized trivalent cation Al. In this oxide, calcium ions are partially substituted for lanthanum ions and, as a result, oxygen ion vacancies are formed to compensate charge neutrality in the crystal [12]; i.e., the composition is expressed as $La_{1-x}Ca_xAlO_{3-\alpha}$. Having studied the behaviors of oxygen concentration cells and fuel cells with $La_{1-x}Ca_xAlO_{3-\alpha}$ ($x = 0.1, 0.2$, and 0.3) ceramics as a solid electrolyte, they confirmed that the conduction was partly oxygen ionic and partly electronic (due to electron holes) in air at elevated temperatures and that, under the fuel cell condition, the conduction is predominantly oxide ionic [13]. The $CaTiO_3$ can take aluminum to form a solid solution $CaTi_{1-x}Al_xO_{3-\alpha}$ ($x \leq 0.5$) in which almost stoichiometric amounts of oxygen vacancies are generated [14]. It was confirmed that this solid solution exhibits conduction behavior similar to that of $La_{1-x}Ca_xAlO_{3-\alpha}$ [15], and that the oxide ion conductivity is rather higher than that of the latter. These studies were reported in a Japanese journal in 1967 and 1979, and these results were summarized in English and published in 1971 [16]. Steele et al. also reported oxygen ionic conduction in $CaTiO_3$-based ceramic [17].

$CaTiO_3$ is a typical 2:4-type perovskite composed of large-sized divalent cation Ca and small-sized tetravalent cation Ti, whereas the aforementioned $LaAlO_3$ is a typical 3:3 type composed of large trivalent cation La and small trivalent cation Al. The excellent high oxide ion conductor that was discovered by Ishihara and reported in 1994 [18] is also based on 3:3-type perovskite $LaGaO_3$, as described in detail in Chapter 4.

Oxide ion conductors with the perovskite structure mentioned above belong to so-called single perovskites, which can be expressed as the simple form, ABO_3. Besides these, there are different types of perovskite-related oxides and some of them are known to show oxide ion conduction. One of them is Brownmillerite, $Ba_2In_2O_5$. This composition can be written as $BaInO_{2.5}$, and it is regarded that the ½O per unit formula of a perovskite is deficient; i.e., it can be written as $BaInO_{2.5}\square_{0.5}$, where □expresses oxygen vacancy. At temperature lower than 930°C, the arrangement of oxygen vacant sites is ordered [19]. Goodenough et al. found that the oxide ion conductivity in this Brownmillerite oxide jumps up by more than one order of magnitude above 930°C, as shown in Fig. 3.3 [20]. Above this temperature, the arrangement of oxide ion vacancies becomes disordered, and oxide ions can move easily in assistance of disordered vacancies.

A characteristic feature of perovskite-type oxide ion conductors is that they are often accompanied with p-type electronic conduction under an oxidizing atmosphere such as air at elevated temperatures. As described in Section 3.2, the contribution of electronic conduction depends on P_{O2} in the atmosphere and temperature. As a typical example, Fig. 3.4 shows the P_{O2} dependence of conductivities of $CaTiO_3$- and $SrTiO_3$-based solid solutions at 800°C [16]. P-type electronic conduction appears in the region of high P_{O2} and n-type one under low oxygen partial pressure, i.e., a reducing atmosphere. The shape of the curve $\ln\sigma \sim \ln P_{O2}$ is essentially the same as that shown schematically in Fig. 3.1. In many fluorite-type oxide ion conductors such as stabilized zirconias and

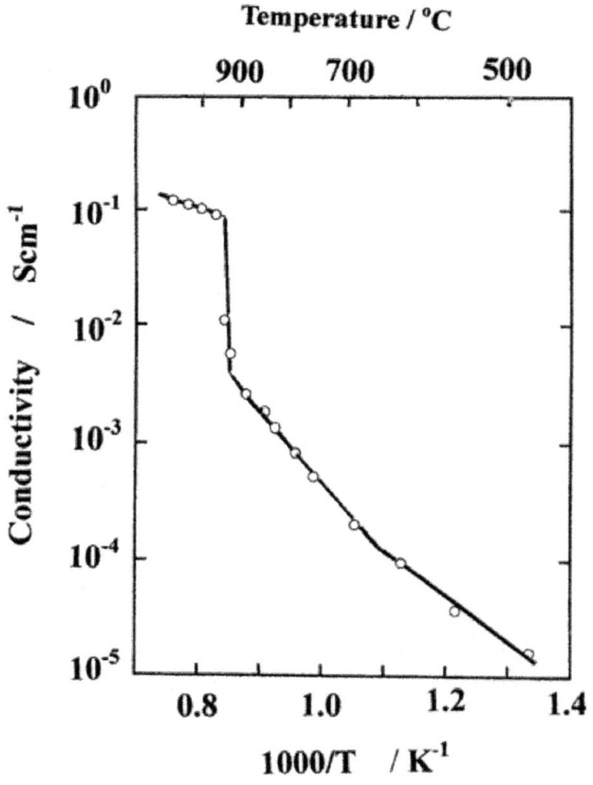

Fig. 3.3 Conductivity of $Ba_2In_2O_5$ as a function of temperature [20]

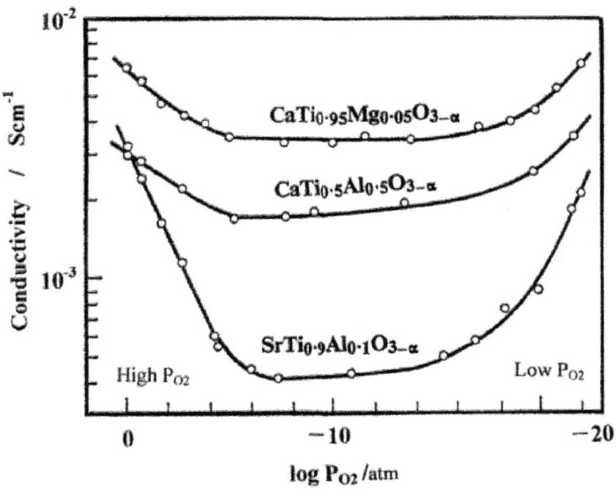

Fig. 3.4 Dependence of conductivities of $CaTiO_3$-based and $SrTiO_3$-based solid solutions on oxygen partial pressure at 800°C [16]

doped cerias, p-type electronic conduction is rarely observed under oxidizing atmosphere even under $p(O_2) = 1$ atm, whereas many perovskite-type oxide ion conductors become the mixed conductors ($O^{2-} + h^+$) under oxidizing atmosphere at elevated temperature. Hole conduction arises from the defect equilibrium with oxygen in gas phase:

$$V_O + 1/2\ O_2 \overset{k_1}{\rightleftarrows} O_O^x + 2\ h^{\cdot} \tag{3.6}$$

In many perovskite-type oxide ion conductors, the equilibrium constant K_I is so large that hole conduction appears even at a relatively weak oxidizing atmosphere such as air, whereas in the case of fluorite-type oxides such as stabilized zirconias, the equilibrium constant K_I is too small to cause p-type electronic conduction in air at elevated temperatures.

Some perovskite-type oxides having transition elements at B sites exhibit mixed conduction at elevated temperatures. A typical example is doped lanthanum cobaltite, in which oxide ions and holes are charge carriers. The electronic conductivity is a few orders of magnitudes higher than that of the oxide-ionic although the ionic conductivity itself is sufficiently high (>10^{-1} S cm^{-1} at several 100°C). This kind of mixed conductor is a promising candidate for the electrode materials of SOFCs and ceramic membrane reactors and is described in Chapters 7 and 8 in detail.

3.5 Proton Conduction

Some perovskite-type oxides exhibit proton conduction under hydrogen-containing atmosphere at elevated temperatures. Cerates or zirconates of alkaline earth elements in which some trivalent cations are partially substituted for cerium or zirconium show such proton-conducting behavior. A typical example is $SrCe_{0.95}Yb_{0.05}O_{3-\alpha}$, a substituted solid solution based on $SrCeO_3$ in which 5% of Ce are replaced by Yb (α is number of oxygen deficiency per unit formula, which depends on concentration of Yb, oxygen partial pressure, and temperature). This oxide has the orthorhombic crystal structure of the perovskite. This material exhibits p-type electronic conduction in dry air free from water vapor. However, if water vapor or hydrogen is introduced into the atmosphere surrounding the ceramic at elevated temperatures, its electronic conduction decreases and protonic conduction appears. When the ceramic is exposed to hydrogen gas, it becomes an almost pure protonic conductor, the conductivity of which is 10^{-3}–10^{-2} S cm^{-1} at 600°–900°C. The proton conduction in these oxides was directly verified by means of electrochemical transport of hydrogen across the oxides [21–23].

This type of proton-conducting cerates was found by the present author's group and the first report was published in 1981 [21] after the preliminary studies on protonic conduction in some oxides, of which conductivities were far lower compared with that of the cerates [24]. Although the existence of hydrogen in the form of protons had been observed in glassy SiO_2 [25], stabilized ZrO_2 [26], $LaAlO_3$ [27], and ThO_2-based solid solutions [28] under hydrogen-containing atmosphere at elevated temperatures, the protonic conduction in these oxides has not been directly confirmed by experiments, probably because of their low conductivity.

Other perovskite-type oxides based on $SrCeO_3$ or $BaCeO_3$, in which trivalent cations are partially substituted to cerium position, are also protonic conductors under the same condition as above [21, 29–31]. The general formula is written as $SrCe_{1-x}M_xO_{3-\alpha}$ or $BaCe_{1-x}M_xO_{3-\alpha}$, where M is some rare earth element, x is less than its upper limit of solid solution formation range (usually less than 0.2), and α is the oxygen deficiency per unit formula, which depends on the concentration of dopant M and surrounding atmosphere. Their conductivities in hydrogen are of the order of 10^{-1} to 10^{-3} S cm^{-1} at 1000°–600°C, as shown in Fig. 3.5.

These oxides are unique solid electrolytes in respect to the fact that they have no host constituents which liberate conducting ions (protons). The oxides take protons from water vapor or hydrogen molecules in ambient gas as a result of equilibria with the defects in the oxide lattice at elevated temperatures. In these oxides, doping by aliovalent cations is indispensable for the appearance of proton conduction. It seems that electron holes and oxygen vacancies formed by doping might play an important role in the formation of protons. For example, substitution of Yb^{3+} for Ce^{4+} in $SrCeO_3$ will provide oxygen vacancies $V^{\cdot\cdot}$ as a result of charge compensation in the stoichiometric condition, and the oxygen vacancies are in equilibrium with the atmosphere at elevated temperature.

The studies of the conductivity as a function of the dopant content or partial pressures of water vapor and oxygen have shown that the following three equilibria are simultaneously established between the defects in the oxide and the atmosphere [32, 33].

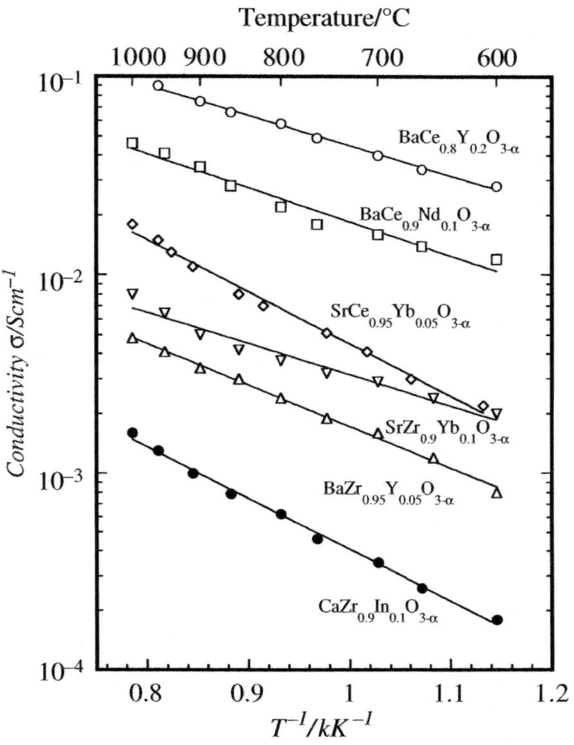

Fig. 3.5 Conductivities of typical proton conducting perovskite-type oxides under H_2-containing atmosphere

$$V_O + 1/2\,O_2 \overset{k_1}{\rightleftarrows} O_O{}^x + 2\,H \tag{3.6}$$

$$H_2O + 2\,H^{\cdot} \overset{k_2}{\rightleftarrows} 2\,H^{\cdot} + 1/2\,O_2 \tag{3.7}$$

$$H_2O + V_O \overset{k_3}{\rightleftarrows} 2\,H^{\cdot} + O_O{}^x \tag{3.8}$$

where, V_O, $O_O{}^x$, H, h˙, and K represent oxygen vacancy, oxide ion at normal lattice site, proton, hole, and equilibrium constant, respectively, and the relationship between the equilibrium constants of each equation is expressed as follows:

$$K_3 = K_1 K_2. \tag{3.9}$$

These equilibria could be qualitatively confirmed by the different kinds of experiments [33, 34]. The protons thus formed is considered to be hydrogen bonding with oxygen ions in the lattice and sometimes the proton is written as

$OH^{\cdot}$ instead of $H^{\cdot}$ in the equilibrium Eqs. (7) and (8). However, the readers should note that it is not $OH^{\cdot}$ but $H^{\cdot}$ that can be mobile.

After the discovery of $SrCeO_3$-based protonic conductors, $KTaO_3$-based oxides [35] and Y_2O_3 ceramic [36] were reported to have protonic conduction at high temperatures, although the conductivities were not as high as those of the cerate-based perovskite-type oxide ceramics. Some doped zirconates based on $CaZrO_3$, $SrZrO_3$, or $BaZrO_3$ [37, 38] and scandates based on $LnScO_3$ (Ln = La, Nd, Sm, or Gd) [39] were also confirmed to exhibit the same behavior as the cerates, although their conductivities were rather low, as shown in Fig. 3.5. Among the oxides described above, the $BaCeO_3$-based oxide shows the highest conductivity. However, the contribution of oxide ions on the conduction increases markedly as the temperature is raised [40]. Although the conductivity of the $SrCeO_3$-based ceramic is rather low, the transport number of protons is higher than that of the $BaCeO_3$-based one. The conductivities of zirconate-based ceramics are lower than those of the cerates, but they are superior proton conductors from the aspect of chemical and mechanical strength. The cerates dissolve easily in the strong acids. For example, $SrCeO_3$-based ceramics dissolve into a concentrated hydrochloric acid liberating chlorine gas. In contrast, zirconates hardly react with acid solution, and they are stable in CO_2 atmosphere, which reacts with cerate ceramics below 800°C to form carbonates [41]. Later, bulk conductivity of the $BaZrO_3$-based solid solution was reported to be as high as that of the $BaCeO_3$-based solid solution [42], but such high conductivity has not been observed in its sintered body, probably because of high resistivity of the grain boundaries.

Besides single perovskites, some complex perovskites are known to exhibit proton conduction. Nowick et al. have reported a series of new protonic conductors [43] in $A_2(B'B'')O_6$ and $A_3(B'B'_2)O_9$ in which the charge of A ions is always 2+ and the B′ and B″ ions have charges 3+ and 5+ in the former case and 2+ and 5+ in the latter. Protonic conduction occurs when the composition of cations is slightly shifted from the stoichiometric one. As an example, $Ba_3(Ca_{1.18}Nb1._{1.82})O_{9-\alpha}$, which is derived from $Ba_3(CaNb_2)O_9$, exhibits a conductivity as high as that of $BaCeO_3$-based ceramics.

Some of the perovskite-related oxides were also reported to show a protonic conduction under similar conditions as above. Indium- or magnesium-doped Sr_2TiO_4 ceramics whose crystal structure belongs to K_2NiF_4 type (a kind of layered perovskite type) show protonic conduction under a hydrogen-containing atmosphere at elevated temperature, although appreciable electronic conduction accompanies this [44]. Similar behavior was observed in $Sr_3Ti_2O_7$-based ceramic, although the conductivity is low compared to that of Sr_2TiO_4-based solid solution [45].

In Table 3.2, typical examples of high temperature-type proton-conducting oxides and their distinct features are listed classifying them as their crystal structure. It is still unclear what is essential for good protonic conduction in this class of oxides. However, empirical requirements for good protonic conductors of this type seem to be (1) high basicity of the constituent cation

Table 3.2 Typical examples of host oxides for proton conducting perovskites and their distinctive features

Crystal structure	Host oxides	Distinctive features
Single perovskites	$SrCeO_3$	High proton transport number Pure protonic conductor under H_2 below 800°C Degradation by CO_2
	$BaCeO_3$	High protonic conductivity Considerable contribution of oxide ion cond'n Marked degradation by CO_2
	$CaZrO_3$	Chemically stable compared to $SrCeO_3$
	$SrZrO_3$	Mechanically strong compared to $SrCeO_3$
	$BaZrO_3$	The conductivities of ceramics are low
	$LaSrO_3$	High proton transport number even at 1000°C Low conductivity
Complex perovskites	$Sr_2(ScNb)O_6$ $Ba_3(CaNb_2)O_9$	High protonic conductivity Low chemical stability Insufficient mechanical strength
Layered perovskites	Sr_2TiO_4 $Sr_3Ti_2O_7$	Large contribution of electronic conduction Low protonic conductivity

in the oxides, (2) partial substitution of aliovalentcation (dopant) for host cation or slight shift of composition from the stoichiometric one, and (3) relatively high content of oxygen per chemical formula.

Although the studies on proton-conducting perovskite-type oxides have been flourishing since the beginning of the 1990s, enough materials for practical use in a coming hydrogen energy system have not yet been established, in contrast to $LaGaO_3$-based ceramics in the case of oxide ion conductors for SOFC. The readers can find recent progress in the science and technology of proton-conducting perovskites in Chapters 11 through 14.

3.6 Lithium Ion Conduction

$La_{2/3}TiO_3$ is a perovskite-type oxide in which one third of the A-site cations is deficient. When lithium is partially substituted for La in A sites of this oxide, lithium ions become mobile [46]. The lithium ion conductivity of $La_{0.51}Li_{0.34}TiO_{2.94}$ is about 10^{-3} S cm^{-1} at room temperature [47]. This value belongs to the highest among lithium ion conductors that are chemically stable in an atmospheric environment. As La and Li ions are randomly distributed in the A-site position in the perovskite-type structure and, therefore, A-site vacancies are also distributed randomly, it is considered that the lithium ions can easily move through the vacancies. The relationship between the conductivity and content of lithium ions obeys so-called percolation theory [48].

Lithium ion conduction has been observed in similar perovskite-type oxides of light rare earths such as Pr, Nd, and Sm; the conductivity decreases with

Fig. 3.6 Lithium ion conductivities of $Ln_xLi_yTiO_3$ (Ln = La, Pr, Nd, or Sm) [47]
1, $La_{0.51}Li_{0.34}TiO_{2.94}$;
2, $Pr_{0.56}Li_{0.34}TiO_{3.01}$;
3, $Nd_{0.55}Li_{0.34}TiO_{3.00}$;
4, $Sm_{0.5}Li_{0.38}TiO_{2.97}$

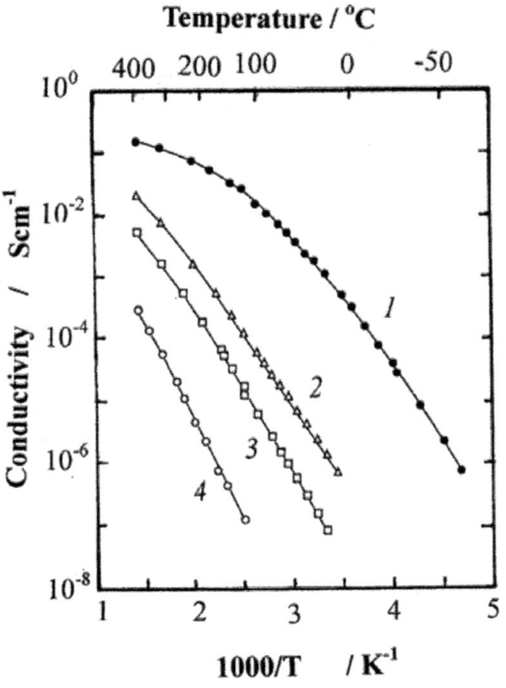

increasing atomic number (Fig. 3.6) [49]. These lithium ion conductors are not durable against reducing conditions, and the oxide is easily reduced on contacting metallic lithium.

Relatively high lithium ion conductivity was observed in perovskite-type $SrVO_{3-\delta}$ in which lithium ions were electrochemically inserted [50]. This material is an electronic conductor and has been studied as a candidate for a high-performance cathode material for lithium ion batteries. The lithium ion conductivity in this oxide is estimated to be about 10^{-5} S cm^{-1} at room temperature.

3.7 Halide Ion Conduction

Different kinds of non-oxide perovskite-type compounds have been known in carbides, halides, nitrides, and hydrides [2]. Conjecturing from the oxide ion conduction in ABO_3, it would be possible to expect anionic conduction, such as halide ionic or nitride ionic, in these non-oxide perovskite compounds ABX_3.

Actually, some double fluorides and chlorides crystallized in the perovskite-type structure show halide ion conduction at elevated temperature. It has been reported that ABX_3-type double halides such as $CsPbCl_3$ and $CsPbBr_3$

exhibited chloride ion conduction and bromide ion conduction, respectively, the conductivities of which were about 1×10^{-3} S cm^{-1} at 500°C [51]. Doping is also effective for the enhancement of the conductivity. For example, by replacing 1% of Pb^{2+} in $CsPbCl_3$ with Na^+, the conductivity increased one order of magnitude. It was reported that, when F in $KaCaF_3$ was partially replaced by oxygen, fluoride ion conductivity increased significantly [52]. Although fluoride ion conductivity of the perovskite-type fluorides is rather low compared to that of lead fluoride belonging to non-perovskite structure, the conductivities of larger halide ion (such as Cl^- and Br^-) conductors are essentially the same order or somewhat higher than that of non-perovskite halides.

3.8 Silver Ion Conduction

Silver iodide sulfide, Ag_3SI, is known to exhibit high silver ion conductivity of 1×10^{-2} S cm^{-1} at room temperature [53]. The crystal structure of this compound has three morphologies: α (>519 K), β (519–157 K), and γ (157 K). Among them, the γ phase crystallizes an anti-perovskite structure in which the anions S^{2-} and I^- occupy A- and B sites, respectively, and silver occupies the O site in ABO_3, as shown in Fig. 3.7(a). In the β phase, which is stable at room temperature, the sublattice of anions is the same as the case of the γ phase, whereas Ag ions do not occupy a strictly face-centered position but occupy, on average, four sites about 0.05 nm apart from the center along the direction of [100], as shown in Fig. 3.7(b) [54, 55]. Because Ag ions can easily move among these clumps of four sites, this compound is a good silver ion conductor at room temperature. Ag_3SBr has a similar crystal structure to this and exhibits silver ion conduction [56].

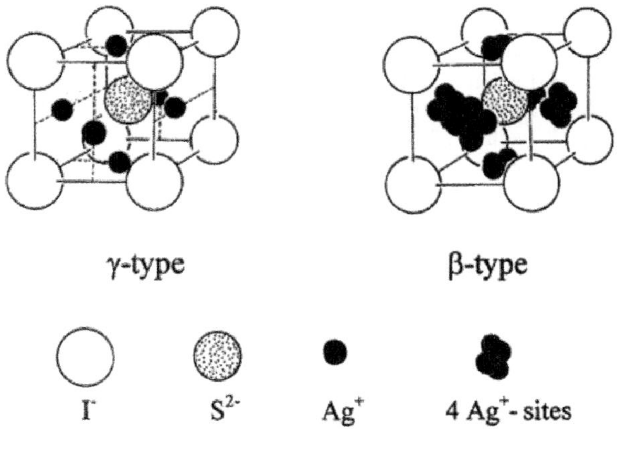

Fig. 3.7 Anti-perovskite structures of γ- and β-Ag_3SI [52]

References

1. F.A. Groger, "The Chemistry of Imperfect Crystals," North Holland, Amsterdam (1964)
2. F.S. Galasso, "Structure, Properties and Preparation of Perovskite Type Compounds," Pergamon Press (1969)
3. S. Swanson, Phys. Rev. 69, 546 (1946)
4. C.V. Stephenson, Bull. Am. Phys. Soc. 3, 299 (1958)
5. C.V. Stephenson, C.E. Franagan, J. Chem. Phys. 34, 2203 (1961)
6. K. Kuikkola, C. Wagner, J. Electrochem. Soc. 104, 379 (1957)
7. R.C. Heckman, D.D. Glower, C.R. Hills, Bull. Am. Phys. Soc., Series 2, 8, 601 (1963)
8. A. Ezis, J.G. Burt, R.A. Krakowski, J. Am. Ceram. Soc. 53, 521 (1970)
9. D.D. Glower, R.C. Heckman, D.L. Hester, Bull. Am. Phys. Soc., Series 2, 8, 601 (1963)]
10. D.D. Glower, R.C. Heckman, J. Chem. Phys. 41, 877 (1964)
11. W. van Gool, Phillips Res. Rep. 20, 81 (1965)
12. F. Forrat, R. Jansen, P. Trevoux, Compt. Rend. 257, 1271 (1963)
13. T. Takahashi, H. Iwahara, Denki Kagaku 35, 433 (1967)
14. R.V. Coates; J. Appl.Chem. 14, 346 (1964)
15. T. Takahashi, H. Iwahara, Denki Kagaku 37, 857 (1969)
16. T. Takahashi, H. Iwahara, Energy Conversion 11, 105–111 (1971)
17. B.H.C. Steele, B.E. Powell, P.M.R. Moody, Proc. Br. Ceram. Soc., No87, (1968)
18. T. Ishihara, H. Matsuda, Y. Takita, J. Am. Chem. Soc. 116, 3801 (1994)
19. K. Mader, H.K. Muller-Bushbaum, J. Less Common Metals 15, 771 (1990)
20. J.B. Goodenough, J.E. Ruiz-Diaz, Y.S. Zhen, Solid State Ionics 44, 21 (1990)
21. H. Iwahara, T. Esaka, H. Uchida, N. Maeda, Solid State Ionics 3/4, 359 (1981)
22. H. Iwahara, H. Uchida, N. Maeda, J. Power Sources 7, 293 (1982)
23. Hi. Iwahara, Solid State Ionics 125, 271 (1999)
24. T. Takahashi, H. Iwahara, Rev. Chim. Minerale 17, 243 (1980)
25. P.J. Jorgensen, F.J. Norton, Phys. Chem. Glasses 10, 23 (1969)
26. S. Stotz, C. Wagner, Ber. Bunsenges. Physik. Chem. 70, 781 (1966)
27. F. Forrat, D. Dauge, P. Trevoux, Compt. Rend. 258, 1271 (1964)
28. D.A. Shores, R.A. Papp, J. Electrochem. Soc. 119, 300 (1972)
29. H. Iwahara, H. Uchida, K. Ono, K. Ogaki. J. Electrochem. Soc. 135, 529 (1988)
30. N. Bonanos, B. Ellis, M.N. Mahmood, Solid State Ionics 44, 305 (1991)
31. H. Iwahara, T. Shimura, H. Matsumoto, Electrochemistry 68, 154 (2000)
32. H. Uchida, N. Maeda, H. Iwahara, Solid State Ionics 11, 117 (1983)
33. H. Uchida, H. Yoshikawa, H. Iwahara, Solid State Ionics 34, 103 (1989)
34. T. Yajima, H. Iwahara, Solid State Ionics 50, 281 (1992)
35. W. Lee, A.S. Nowick, L.A. Boaturmer, Solid State Ionics, 18/19, 989 (1986)
36. T. Norby, P. Kofstad, J. Am. Ceram. Soc. 67, 786 (1984)
37. T. Yajima, H. Kazeoka, T. Yogo, H. Iwahara, Solid State Ionics 47, 271 (1991)
38. H. Iwahara, T. Yajima, T. Hibino, K. Ozaki, H. Suzuki, Solid State Ionics 61, 65 (1993)
39. H. Fujii, K. Katayama, T. Shimura, H. Iwahara, J. Electroceramics 2, 199 (1998)
40. H. Iwahara, T. Yajima, T. Hibino, H. Ushida, J. Electrochem. Soc. 140, 1687 (1993)
41. T. Yajima, H. Suzuki, T. Yogo, H. Iwahara, Solid State Ionics 51, 101 (1992)
42. K.D. Kreuer, S. Adams, W. Munch, A. Fuchs, U. Klock, J. Maier, Solid State Ionics 145, 295 (2001)
43. A.S. Nowick, Y. Du, Solid State Ionics 77, 137 (1995)
44. T. Shimura, K. Suzuki, H. Iwahara, Solid State Ionics 104, 79 (1997)
45. T. Shimura, K. Suzuki, H. Iwahara, Solid State Ionics 113, 355 (1998)
46. L. Latie, G. Villenur, D. Conte, G.L. Flem, J. Solid State Chem. 51, 293 (1984)
47. Y. Inaguma, C. Liquan, M. Itoh, T. Nakamura, T. Uchida, H. Ikuta, M. Wakihara, Solid State Commun. 86, 689 (1993)
48. Y. Inaguma, M. Ito, Solid State Ionics 86, 257 (1996)

49. M. Ito, Y. Inaguma, W.H. Yong, L. Chen, T. Nakamura., Solid State Ionics 70, 203 (1994)
50. Y.J. Shan, L. Chen, Y. Inaguma, M. Shikano, M. Itoh, T. Nakamura, Solid State Ionics 70, 409 (1994)
51. J. Mizusaki, K. Arai, K. Fueki, Solid State Ionics 11, 203 (1983)
52. A.V. Chadwick, J.H. Strange, G. A M. Terenzi, Solid State Ionics 9, 555 (1983)
53. T. Takahashi, O. Yamamoto, Electrochim. Acta 11, 911 (1966)
54. S. Hoshino, T. Sakuma, Y. Fujii, J. Phys. Soc. Jpn. 47, 1252 (1979)
55. F. Billi, H.E. Roman, Dietrich, Solid State, Ionics 28, 58 (1988)
56. B. Reuter, K. Hardel, Z. Anorg. Allgem. Chem. 340, 168 (1965)

Chapter 4
Oxide Ion Conductivity in Perovskite Oxide for SOFC Electrolyte

Tatsumi Ishihara

4.1 Introduction

The electrolyte used in a solid oxide fuel cell (SOFC) must be stable in both reducing and oxidizing environments and must have sufficiently high ionic conductivity but low electronic conductivity at the cell operation temperature. If a small amount of electronic conductivity appears, particularly under a reducing atmosphere, then the energy conversion efficiency will decrease because of the consumption of fuel by chemically leaked oxygen. SOFCs have commonly used the fluorite structure stabilized zirconia, especially yttria-stabilized zirconia, as the electrolyte. Other oxide ion conductors, such as doped ceria and perovskite oxides, have also been proposed as the electrolyte materials for SOFCs, especially for reduced-temperature operation (873–1073 K). Either oxide ion- or proton-conducting oxides can be used for the electrolyte of fuel cells; therefore, development of high oxide ion- or proton-conducting ceramics is also an important subject for achieving high performance for SOFCs. At present, oxide ion conductors are mainly used for the electrolyte of solid oxide fuel cells (SOFCs). For application as the electrolyte of an SOFC, perovskite oxides of $LaGaO_3$ and $CaTiO_3$ (oxide ion conductors), $BaCeO_3$ and $BaZrO_3$ (proton conductors), are also an important group. The electrolyte for SOFCs must meet the following requirements: (1) high ionic conductivity, (2) low electronic conductivity, (3) chemical and physical stability under reducing and oxidizing atmospheres at the operating temperature, and (4) ease of preparation in the form of a dense film. In this chapter, oxide ion conductivity in perovskite oxides, in particular, $LaGaO_3$-based oxides, is introduced. The history of ionic conductors with perovskite structure is also overviewed in Chapter 3.

T. Ishihara (✉)
Department of Applied Chemistry, Faculty of Engineering, Kyushu University,
Motooka 744, Nishi-Ku, Fukuoka, 819-0395, Japan
e-mail: ishihara@cstf.kyushu-u.ac.jp

T. Ishihara (ed.), *Perovskite Oxide for Solid Oxide Fuel Cells*,
Fuel Cells and Hydrogen Energy, DOI 10.1007/978-0-387-77708-5_4,

4.2 Oxide Ion Conductivity in Oxide

Oxide ion conductivity was first found in ZrO_2 with 15 wt% Y_2O_3 (stabilized zirconia denoted as YSZ) by Nernst [1] as early as 1899, and so the history of the oxide ion conductor covers more than a century. In the history of oxide ion conductors, fluorite structure oxides consisting of tetravalent cations have been widely studied. The fluorite-type structure is a face-centered cubic arrangement of cations with anions occupying all the tetrahedral sites. It has a large number of vacant octahedral interstitial sites; thus, this structure is a rather open structure and rapid ion diffusion is achieved. In the literature, there are many reports suggesting that the ionic size of the dopant is highly important in determining the oxide ion conductivity. In this section, the effects of dopant on oxide ion conductivity are briefly introduced.

To achieve high oxide ion conductivity, the introduction of oxygen vacancies by the substitution of a lower valence cation is essential. In the case of zirconia-based oxides, the dissolution of yttria into the fluorite phase of ZrO_2 can be written by the following defect equation in Kroger–Vink notation:

$$Y_2O_3(ZrO_2) \rightarrow 2Y'_{Zr} + 3O_o^{x} + V_o^{\bullet\bullet} \tag{4.1}$$

The concentration of the vacancies is given simply by the electron neutrality condition. In this case, therefore, $2[Y'_{Zr}] = [V_o^{\bullet\bullet}]$. Here, it is well known that the ionic conductivity, σ, can be expressed by Eq. (4.2):

$$\sigma = en\mu \tag{4.2}$$

where n is the number of mobile oxide ion vacancies, μ their mobility, and e the charge. Therefore, the vacancy concentration is linearly dependent upon the dopant level. However, this is not entirely true, and the higher doping concentrations lead to the formation of vacancy and dopant cluster resulting in the decreased oxide ion conductivity. In fact, as shown in Fig. 4.1, the conductivities

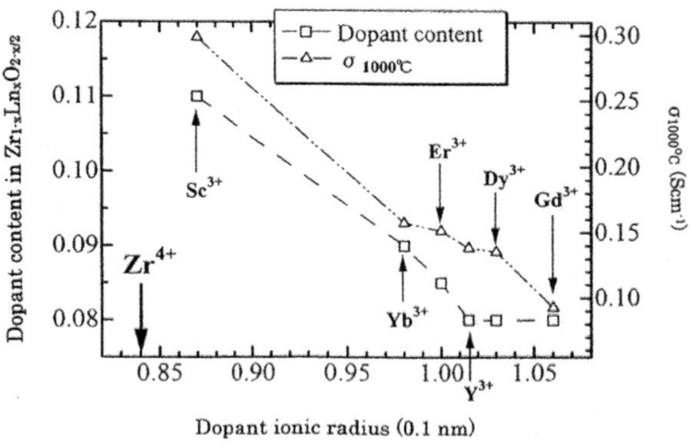

Fig. 4.1 The maximum dopant content and conductivities at 1273k as a function of dopant ionic radii in doped zirconia

of doped zirconia show a maximum at a specific concentration of dopant, which was reported by Arachi et al. [2] for the ZrO_2–Ln_2O_3 (Ln = lanthanide) system. It is obvious that the electrical conductivity in ZrO_2 is strongly dependent on the dopant element and its concentration. For a low concentration of dopant, the conductivity monotonically increases with dopant amount, as is expected from theory, and evidently the defects behave as a point defect. Therefore, conductivity is mainly determined by the amount of oxygen vacancy, namely, the amount of dopant. On the other hand, conductivity as well as activation energy for conduction are strongly affected by dopant ionic size. It is evident that the conductivity increases with decreasing ionic size of doped cations. Explanations for this conductivity behavior have been tried based on structural effects. The content of dopant with the highest conductivity in the ZrO_2–Ln_2O_3 system decreases with increasing dopant ionic radius. The dopants Dy^{3+} and Gd^{+3} with larger ionic radii show a limiting value of 8 mol%. The dopant Sc^{3+}, which has the closest ionic radius to the host ion, Zr^{4+}, shows the highest conductivity and the highest dopant content at which cluster formation starts. Similar conductivity dependence on the dopant level was observed in the CeO_2 system. Figure 4.2 shows the conductivity of CeO_2 as a function of ionic size of Ln_2O_3. The highest conductivity was found at 10 mol% for the Sm_2O_3 dopant and at 4 mol% for Y_2O_3. The diffusion of oxide ion vacancies is affected by the local strain energy, which is related to the mismatch between the host and dopant cation size [3]. Therefore, not only dopant concentration but also ionic size is an important factor for achieving high oxide ion conductivity. Recent studies on oxide ion conductivity reveal that the clusters form at a dopant concentration much lower than that considered previously [4]. Therefore, to achieve high oxide ion conductivity, design of the dopant and its concentration is highly important.

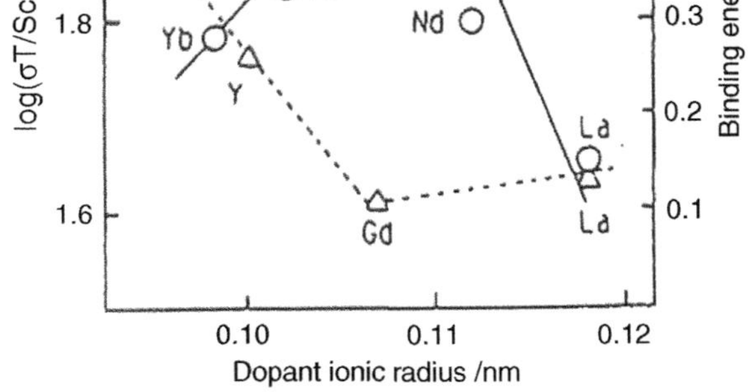

Fig. 4.2 Conductivity and binding energies of rare earth dopants in CeO_2 as a function of ionic size

Although at present YSZ is most commonly used for the electrolyte in a SOFC, higher oxide ion conductivity is a major requirement for SOFC operation at lower temperatures. However, in the case of tetravalent oxides with fluorite structure, up to now a higher oxide ion conductor is only achieved with lower chemical stability in reducing atmosphere and so the alternative candidates to YSZ are quite limited: only Sc_2O_3–ZrO_2 doped with 1 mol% CeO_2, or Sm_2O_3 or Gd_2O_3 doped with CeO_2. Similar to the fluorite oxides, the perovskite structure also has a large free volume, and so oxide ion conductivity in the perovskite oxide presents an alternative to the fluorites for use as an SOFC electrolyte.

4.3 Oxide Ion Conductivity in Perovskite Oxides

Although oxides with perovskite structure are predicted to be good oxide ion conductors, typical perovskite oxides such as $LaCoO_3$ and $LaFeO_3$ are widely known as mixed electronic and oxide ionic conductors, which can be used as cathodes in SOFCs. Therefore, these mixed conducting perovskite oxides present a promising material group for cathode catalysts for SOFC or oxygen-permeating membranes. The large majority of perovskite oxides exhibiting oxide ion conduction are classified as mixed conductors, which show both electronic and oxide ionic conduction and thus cannot be used as the electrolyte of an SOFC.

Takahashi and Iwahara have done pioneering work on perovskite oxide ion conductors [5]. They reported fast oxide ion conductivity in Ti- and Al-based perovskite oxides, and it is evident that Al- or Mg-doped $CaTiO_3$ exhibits high conductivity, but it is still lower than that of YSZ. Iwahara and Takahashi investigated the oxide ion conductivity in $CaTiO_3$ in detail [5]. On the other hand, although the oxide ion exhibits a high transport number in $CaTi_{0.95}Mg_{0.05}O_3$ at intermediate temperatures, Ca-doped $LaAlO_3$ is another attractive candidate as an oxide ion conductor, since no electronic conduction appears in a reducing atmosphere and transport number is higher than 0.9 over the entire temperature range.

After the report of Takahashi and Iwahara [5], many researchers investigated the oxide ion conductivity in $LaAlO_3$-based oxide. However, the reported oxide ion conductors with the perovskite structure exhibited lower ionic conductivity than that of Y_2O_3–ZrO_2. In the conventional study of perovskite oxides, ABO_3, it was believed that the electric or dielectric properties were strongly dependent on B-site cations. However, a migrating oxide ion has to pass through the triangular orifice consisting of two large A- and one small B-site cations in the crystal lattice. Therefore, the ionic size of the A-site cation seems to influence greatly the oxide ion conductivity. Effects of A-site cations on oxide ion conductivity in $LnAlO_3$-based perovskite oxide were reported. Figure 4.3 shows the electrical conductivity in $LnAlO_3$-based oxides [6]. Electrical conductivity of Al-based perovskite oxides increased with increasing the ionic size of A site cations, which suggests that the

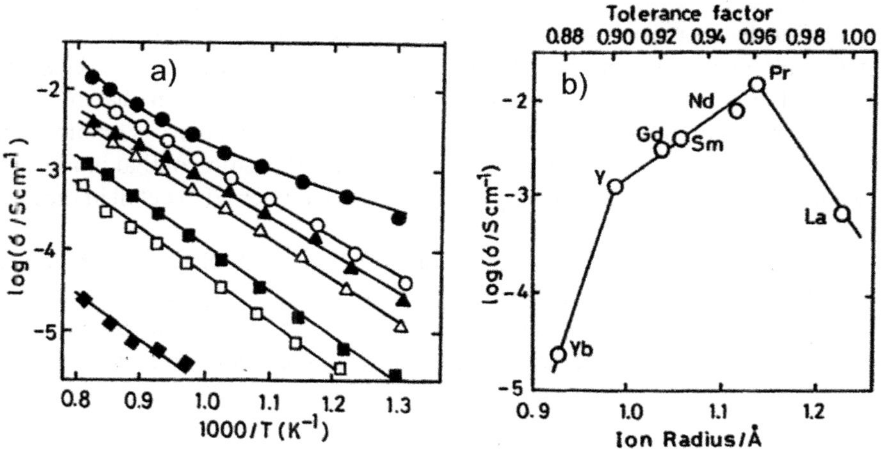

Fig. 4.3 Arrhenius plot of the electrical conductivity of Ca-doped $LnAlO_3$ [Ln = La (□), Pr (●), Nd (○), Sm (▲),Gd (△), Y (■), Yb (◆)] **(a)** and electrical conductivity at 1223 K as a function of A-site ion size **(b)**

larger unit cell volume is important for higher oxide ion conductivity because of the larger free volume. Therefore, doping larger cations for the B site is also important for achieving high oxide ion conductivity. The oxide ion conductivity in $NdAlO_3$ doped with Ga at Al sites was studied [6]. The addition of Ga^{3+}, a larger ion than Al^{3+}, to B sites of $NdAlO_3$ is effective for improving oxide ion conductivity [7]. Figure 4.4 shows the oxide ion conductivity at 1123 K as a function of Ga content. In accordance with the prediction, oxide ion conductivity increased with an increasing amount of Ga, and it attained the maximum (log σ/Scm^{-1}) = −1.5 at 1223 K when 50 mol% Ga was doped at Al sites in $LaAlO_3$. Since no oxygen vacancy is formed by doping Ga^{3+} because of the same valence

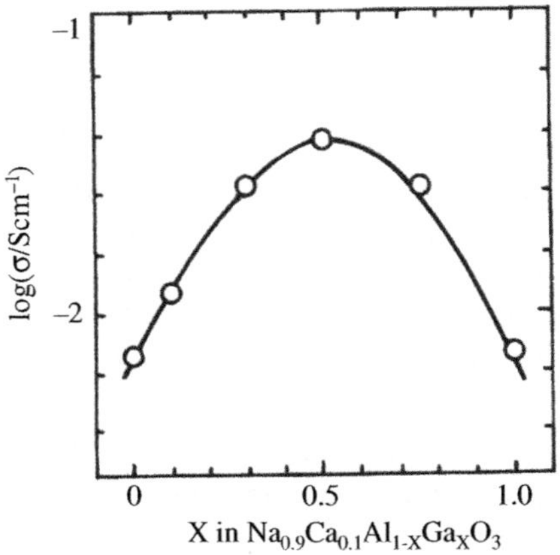

Fig. 4.4 Oxide ion conductivity in $Na_{0.9}Ca_{0.1}Al_{1-x}Ga_xO_3$ at 1123 K as a function of Ga amount

as Al, higher oxide ion conductivity is brought about by the improved mobility of oxide ion by increasing the unit cell volume, or free volume in the unit cell. Although high oxide ion conductivity was obtained on $Nd_{0.9}Ca_{0.1}Al_{0.5}Ga_{0.5}O_3$, the oxide ion conductivity of this compound is still lower than that of YSZ. However, it is of great interest that higher oxide ion conductivity is reported for $LaGaO_3$-based perovskite oxide, which is the end composition of that in Fig. 4.4. Oxide ion conductivity in $LaGaO_3$-based oxide is overviewed in detail in the next section.

Another type of perovskite oxide ion conductor is $LaScO_3$, which is also reported as a high-temperature proton conductor [8]. Figure 4.5 shows the temperature dependence of four different perovskite oxides with similar composition [9]. In spite of the similar compositions, the oxide ion conductivity is

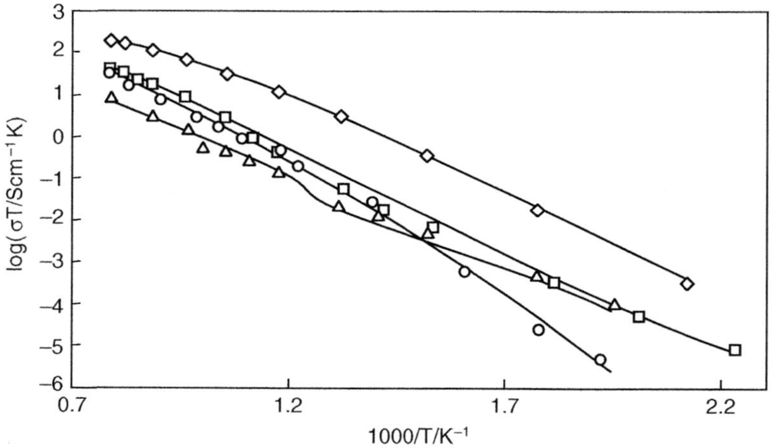

Fig. 4.5 Arrhenius plot of four different perovskite oxide in air. $La_{0.9}Sr_{0.1}Ga_{0.9}Mg_{0.1}O_3$ (◇); $La_{0.9}Sr_{0.1}Sc_{0.9}Mg_{0.1}O_3$ (□); $La_{0.9}Sr_{0.1}Al_{0.9}Mg_{0.1}O_3$ (○); $La_{0.9}Sr_{0.1}In_{0.9}Mg_{0.1}O_3$ (△) (Cited from Ref. (9))

very different, i.e., higher for $LaGaO_3$. However, $LaAlO_3$, $LaScO_3$, and $LaInO_3$ show lower oxide ion conduction with hole conduction in the higher P_{O2} range. A similar study was performed by Nomura et al., and the order of oxide ion conductivity in these perovskite oxides is not simply explained by free volume size, but by size matching of dopant, in particular, Mg to B-site cations. Mg is too large as a dopant on the B site of these four perovskite oxides; however, it is the one closest to Ga in size [10]. Comparison of the defect association energy is further required, as discussed later; however, it is evident that $LaGaO_3$ is a promising oxide ion conductor over a wide P_{O2} range [11]. Oxide ion conductivity in $LaGaO_3$ is discussed in the next section.

4.4 $LaGaO_3$-Based Oxide Doped with Sr and Mg (LSGM) as a New Oxide Ion Conductor

4.4.1 *Effects of Dopant for La and Ga Site*

The high oxide ion conductivity in $LaGaO_3$-based perovskites, which is a pure oxide ion conductor, was first reported in 1994 [12]. The high oxide ion conductivity in this oxide is achieved by double doping of a lower valence cation into both the A- and the B sites of perovskite oxide, ABO_3. It is obvious that the oxide ion conductivity strongly depends on the cations on the A site, which is similar to the Al-based oxide, and the highest conductivity is achieved for $LaGaO_3$, which has also the largest unit cell volume of the Ga-based perovskites. The electrical conductivity of all the Ga-based perovskite oxides is almost independent of the oxygen partial pressure, and, therefore, it is expected that the oxide ion conduction will dominate in all Ga-based perovskite oxides.

Doping with a lower valence cation generally forms oxygen vacancies due to the requirement for electric neutrality; the oxide ion conductivity increases with increasing the amount of oxygen vacancies. Therefore, doping alkaline earth cations onto La sites was investigated, and the oxide ion conductivity is shown in Fig. 4.6 [12]. The electrical conductivity of $LaGaO_3$ depends strongly on the alkaline earth cations and also increase in the following order Sr > Ba > Ca.

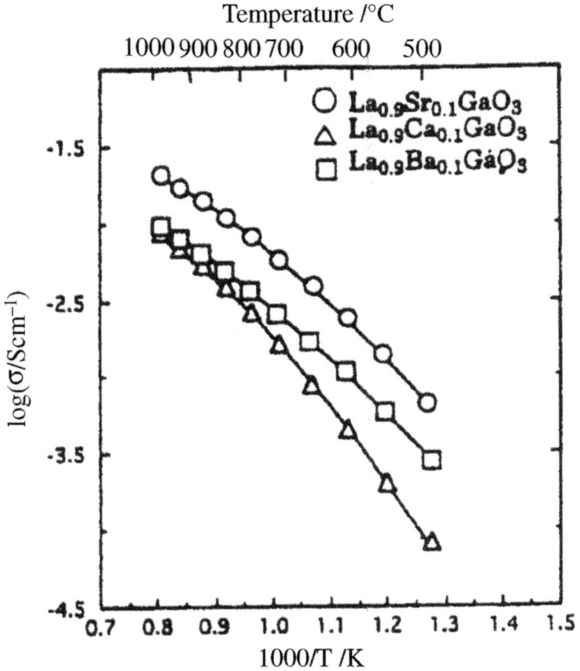

Fig. 4.6 Effects of the nature of alkaline earth cations on the La site on oxide ion conductivity in $LaGaO_3$

Therefore, strontium, of which the ionic size is almost the same as that of La^{3+}, is the most suitable dopant for the La sites in $LaGaO_3$. Theoretically, increasing the amount of Sr will increase the amount of oxygen vacancies and hence the oxide ion conductivity. However, solid solubility of Sr into La sites of $LaGaO_3$ is poor, and the secondary phases, $SrGaO_3$ or La_4SrO_7, form when the amount of Sr becomes higher than 10 mol%. Thus, the concentration of oxygen vacancies introduced by La site doping is not large.

The effects of dopant on Ga sites of $La_{0.9}Sr_{0.1}GaO_3$ were also studied for further improvement of electrical conductivity. It is found that doping with Mg is very effective at increasing the conductivity because additional oxide ion vacancies are formed. The oxide ion conductivity is further increased by increasing the amount of Mg added; the maximum conductivity is attained at 20 mol% Mg doped for Ga sites. The lattice parameter also increases by doping Mg for Ga sites as the ionic radius of Mg is larger than that of Ga. The solid solubility of Sr into the $LaGaO_3$ lattice seems to reach a limit around 10 mol% without Mg; however, it increases up to 20 mol% by doping Mg for Ga. This enlargement in the limit of Sr solid solution was also reported by Majewski et al. [13], which seems to be a result of the enlarged crystal lattice. In any case, the highest oxide ion conductivity in the $LaGaO_3$-based oxide is reported at the composition of $La_{0.8}Sr_{0.2}Ga_{0.8}Mg_{0.2}O_3$ [14].

Because this oxide consists of four elements, the optimum composition varies slightly from group to group. Oxide ion conductivity in $LaGaO_3$-based oxide was investigated by several groups [15, 16], and various cations were examined as a dopant for $LaGaO_3$-based oxides. Huang and Petric investigated the oxide ion conductivity of various compositions [16] and expressed the oxide ion conductivity in contour maps [17] (Fig. 4.7), in which the optimum composition reported by two other groups is shown. Huang et al. reported that the highest oxide ion conductivity was obtained at the composition of $La_{0.8}Sr_{0.2}$-$Ga_{0.85}Mg_{0.15}O_3$ [17] On the other hand, Huang et al. and Huang and Goodenough reported the optimized composition in $La_{1-X}Sr_XGa_{1-Y}Mg_YO_3$ at $X = 0.2$, $Y = 0.17$ [17, 18]. However, the optimized composition among the three groups is close to each other, and the optimized composition in Sr- and Mg-doped $LaGaO_3$ exists between $Y = 0.15$ to 0.2 in $La_{0.8}Sr_{0.2}Ga_{1-Y}Mg_YO_3$. The difference may come from the uniformity of composition and also grain size.

Figure 4.8 shows the comparison of oxide ion conductivity of doubly doped $LaGaO_3$ with the conventional fluorite oxide ion conductors. It is obvious that the oxide ion conductivity in $La_{0.8}Sr_{0.2}Ga_{0.8}Mg_{0.2}O_3$ is higher than the typical conductivity of ZrO_2- or CeO_2-based oxides and somewhat lower than those of Bi_2O_3-based oxides. It is well known that electronic conduction is dominant in CeO_2- or Bi_2O_3-based oxides under a reducing atmosphere; furthermore, thermal stability is not satisfactory in Bi_2O_3-based oxides. In contrast, $La_{0.8}Sr_{0.2}Ga_{0.8}Mg_{0.2}O_3$ exhibits wholly ionic conduction from $P_{O2} = 10^{-20}$ to 1 atm. Therefore, doubly doped $LaGaO_3$ perovskite oxide shows great promise as the solid electrolyte for fuel cell and oxygen sensor.

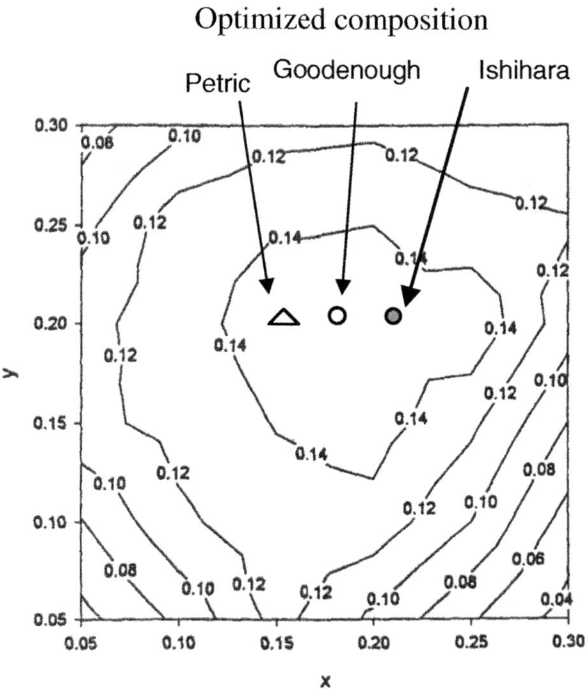

Fig. 4.7 Contour plot of conductivity at 1073 K as a function of x and y in $La_{1-X}Sr_XGa_{1-Y}Mg_YO_3$

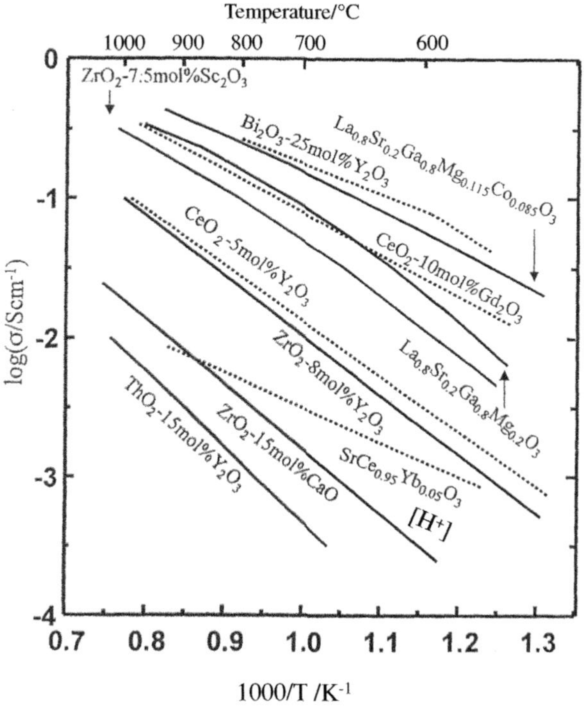

Fig. 4.8 Comparison of oxide ion conductivity in doubly doped $LaGaO_3$ with conventional oxide ion conductors

4.4.2 Transition Metal Doping Effects on Oxide Ion Conductivity in LSGM

To improve oxide ion conductivity, several groups have already investigated the effects of the various cation dopants on oxide ion conductivity in $LaGaO_3$-based oxides [19]. Doping transition metal cations such as Co, Ni, or Fe is not preferable for a solid electrolyte due to the formation of partial electronic conduction. However, it is reported that oxide ion conductivity is increased by doping Co without decreasing the transport number of oxide ions [20], if the amount of Co is smaller than 10 mol%. In this section, the effect of transition metal doping on oxide ion conductivity is briefly summarized. As shown in Fig. 4.8, the Arrhenius plot for LSGM is slightly bent around 1000 K, suggesting a change in conduction mechanisms. Detailed crystal structure analysis by neutron diffraction suggests that the crystal phase changes from triclinic to pseudo-cubic lattice. This phase change might be related to the mismatch of the Mg^{2+} ionic size with that of Ga^{3+}. Doping with a trivalent cation similar in size to that of Mg^{2+} might be effective in stabilizing the high-temperature cubic phase. Considering the ionic size of trivalent cation with a 6-coordinated number, it seems that Fe, Co, and Ni are candidate ions.

Baker et al. investigated the effects of doping with transition metals Cr and Fe on the oxide ion conductivity of $LaGaO_3$-based oxide [21]. It was reported that doping with Cr or Fe on Ga sites induces hole conduction in $LaGaO_3$-based oxides, resulting in decreased stability against reduction. Figure 4.9 shows an Arrhenius plot of electrical conductivities of the $LaGaO_3$-based oxides doped with some transition metal cations on Ga sites [20]. It was observed

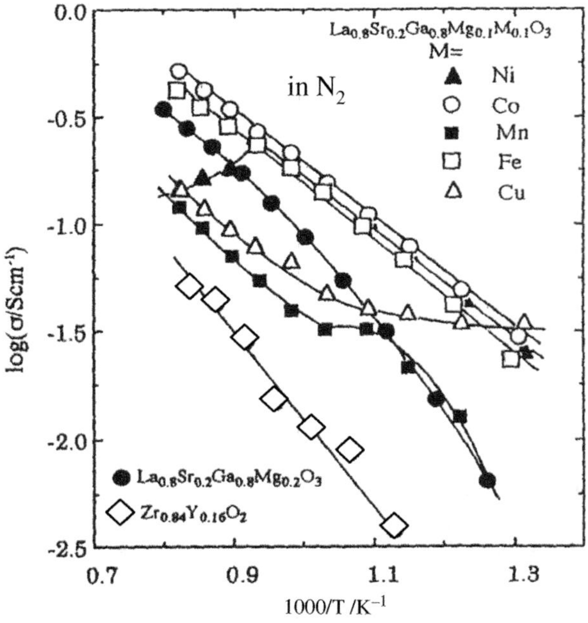

Fig. 4.9 Arrhenius plot of electrical conductivities of the $LaGaO_3$-based oxides doped with some transition metal cations for Ga sites

that the conductivity is improved by doping with Co and Fe and is lowered by doping with Cu and Mn. In the case of Ni, the conductivity decreased with increasing temperature above 1073 K, whereas below 973 K, it increased with increasing temperature. Such a decrease in conductivity in spite of increasing temperature may result from the significant electronic conductivity caused by the thermal reduction of Ni. From the P_{O2} dependence of the electrical conductivity, n-type conductivity is greatly enhanced by doping with Mn and Ni, and p-type conduction increases by doping with Cu. Kharton et al. also investigated the effects of transition metal dopant on oxide ion conductivity in $LaGa_{0.8}Mg_{0.2}O_3$ [22]. Although the amount of doped transition metal is much larger, i.e., 40 mol% for Ga sites, they also reported that doping with Mn and Cr decreases oxide ion conductivity. However, Thangaduari et al. reported that the $La_{0.9}Sr_{0.1}$-$Ga_{0.8}Mn_{0.2}O_3$ exhibits an oxide ionic conductivity that is comparable with that of $La_{0.9}Sr_{0.1}Ga_{0.8}Mg_{0.2}O_3$ [23]. In addition, the activation energy for ion conductivity in the Mn-doped sample is much smaller than that of Mg-doped ones. However, the small activation energy may suggest dominant electronic conduction in this oxide. In contrast, the total conductivity of Fe- or Co-doped specimens is almost independent of the oxygen partial pressure, which suggests that oxide ion conductivity increases by doping Co or Fe [24]. Therefore, oxide ion conductivity in Co-doped $LaGaO_3$-based oxide will be important. On the other hand, it is reported that the higher Fe-doped $La(Sr)GaO_3$ exhibits mixed conductivity with both hole and oxide ions contributions. This material shows large oxygen permeation flux when used as an oxygen separation membrane.

Figure 4.10 shows the electrical conductivity of Co-doped LSGM at 1273 K, $P_{O2} = 10^{-5}$ atm and the transport number of the oxide ion at 1273 K as a

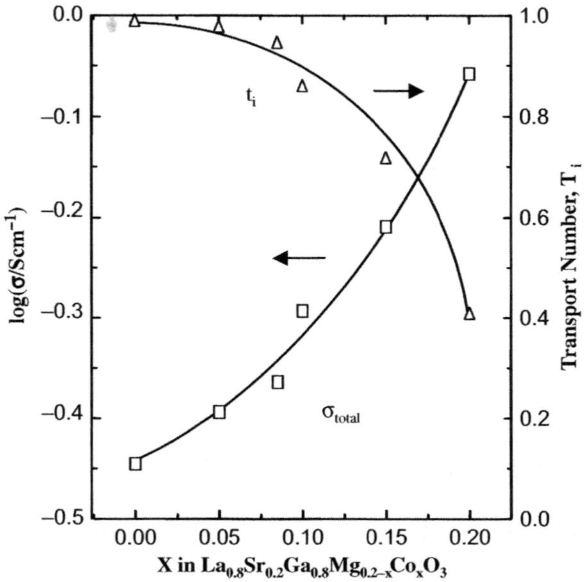

Fig. 4.10 Electrical conductivity of Co-doped LSGM at 1273 K, $P_{O2} = 10^{-5}$ atm and the transport number of oxide ion at 1273 K as a function of Co content

Fig. 4.11 Oxide ion conductivity estimated from the transport number and the total conductivity in LSGM

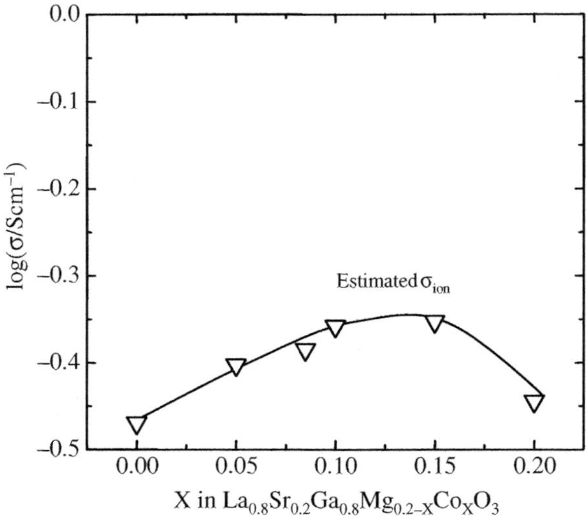

function of Co content [20]. The electrical conductivity increases, whereas the transport number of the oxide ion decreases with an increasing amount of Co. The oxide ion conductivity values estimated from the transport number and the total conductivity are shown in Fig. 4.11. The electrical conductivity becomes higher with increasing the amount of Co and attains a maximum value at around 10 mol%. The apparent activation energy for the electronic conduction monotonically decreased with increasing the Co concentration and reaches a value of 0.45 eV for 10 mol% Co, which is almost half of the value reported for YSZ. Although the highest oxide ion conductivity is obtained at $X = 0.1$, the transport number of oxide ion becomes smaller than 0.9. Since the decreased transport number of the oxide ion leads to a decrease in the energy conversion efficiency of a SOFC, it is considered that the desirable composition as the electrolyte for SOFC is $La_{0.8}Sr_{0.2}Ga_{0.8}Mg_{0.115}Co_{0.085}O_3$ (denoted as LSGMC-8.5) or one with even lower Co content.

In Fig. 4.8, the temperature dependence of oxide ion conductivity in Co- doped $LaGaO_3$-based oxide is also shown. It is seen that Co-doped $LaGaO_3$-based oxides exhibit even higher conductivity than that of LSGM and Gd-doped CeO_2. Another interesting point is the disappearance of the slope change around 1000 K, suggesting that the high-temperature cubic phase is stabilized. The conductivity value of this Co-doped LSGM is close to that of Bi_2O_3-based oxide, which exhibits pure oxide ion conduction in a limited P_{O2} range. Therefore, Co-doped $LaGaO_3$, in particular, 8.5 mol% Co-doped LSGM, is a good choice to use as the electrolyte of a solid oxide fuel cell operable at intermediate temperatures.

4.5 Basic Properties of the LSGM Electrolyte System

4.5.1 Phase Diagram of La-Sr-Ga-Mg-O

The phase diagram of the quaternary system $LaO_{1.5}$–SrO–$GaO_{1.5}$–MgO has been reported for $LaGaO_3$-based oxides (Fig. 4.12) [18, 25, 26]. Several impurity phases, i.e., $LaSrGa_3O_7$ (237), $LaSrGaO_4$ (214) phases, are reported for this $LaGaO_3$ perovskite oxide, and the single, two-phase, and three-phase regions appeared in phase diagrams. However, no phase containing Mg was found in the compositional range of Fig. 4.12, which implies a higher solubility of Mg in

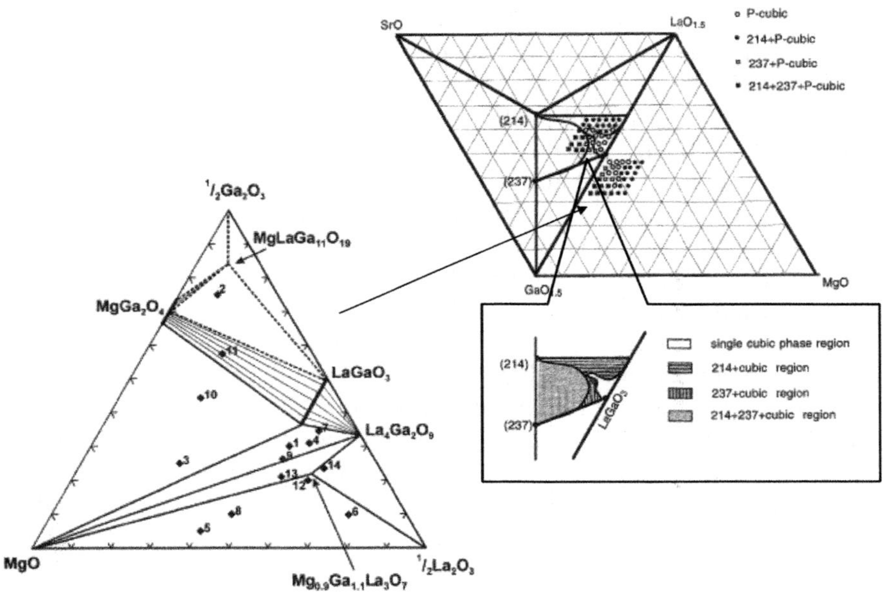

Fig. 4.12 Phase diagram of pseudo-quaternary $LaO_{1.5}$–SrO–$GaO_{1.5}$–MgO system

the perovskite phase and related compound. As already discussed in a previous section, doping with Mg is also effective for expanding the solubility of Sr in the La site and expands the perovskite regions compared to the non-doped La_2O_3–Ga_2O_3 binary phase diagrams; thus, doping Mg is effective not only for the introduction of vacancies but also in expanding the perovskite regions.

4.5.2 Reactivity with SOFC Component

The reactivity of this $LaGaO_3$ oxide has been also investigated by several groups. The reactivity of $LaGaO_3$-based oxide [27] with $La(Sr)CoO_3$ perovskite oxide or a Pt electrode [28] is important. In particular, platinum seems to react easily with gallium oxide to reduce Ga^{3+} to Ga^{+} to form Ga_2O, which is volatile [28]. Therefore, for the practical application of this material to the SOFC, one should pay attention to the choice of the electrode material and/or its conditions of use, such as temperature or atmosphere. Another undesirable reaction of LSGM for the electrolyte of a SOFC is that with an Ni-based anode [27]. Because $LaNiO_3$ is one of the typical perovskite oxides, Ni is easily substituted with Ga on the B site of the perovskite phase to form a highly resistive phase between the LSGM electrolyte and NiO anode during cell preparation [29]. To prevent the reaction between the components, many buffer layers are used for the current SOFC, even for the case of YSZ electrolyte. In the case of the LSGM perovskite electrolyte, it is reported that CeO_2 doped with La (LDC) shows low reactivity when the amount of La is in a narrow range, around 40 mol%. Therefore, by insertion of an LDC layer between an NiO anode and an LSGM electrolyte, a SOFC with high power density can be achieved [30]. It is also reported that insertion of an LDC layer is effective for preparing an LSGM thin film by a conventional method such as slip casting. However, sintering LDC is rather difficult, and also the electrical conductivity of this compound is low. Therefore, this LDC buffer layer makes a significant contribution to the total resistance of the cell. Further suitable buffer layer compound for LSGM electrolyte is still needed to prevent the reaction between NiO and LSGM. One of the useful compounds is Sm-doped CeO_2 (SDM), which also exhibits high oxide ion conduction.

In contrast, the reactivity of the LSGM electrolyte with the cathode perovskite oxide is not extensive. When YSZ is used for the electrolyte, Co-based perovskite oxides such as $LaCoO_3$ cannot be used, in spite of the high surface activity to oxygen dissociation; this is because the reaction between YSZ and $LaCoO_3$ forms the $La_2Zr_2O_7$ pyrochlore phase with high resistivity [31]. However, in case of LSGM electrolyte, compatibility with $LaCoO_3$ is high enough to use it as a cathode of SOFCs. Horita et al. reported that no reaction products were observed after a reasonably long period of operation [32]. As a result, Co-based perovskites can be used as cathodes with low values of cathodic overpotentials. High compatibility with Co-based perovskites is one of the main advantages of $LaGaO_3$ perovskites as the electrolyte of SOFCs. It is also noted that La_2NiO_4 has low reactivity toward LSGM; however, in the case of $LaMnO_3$, which is the most popular cathode material, some interactions between La_2NiO_4 and $LaMnO_3$ were observed with the formation of the insulator layer.

4.5.3 Thermal Expansion Behavior and Other Properties

Thermal expansion is another important property for the application of materials to SOFC. The thermal expansion increased with the increase in the dopant concentration. Anomalies in thermal expansion behavior for $LaGaO_3$ and

$La_{0.9}Sr_{0.1}GaO_3$ were observed around 400 K and were assigned to the phase transition from orthorhombic to rhombohedral structure. On the other hand, Sr- and Mg-doped materials show a monotonic expansion; the average thermal expansion coefficient measured was around $11.5 \times 10^{-6}\ K^{-1}$ within the temperature range from 298 to 1273 K. Therefore, the average thermal expansion coefficient is slightly larger than but close to that of Y_2O_3-stabilized ZrO_2.

Thermal conductivity of this material has also been studied. Table 4.1 summarizes the thermal conductivity, specific heat, and fracture strength of $LaGaO_3$ perovskite. The average thermal conductivity of this LSGM is slightly smaller than that of YSZ. Therefore, SOFCs using $LaGaO_3$-based perovskites are more desirable from the point of view of uniform temperature distribution.

Table 4.1 Thermal properties of $LaGaO_3$ oxide

Temperature (K)	Specific heat (J/g k)	Thermal conductivity (W/m K)	Fracture strength (MPa)
298	0.410	1.55	220
673	0.464	1.55	180
1073	0.556	1.77	136

4.5.4 Behavior of Minor Carrier

Since the concentration of the minor carriers (electrons and/or holes) determines the chemical leakage of oxygen when oxide ion conductor is used as the electrolyte in SOFCs [33], analysis of the performance of electron and hole conduction is an important subject for the electrolyte materials. Partial electronic conduction is commonly analyzed by the ion blocking method, the so-called Wagner polarization method. Partial electronic conductivity is the sum of electronic and hole contribution to the total conductivity, and each conductivity is proportional to a carrier density. Therefore, the total electronic conductivity can be expressed as follows:

$$\sigma = ILF/RT = \sigma_n + \sigma_p = n^0\{1 - \exp(FE(L)/RT)\} + \sigma p^0\{\exp(FE(L)/RT) - 1\} \quad (4.3)$$

Here, I, L, E(L),F, R, and T are current, length of the sample, applied voltage, the Faraday constant, the gas constant, and temperature, respectively. When the hole conduction is dominant, the second term in the above equation is dominant and the current increases exponentially with applied potential. Since the predominant charge carriers change from holes to electrons with decreasing P_{O2} and p–n transition occurs at intermediate P_{O2}, current, I, shows a typical "S"-shaped curve against potential, E(L). Figure 4.13 shows the typical I–E(L) curve observed in Ni-doped LSGM by the ion blocking method [34].

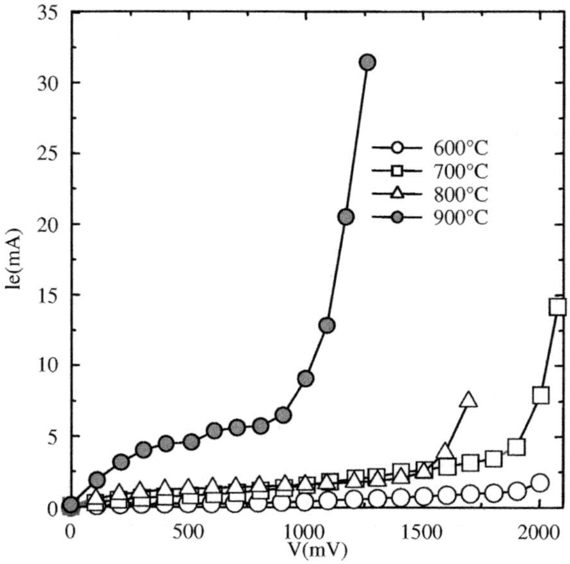

Fig. 4.13 Typical I–E (L) curves observed in Ni-doped LSGM by the ion blocking method

Differential of the observed current with respect to potential, which corresponds to the P_{O2} in the sample, gives the dependence of the partial electronic conductivity on P_{O2}.

Determination of hole and electron conductivities and transport numbers of oxide ion in $LaGaO_3$-based oxides were performed by the polarization method by Baker et al. [21], Yamaji et al. [35], and Kim and Yoo [36]. Kim et al. reported that P_{O2} dependence of hole and electron conductivity is proportional to $P_{O2}{}^{1/4}$ and $P_{O2}{}^{-1/4}$, respectively, and well obeys the Hebb–Wagner theory. The results of the polarization method clearly indicate that $LaGaO_3$-based oxides exhibit almost pure oxide ion conductivity over a wide oxygen partial pressure range ($10^5 > P_{O2} > 10^{-25}$ atm). Compared with CeO_2-based oxides or Bi_2O_3 oxide, this is a major advantage of the $LaGaO_3$-based oxides as well as the redox stability comparable to that of ZrO_2-based oxide. Consequently, $LaGaO_3$-based oxides are highly promising as an electrolyte for SOFCs, particularly when compared with ceria-based oxides. Kim et al. [36] also investigated the temperature dependence of hole and electronic conductivity in Mg-doped gallate, $La_{0.9}Sr_{0.1}Ga_{0.8}Mg_{0.2}O_3$, with the polarization method. Figure 4.14 shows the evaluated boundaries of the electrolytic domain in $La_{0.9}Sr_{0.1}$-$Ga_{0.8}Mg_{0.2}O_3$ plotted in the axis of log (P_{O2}/atm) versus reciprocal temperature. The lower boundary of the electrolytic domain (defined as $t_{ion} > 0.99$) for LSGM is 10^{-23} atm at 1273 K. This pressure is even lower than that of CaO-stabilized ZrO_2 and that of YSZ, which is also plotted in Fig. 4.14. Consequently, it is clear that the electrolytic domain covers the P_{O2} range required for the operation of SOFCs and that the LSGM can be successfully used as electrolyte in SOFCs.

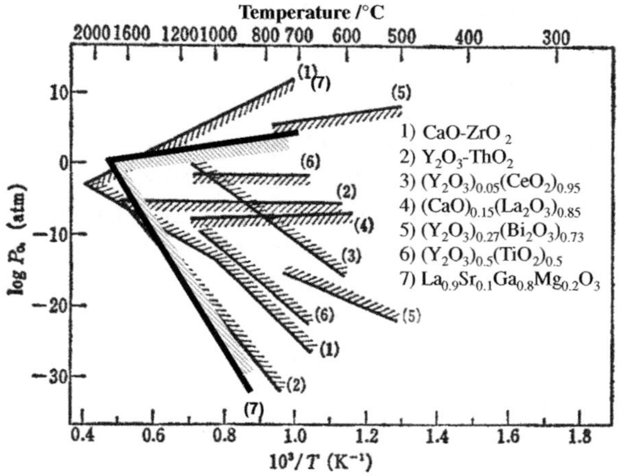

Fig. 4.14 Boundaries of the electrolytic domain of $La_{0.9}Sr_{0.1}Ga_{0.8}Mg_{0.2}O_3$ in the plane of log (P_{O2}/atm) versus reciprocal temperature

On the other hand, doping by a small amount of Co is effective in increasing the electrical conductivity, as discussed in Section 4.2. However, with the increasing Co concentration, the partial hole conductivity also increases. As a result, hole, electron, and oxide ion conductivities in Co-doped LSGM have also been studied by the polarization method [34], and the estimated conductivities at 1073 K as a function of Co content are shown in Fig. 4.15. It is seen that both the electronic and hole conductivities become significant with the increasing Co concentration; however, at Co amount less than 5 mol% on the Ga site, the oxide ion conductivity is much higher than partial hole conductivity. Furthermore, the oxide ion conductivity also increases as Co amount increases. Although the hole conduction becomes significant and the oxide becomes a mixed electronic and ionic conductor when the amount of Co is excess, doping Co is preferable for improving the oxide ion conductivity in $LaGaO_3$-based oxide.

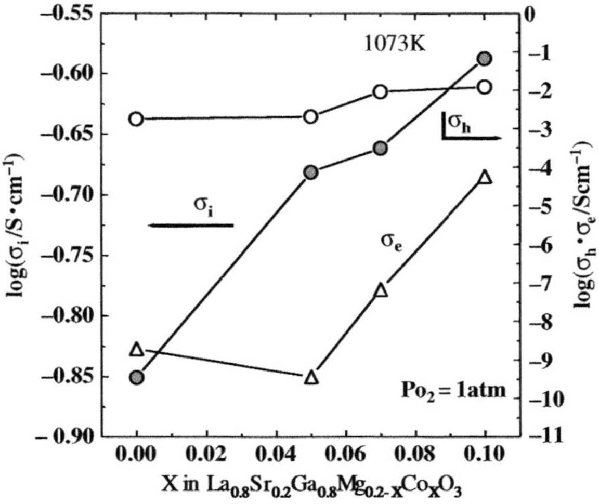

Fig. 4.15 Estimated ionic and electronic conductivity in LSGM at 1073 K as a function of Co content under $P_{O2} = 10^{-5}$ atm

4.5.5 Diffusivity of Oxide Ion

The diffusivity of oxide ions in LSGM was further studied by ^{18}O tracer diffusion measurements [37]. Diffusion of oxide ion in perovskite oxide is explained in detail in Chapter 5. LSGM exhibits large values of diffusion coefficient, and the observed fast diffusion in LSGM originates from the higher mobility of oxide ions in the perovskite structure as compared with the fluorite structure (Table 4.2), presumably due to a large free volume in the lattice.

Recently, atomic simulation of the oxygen transport in the perovskites, in particular, $LaGaO_3$, were performed based on quantum chemistry [38, 39]. In the case of perovskite oxides, the migrating oxygen ion must pass through the triangular orifice defined by two A-site (La^{3+}) ions and one B-site ion. As a result of lattice relaxation during oxygen ion migration, it was suggested that there is a small deviation from the direct path for oxygen migration, as illustrated schematically in Fig. 4.16. Indeed the calculations predicted a curved path around the

Table 4.2 Comparison of mobility of oxide ion in selected fluorite and LSGM oxide at 1073 K (Cited from Ref. 54)

	Dt /em^2/s	Ea /eV	δ	$[V_o^{\cdot\cdot}]$ /cm^3	D cm^2/s	μ cm^2/Vs
$Zr_{0.81}Y_{0.19}O_{2-\delta}$	6.2x10^{-8}	1.0	0.10	2.95x10^{21}	1.31x10^{-6}	1.41x10^{-5}
$Zr_{0.858}Ca_{0.142}O_{2-\delta}$	7.54x10^{-9}	1.53	0.142	4.19x10^{21}	1.06x10^{-7}	1.15x10^{-6}
$Zr_{0.85}Ca_{0.15}O_{2-\delta}$	1.87x10^{-8}	1.22	0.15	4.43x10^{21}	2.49x10^{-7}	2.69x10^{-6}
$Ce_{0.9}Gd_{0.1}O_{2-\delta}$	2.70x10^{-8}	0.9	0.05	1.26x10^{21}	1.08x10^{-6}	1.17x10^{-5}
$La_{0.9}Sr_{0.1}Ga_{0.8}Mg_{0.2}O_{3-\delta}$	3.24x10^{-7}	0.74	0.15	2.53x10^{21}	6.4x10^{-6}	6.93x10^{-5}
$La_{0.8}Sr_{0.2}Ga_{0.8}Mg_{0.2}O_{3-\delta}$	4.13x10^{-7}	0.63	0.20	3.34x10^{21}	6.12x10^{-6}	6.62x10^{-5}
$La_{0.8}Sr_{0.2}Ga_{0.8}Mg_{0.125}Co_{0.085}O_{3-\delta}$	4.50x10^{-7}	0.42	0.1645	2.78x10^{21}	8.21x10^{-6}	8.89x10^{-5}

D_t: Tracer diffusion coefficient, D: Self diffusion coefficient

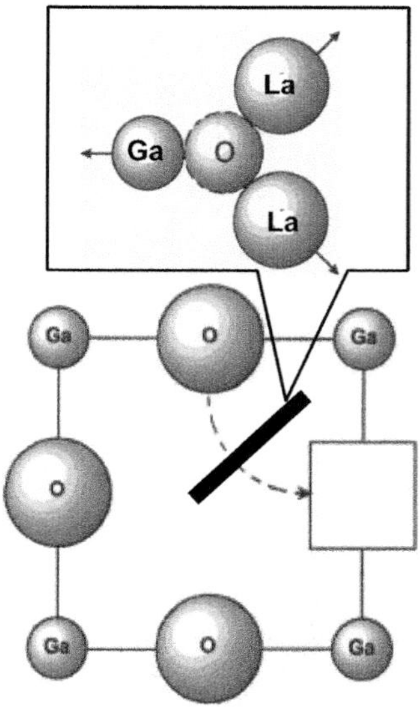

Fig. 4.16 Calculated path for oxygen vacancy migration. □: Vacancy

octahedron edge with the saddle point located away from the adjacent B-site cation. Similar experimental results are discussed in Chapter 6. Further detailed molecular dynamic calculations were also done for the LSGM system. Figure 4.17 shows the calculated mean square displacement of constituent atoms as a function of time. It is evident that the mean displacement between O–O ions expanded with time; however, the positions of other ions remain essentially constant, which suggests only diffusion of oxygen but not of cations in LSGM. The diffusion coefficients were calculated according to the random walk theory by using the slope in Fig. 4.17, and the results are compared in Fig. 4.18 with the diffusion

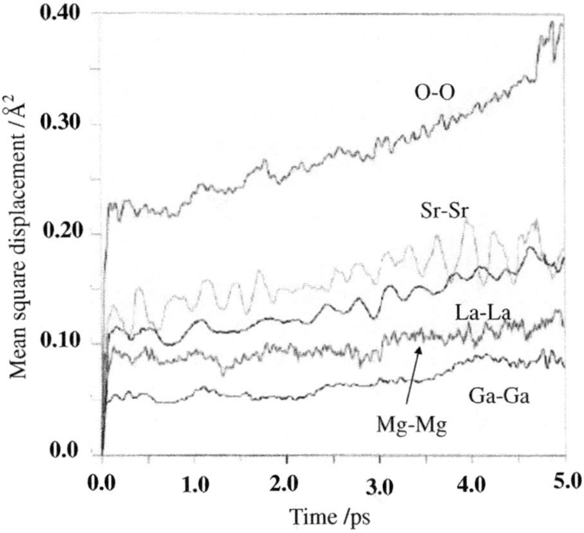

Fig. 4.17 Mean displacement of constituent atoms as a function of time in LSGM

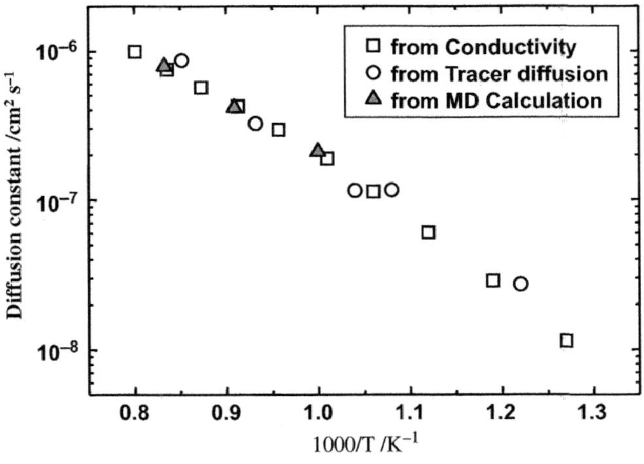

Fig. 4.18 Comparison of diffusion coefficient in LSGM estimated from ionic conductivity and tracer diffusion measurements and calculated by molecular dynamic calculations

coefficients measured during the tracer diffusion experiments. A good agreement of the values of oxygen diffusion coefficients estimated from the results of the computer simulation and experimentally measured in tracer diffusion measurements is observed, suggesting that the mobile oxygen vacancies and substituted ions in this oxide behave as ideal noninteracting point defects.

Furthermore, the defect binding energy in $LaGaO_3$ was also calculated by Islam et al. [38, 39], and Table 4.3 summarizes the calculated cluster binding

Table 4.3 Calculated cluster binding energies (E_b) for some selected dopant on the La and Ga sites in $LaGaO_3$

Location	Cluster pair	Cluster binding energy (E_b/eV/defect)
La site	Sr^{2+} -oxygen vacancy	–0.01
	Ca^{2+} -oxygen vacancy	–0.10
Ga site	Mg^{2+}-oxygen vacancy	–0.90
	Co^{2+} -oxygen vacancy	–0.87
	Ni^{2+} -oxygen vacancy	–0.91
	Cu^{2+} -oxygen vacancy	–0.65

energies (E_b) for some selected dopants on the La and Ga sites. The low binding energy calculated for the Sr substitution on the La site could be assigned to the similarity in ionic size of Sr with that of La, leading to smaller local stresses in the lattice. In contrast, rather large cluster binding energies of about 0.9 eV are estimated for B-site dopants, suggesting that the oxygen vacancies tend to be trapped around dopants in the Ga site. Although dopants on the Ga sites are effective for expanding the unit cell volume and improving the solubility of Sr on the La site, observed binding of oxygen vacancies may decrease the oxygen vacancy diffusivity. The results of computer calculations presented in Table 4.3 suggest that doping of Cu on the Ga site has the smallest binding energy; however, the electrical conductivity of LSGM decreased when doped with Cu^{2+}, presumably because of the low chemical stability of Cu^{2+}. Consequently, from the aspect of chemical stability, Co^{2+}/Co^{3+} doping is more desirable experimentally. In any case, due to the high crystal symmetry, it is evident that the perovskites are interesting subjects for a computer simulation of ion conductivity, in particular, oxygen ion diffusivity.

4.6 Performance of a Single Cell Using LSGM Electrolyte

The applications of LSGM as the electrolyte in fuel cells are now commonly investigated. Figure 4.19 shows the temperature dependence of the maximum power density and the open circuit potential (OCV) of the cell with $Sm_{0.5}Sr_{0.5}CoO_3$ cathode and Ni anode [40]. It is seen that open circuit potential (OCV) increased with the decrease in the operating temperature; the results are in good

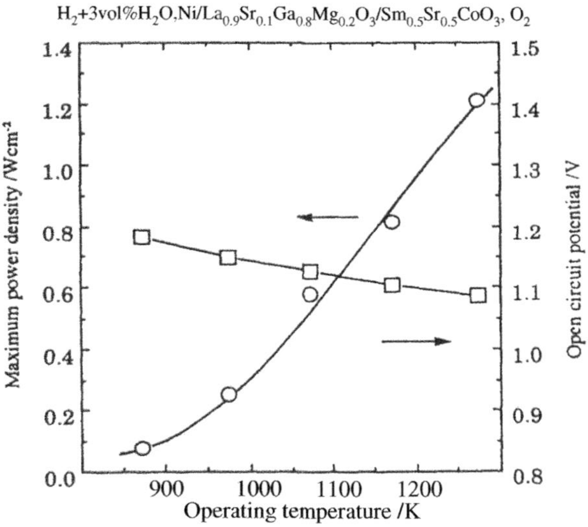

Fig. 4.19 Temperature dependence of the maximum power density and the open circuit potential (OCV) of the cell with $Sm_{0.5}Sr_{0.5}CoO_3$ cathode o and Ni anode. Thickness of electrolyte was 0.5 mm

agreement with the theoretical values estimated from the Nernst equation. Furthermore, the power density was improved by using $Sm_{0.5}Sr_{0.5}CoO_3$ as cathode at all temperatures studied, comparing with that of $LaCoO_3$-based conventional cathode. The maximum power density was higher than 1.0 and 0.1 W/cm^2 at 1273 and 873 K, respectively, in spite of the 0.5-mm thickness of the electrolyte [40]. In comparison with the power densities of the cell utilizing YSZ electrolyte, a reasonably high power density was achieved at 873 K. Goodenough and coworkers also investigated the application of $LaGaO_3$-based oxides as the electrolyte in fuel cells [41]. Similar large values of power density were reported at intermediate temperatures with $La_{0.6}Sr_{0.4}CoO_3$ cathode and Ni-La-doped CeO_2 cermet anode [42]. Due to high power density, LSGM has been attracting much interest as a promising SOFC electrolyte for intermediate temperature SOFC.

Oxide ion conductivity in LSGM is further improved by doping Co to Ga site, albeit the hole conductivity increases. Figure 4.20 shows the open circuit potential as well as the maximum power density at 1073 K as a function of Co content in $LaGaO_3$-based oxide electrolyte [40]. The open circuit potential (OCV) decreases monotonically with increasing Co concentration. In particular, the decrease in OCV is significant when the Co content on the Ga site is higher than 10 mol%; this is caused by the increased hole conductivity due to electronic compensation of doped Co ions. The dependence of OCV on the amount of doped Co is in good agreement with that of the transport number of oxide ion (see Fig. 4.10). On the other hand, the power density increases with increasing Co content and attains a maximum value when 8.5 mol% Co is doped on the Ga sites. Improvement in the power density is simply explained by the enhanced oxide ion conductivity resulting from Co doping. When the amount of doped Co further increases, the hole conduction becomes significant, short-circuiting the

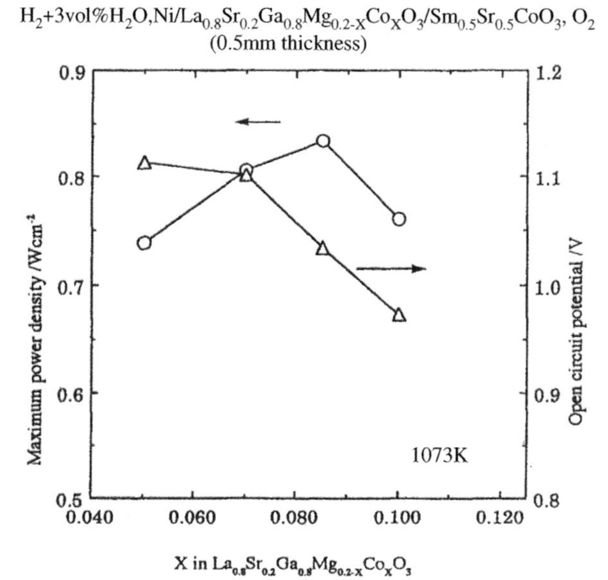

Fig. 4.20 Open circuit potential as well as the maximum power density at 1073 K as a function of Co content in $LaGaO_3$-based oxide electrolyte

cell and decreasing the power density. Consequently, the maximum power density is obtained when 8.5 mol% Co-doped LSGM electrolyte is used.

On the other hand, it is expected that the power density of the cell increased with decreasing thickness of the electrolyte, since the main internal resistance is due to IR losses. Figure 4.21 shows the power density of an H_2-O_2 cell with 0.18-mm-thick LSGMC-8.5 electrolyte at 1073 and 873 K. As expected, the power density of the cell increases with decreasing the thickness of the LSGMC electrolyte. However, the open circuit potential exhibited a tendency

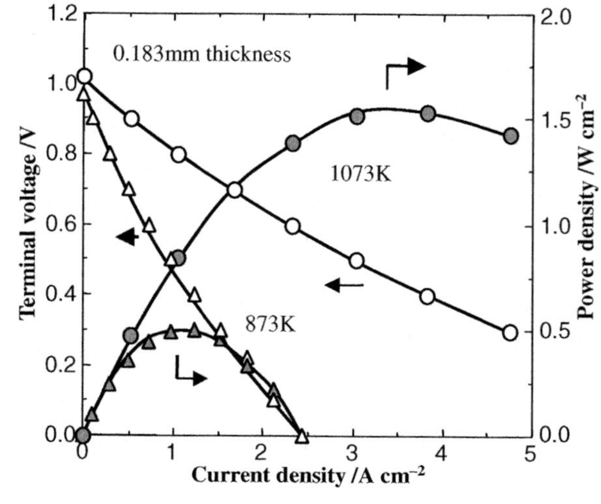

Fig. 4.21 Power-generating property of H_2-O_2 cell at 1073 and 873 K using 0.18-mm thickness LSGMC-8.5 as the electrolyte

to decrease with decreasing thickness of the electrolyte, which is explained by the increased amount of oxygen leakage with the decreasing thickness, as LSGMC exhibits a small hole conductivity. At 0.18-mm electrolyte thickness (as shown in Fig. 4.21), the open circuit potential decreased to 0.94 V at 873 K. Therefore, from the point of view of the conversion efficiency, it is expected that the optimal thickness of electrolyte will exist. However, it is obvious that an extremely large power density is attained for thinner electrolytes. The maximum power densities were 1.58 and 0.50 W/cm^2 at 1073 and 873 K, respectively (Fig. 4.21). The observed value of the power density at 873 K suggests that an SOFC operable at less than 873 K can be constructed using a LSGMC thin-film electrolyte. The power density of an even larger-sized cell (φ 150 mm) using $La_{0.8}Sr_{0.2}Ga_{0.8}Mg_{0.15}Co_{0.05}O_3$ has also been reported recently [42, 43]. Based on the high values of power density of the cell using LSGMC for an electrolyte, stacks of cells containing LSGMC electrolyte were developed under collaboration with Mitubishi Materials and Kansai Electronic Power Co, for which results are presented in Chapter 9.

4.7 Preparation of $LaGaO_3$ Thin-Film Electrolytes for Application at Temperatures Lower Than 773 K

Operating temperatures below 973 K are preferred to use metallic SOFC components (e.g., interconnects) and reduce the start-up time. Decrease of the operating temperature can be achieved by using $LaGaO_3$ thin films. In this section, the preparation of LSGM films is briefly described, and performance of SOFC cells using the LSGM thin-film electrolytes is discussed.

Much effort has been recently made to fabricate a SOFC single cell with LSGM thin film as electrolyte [44–46]. Because of partial electronic conductivity, LSGM is more suitable as an electrolyte film than LSGMC. Because of the reaction between LSGM and NiO during the fabrication of the cell, the high performance of SOFC as expected is not achieved in the cell using LSGM film [45, 46]. However, as already discussed, the addition of an La-doped CeO_2 buffer layer was found to be effective in preventing the reaction. On the other hand, in SOFC applications, the electrolyte films are normally deposited directly on the porous substrates. If a porous substrate is employed, the thickness of the electrolyte film will have some limitation, which is a relatively large value around 30–50 μm. However, a much thinner electrolyte film ($<$5 μm) can be easily fabricated on dense substrates. As a result, a new method for deposition of thin $LaGaO_3$-based oxide film for the intermediate-temperature SOFC was proposed [47]. The dense anode layer was used as a substrate for deposition of a thin LSGM electrolyte film by pulsed laser deposition (PLD). To prevent the reaction between the substrate and electrolyte, Sm-doped CeO_2 thin film was used as a buffer layer instead of La-doped CeO_2 because of the higher oxide ion conductivity.

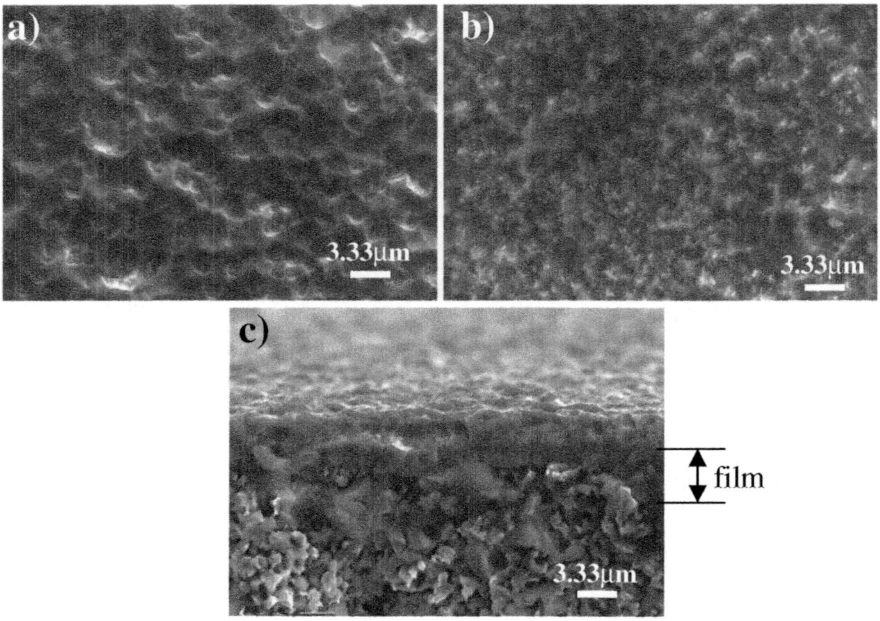

Fig. 4.22 Surface morphology and cross-sectional view of LSGM/SDC composite film prepared by pulsed laser deposition method: (**a**) as deposited; (**b**) after post annealing; (**c**) cross-section view after NiO reduction

Figure 4.22 shows the surface morphology and cross-sectional view of LSGM/SDC composite film before and after heat treatment. No pinholes or cracks were observed on the surface of the as-deposited and post-annealed films. The LSGM/SDC composite film was dense and uniform in thickness, as shown in Fig. 4.22 (b). Although the substrate became porous after the in situ reduction of NiO anode substrate, the electrolyte film was still gas tight, indicating that the strength of the LSGM/SDC film is high and that the dimensional changes of the anode substrate are not large during the reduction.

The power density of the SOFC single cell using $Sm_{0.5}Sr_{0.5}CoO_3$ as cathode is shown in Fig. 4.23. Because LSGM/SDC thin film was utilized as electrolyte in this cell, the high value of power density was expected even though the cell was only operated within the relatively low temperature range (673–973 K). The open circuit voltage (OCV) at 973 K reached a value of 1.08 V, which is close to the theoretical value of the electromotive force (1.15 V) for a SOFC operating under identical conditions and fed by H_2 and O_2 as fuel and oxidant, respectively. This result confirms that the LSGM/SDC composite film was gas tight. Upon lowering the operation temperature, the OCV gradually decreased, which could be ascribed to the enlarged gas leakage from the molten Pyrex glass seal. Although the porosity of the anode substrate was lower when compared with that of the typical anode deposited on an electrolyte, it is interesting that no concentration polarization was observed on the I–V curves even at very high

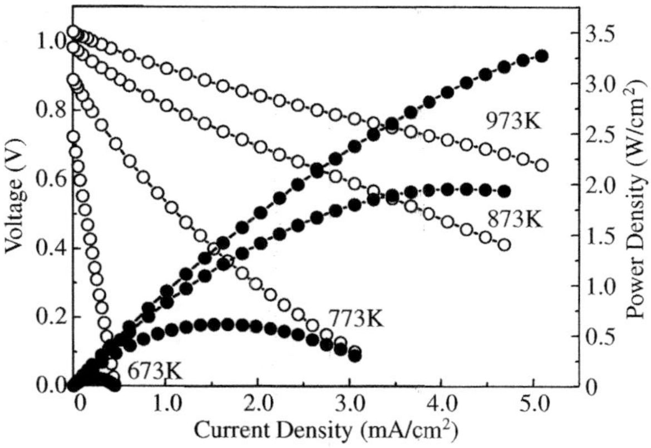

Fig. 4.23 Power-generating property of the SOFC single cell using $Sm_{0.5}Sr_{0.5}CoO_3$ as cathode

current densities; this means H_2 easily permeates through the anode substrate. The maximum power density of this cell reached the values of 3.27, 1.95, 0.61, and 0.08 W/cm^2 at 973, 873, 773, and 673 K, respectively. These values indicate that the thin-film SOFC can operate at temperatures lower than 773 K with a reasonably high power density. Recently, there were several attempts to prepare an LSGM thin film using conventional methods such as slurry coating. Although LSGM reacts easily with NiO, La-doped CeO_2 was found to be effective as the buffer layer, and reasonably high power densities were obtained on the cell using LSGM electrolyte. A successful operation of the cell using LSGM film at intermediate temperatures (673–873 K) may expand the possibility of SOFC operation at further decreased temperature.

4.8 Oxide Ion Conductivity in the Perovskite-Related Oxides

In this section, oxide ion conductivity in the oxides with perovskite-related structure is discussed. In particular, $Ba_2In_2O_5$-based oxide has been a subject of extensive research. Brownmillerite ($A_2B_2O_5$) is a perovskite-related structure in which one-sixth of the oxygen sites in the unit cell is vacant (see Chapter 1). In this oxide, the oxygen vacancies are ordered in the (101) direction at low temperature but became disordered at higher temperatures. The high oxide ion conductivity in $Ba_2In_2O_5$-based oxides was first reported by Goodenough in 1990 (Fig. 4.24) [48]. The oxide ion conductivity in this disordered $Ba_2In_2O_5$ phase is higher than that of YSZ, and several studies have been carried out to elucidate the effect of doping on ionic conductivity in this oxide [49, 50] Similar to ZrO_2, a high-temperature cubic phase can be stabilized at lower temperatures by aliovalent doping on the cation sites. Thus doping with Zr by Goodenough et al. [48] and Ga by Yao et al. [49] on the In site was found to be efficient in stabilizing the high-temperature cubic phase. The formation of the cubic phase resulted in enhancement of ionic conductivity at low temperatures and the disappearance of the discontinuity present in the Arrhenius plot for

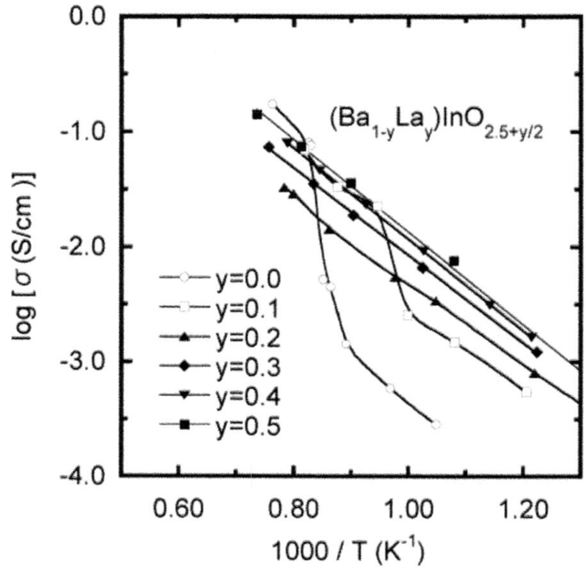

Fig. 4.24 Oxide ion conductivity in $Ba_2In_2O_5$-based oxide

$Ba_2In_2O_5$ (Fig. 4.24). Although the low-temperature conductivity was significantly improved by doping (as shown in Fig. 4.24), conductivity at high temperatures did not increase and even showed a slight reduction. Therefore, the high-temperature conductivity of doped $Ba_2In_2O_5$ is almost the same as that of YSZ.

Recently, Kakinuma et al. studied the effects of La^{3+} doping on the Ba site on the oxide ion conductivity in $Ba_2In_2O_5$ [51]. In the typical perovskite oxide ion conductor, the formation of oxygen vacancy is achieved by doping with lower valence cations; however, in the case of $Ba_2In_2O_5$, doping with the higher-valence La^{3+} cation on the Ba site resulted in the increase of oxide ion conductivity. Figure 4.25 shows oxide ion conductivity as a function of oxygen content. It is of interest to note that oxide ion conductivity increases almost monotonically with oxygen content, which can be explained by decreasing the degree of vacancies ordering resulted from the introduction of excess oxygen. Additionally, the introduction of Sr^{2+} decreases the unit cell volume and enhances oxide ion conductivity in this oxide. Despite the fact that isovalent doping is not expected to change the oxygen content, the substitution of Sr for the Ba site in $Ba_2In_2O_5$ was effective in increasing oxide ion conductivity. It is thought that the free volume plays an important role in improving the oxide ion conductivity, which is also suggested for perovskite oxide. The highest oxide ion conductivity was reported for $(Ba_{0.5}Sr_{0.2}La_{0.3})InO_{2.85}$ [49].

The application of Sr- and La-doped $Ba_2In_2O_5$ as the electrolyte in SOFCs was also reported [49]. Figure 4.25 shows the power density curves of the cell constructed of $(Ba_{0.5}Sr_{0.2}La_{0.3})InO_{2.85}$ electrolyte and $La(Sr)Mn(Fe)O_3$

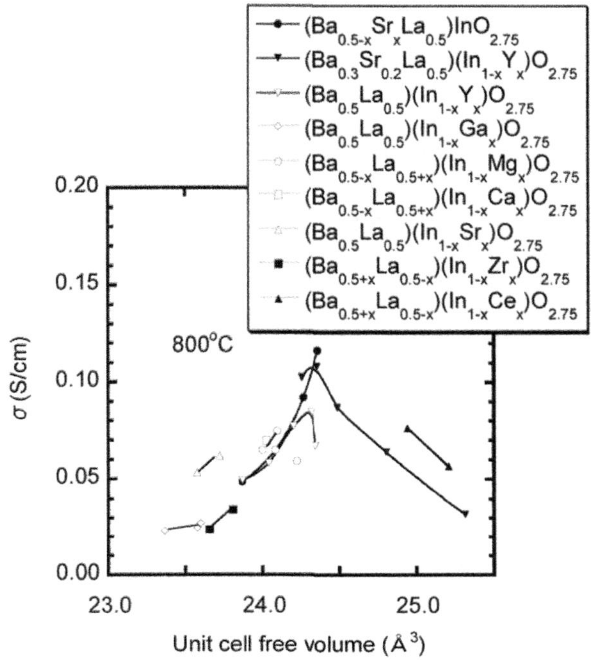

Fig. 4.25 Oxide ion conductivity as a function of oxygen content in $Ba_2In_2O_5$

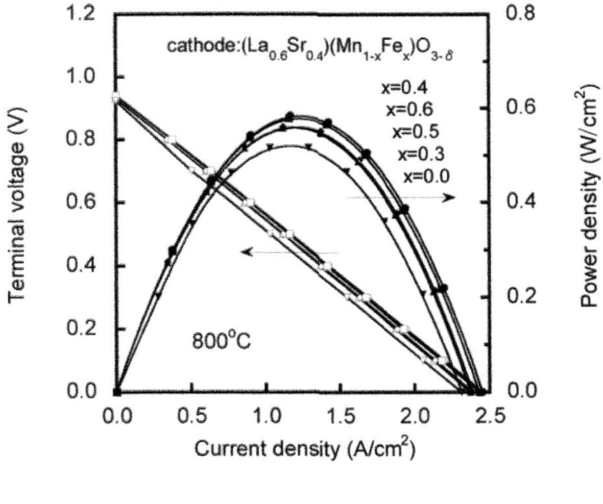

Fig. 4.26 Power generation curves of the cell utilizing $(Ba_{0.5}Sr_{0.2}La_{0.3})InO_{2.85}$ electrolyte and $La(Sr)Mn(Fe)O_3$ cathode (Cited from Ref. 49)

cathode. Open circuit potential was 0.93 V, which was slightly lower than the theoretical value due to the hole conduction at high P_{O2} range. Nevertheless, a reasonably high value of the maximum power density of almost 0.6 W/cm^2 was obtained at 1073 K. This power density suggests that $Ba_2In_2O_5$ perovskite-related oxides are also promising electrolytes in SOFC.

4.9 Summary

In this chapter, the current status of research of promising perovskite or perovskite-related oxide ion conductors for the electrolyte of SOFC is discussed. A large number of such non-fluorite oxide ion conductors have the cubic-like perovskite or perovskite-related structure. Recently, there were some reports of a new group of oxide ion conductors with non-fluorite structure. Some of the non-cubic oxide conductors such as $La_{10}Si_6O_{27}$ or La_2GeO_5 are highly interesting [52, 53]. Therefore, in the near future, there is high possibility that a new non-cubic oxide exhibiting extremely fast ion conductivity will be found. At present, the typical electrolyte for high-temperature SOFC, which operate at around 1273 K, seems to be predominantly Y_2O_3-stabilized ZrO_2; however, the preferred electrolyte material for intermediate temperatures has not yet been determined. The most promising candidate is considered to be an $LaGaO_3$-based oxide. If the operating temperature can be decreased to the temperature range of 673–873 K, the development of SOFCs for practical use would be greatly accelerated. In comparison with the proton-conducting electrolyte, one of the great advantages of SOFC using oxide ion-conducting electrolyte is the direct usage of hydrocarbon-containing fuels. Therefore, a simple fuel reforming system is enough for an SOFC system using an oxide ion conductor as electrolyte. The day when SOFCs will operate at such low temperatures is not far in the future. Perovskite oxides have an extremely high potential as an alternative oxide ion conductor to YSZ, and future development of oxide ion-conducting perovskite materials is greatly expected.

References

1. W. Nernst, Z. Elektrochem 6, 41 (1899)
2. Y. Arachi, H. Sakai, O. Yamamoto, Y. Takeda, N. Imanishi, Solid State Ionics, 121, 133 (1999)
3. J.A. Kilner, R.J. Brook, Solid State Ionics 6, 237 (1982)
4. A. Navrotsky, P. Simoncic, H. Yokokawa, W. Chen, T. Lee, Faraday Discussions, 134, 171 (2007)
5. T. Takahashi, H. Iwahara, Energy Conversion 11, 105–111 (1971)
6. T. Ishihara, H. Matsuda, Y. Takita, J. Electrochem. Soc. 141, 3444 (1994)
7. T. Ishihara, H. Matsuda, Y. Mizuhara, Y. Takita, Solid State Ionics 70/71, 234 (1994)
8. K. Nomura, T. Takeuchi, S. Kamo, H. Kageyama, Y. Miyazaki, Solid State Ionics 175, 553 (2004)
9. D. Lybye, F.W. Poulsen, M. Mogensen, Solid State Ionics 128, 91 (2000)
10. K. Nomura, S. Tanase, Solid State Ionics 98, 229 (1997)
11. M. Mogensen, D. Lybye, N. Bonanos, P.V. Hendriksen, F.W. Poulsen, Solid State Ionics 174, 279 (2004)
12. T. Ishihara, H. Matsuda, Y. Takita, J. Am. Chem. Soc. 116, 3801 (1994)
13. P. Majewski, M. Rozumek, F. Aldinger, J. Alloys Compd. 329, 253–258 (2001)
14. T. Ishihara, H. Matsuda, Y. Takita, Solid State Ionics 79, 147 (1995)
15. M. Feng, J.B. Goodenough, Eur. J. Solid State Inorg. Chem., 31, 663 (1994)
16. P.N. Huang, P. Petric, J. Electrochem. Soc. 143, 1644 (1996)

17. K. Huang, R. Tichy, J.B. Goodenough, J. Am. Ceram. Soc. 81, 2565 (1998)
18. K. Huang, J.B. Goodenough, J. Alloys Compd. 303–304, 454 (2000)
19. R.T. Baker, B. Gharbage, F.M.B. Marques, J. Electrochem. Soc. 144, 3130 (1997)
T. Ishihara, H. Furutani, M. Honda, T. Yamada, T. Shibayama, T. Akbay, N. Sakai,
20. H. Yokokawa, Y. Takita, Chem. Mater. 11, 2081–2088 (1999)
21. R.T. Baker, B. Gharbage, F.M.B. Marques, J. Electrochem. Soc. 144, 3130 (1997)
22. V.V. Kharton, A.A. Yaremchenko, A.V. Kovalevsky, A.P. Viskup, E.N. Naumovich, P.E. Kerko, J. Membr. Sci. 163, 307 (1999)
23. V. Thangadurai, A.K. Shukla, J. Gopalakrishnan, Chem. Commun. 2647 (1998)
24. T. Ishihara, T. Shibayama, M. Honda, H. Nishiguchi, Y. Takita, J. Electrochem. Soc. 147, 1322 (2000)
25. P. Datta, P. Majewski, F. Aldinger, J. Alloys Compd. 438, 232 (2007)
26. A. Matraszek, L. Singheiser, D. Kobertza, K. Hilpert, M. Miller, O. Schulz, M. Martin, Solid State Ionics 166, 343 (2004)
27. K. Huang, M. Feng, J.B. Goodenough, M. Schmerling, J. Electrochem. Soc. 143, 3630 (1996)
28. K. Yamaji, T. Horita, M. Ishikawa, N. Sakai, H. Yokokawa, Solid State Ionics 108, 415 (1998)
29. K. Huang, R. Tichy, J.B. Goodenough, C. Milliken, J. Am. Ceram. Soc. 81, 2581 (1998)
30. K. Huang, J.H. Wan, J.B. Goodenough, J. Electrochem. Soc. 148, A788 (2001)
31. H. Yokokawa, N. Sakai, T. Kawada, M. Dokiya, J. Elecrochem. Soc. 138, 2719 (1991)
32. T. Horita, K. Yamaji, N. Sakai, H. Yokokawa, A. Weber, E. Ivers-Tiffée, Solid State Ionics 133, 143 (2000)
33. H. Yokokawa, N. Sakai, T. Horita, K. Yamaji, Fuel Cell 1, 117 (2001)
34. T. Ishihara, S. Ishikawa, K. Hosoi, H. Nishiguchi, Y. Takita, Solid State Ionics, 175, 319 (2004)
35. K. Yamaji, T. Horita, M. Ishikawa, N. Sakai, H. Yokokawa, M. Dokiya, SOFC V, Electrochem. Soc., Proc. 97–18, 1041 (1997)
36. J.H. Kim, H.I. Yoo, Solid State Ionics 140, 105 (2001)
37. T. Ishihara, J.A. Kilner, M. Honda, Y. Takita, J. Am. Chem. Soc. 119, 2747 (1997)
38. M.S. Islam, Solid State Ionics 154–155, 75–85 (2002)
39. M.S. Khan, M.S. Islam, D.R. Bates, J. Phys. Chem. B. 102, 3099 (1998)
40. T. Ishihara, T. Shibayama, M. Honda, H. Nishiguchi, Y. Takita, Chem. Commun. 1227 (1999)
41. K. Huang, J.H. Wan, J.B. Goodenough, J. Electrochem. Soc. 148, A788 (2001)
42. J. Akikusa, K. Adachi, K. Hoshino, T. Ishihara, Y. Takita, J. Electrochem. Soc. 148, A1275–1278 (2001)
43. T. Yamada, N. Chitose, J. Akikusa, N. Murakami, T. Akbay, T. Miyazawa, K. Adachi, A. Hasegawa, M. Yamada, K. Hoshino, K. Hosoi, N. Komada, H. Yoshida, M. Kawano, T. Sasaki, T. Inagaki, K. Miura, T. Ishihara, Y. Takita, J. Electrochem. Soc. 148, A1712–A1714 (2004)
44. M. Sahibzada, B.C.H. Steele, D. Barth, R.A. Rudkin, I.S. Metcalfe, Fuel 78, 639 (1999)
45. J.W. Yan, Z.G. Lu, Y. Jiang, Y.L. Dong, Y.C. Yu, W.Z. Li, J. Electrochem. Soc. 149 (2002) A1132
46. Z.H. Bi, B.L. Yi, Z.W. Wang, Y.L. Dong, H.J. Wu, Y.C. She, M.J. Cheng, Electrochem. Solid-State Lett. 7 (2004) A105
47. J. Yan, H. Matsumoto, M. Enoki, T. Ishihara, Electrochem. Solid State Lett. 8, A389 (2005)
48. J.B. Goodenough, J.E. Ruiz-Diaz, Y.S. Zhen, Solid State Ionics 44, 21 (2000)
49. T. Yao, Y. Uchimoto, M. Kinuhata, T. Inagaki, H. Yoshida Solid State Ionics 132, 189 (2000)
50. V.V. Kharton, F.M.B. Marques, A. Atkinson, Solid State Ionics 174, 135 (2004)
51. K. Kakinuma, H. Yamamura, T. Atake. Defect Diffusion Forum 159, 242–244 (2005)
52. S. Nakayama, M. Sakamoto, J. Eur. Ceram. Soc. 18, 1413 (1998)
53. T. Ishihara, H. Arikawa, T. Akbay, H. Nishiguchi, Y. Takita, J. Am. Chem. Soc. 123, 203 (2001)
54. T. Ishihara, S. Ishikawa, M. Ando, H. Nishiguchi, Y. Takita, Solid State Ionics, 173, 9 (2004)

Chapter 5
Diffusivity of the Oxide Ion in Perovskite Oxides

J.A. Kilner, A. Berenov, and J. Rossiny

5.1 Introduction

There are a large number of complex metal oxides, with the general formula ABO_3, that form into the many and closely related perovskite-type structures. These materials show a very wide range of valuable physicochemical properties including ferromagnetism, catalytic activity, ferroelectricity, giant magnetoresistive effect, and ionic and mixed conductivity. As just mentioned, there are a number of structure types that fall under the collective name of perovskite, including the ideal cubic structure, rhombohedral, tetragonal, and orthorhombic distortions, and the hexagonal $GdFeO_3$ types. In this chapter, we present a review of the literature on the subject of oxygen ion diffusion in these materials, as this is fundamental to an understanding of the other physical properties. This is a limited review, as the literature on this subject is extensive, and the aim here is to present the trends that will aid in understanding the changes in oxygen transport between different compositions and materials under different conditions. To this end, the structural differences in the materials will be ignored to a first approximation, and this must be remembered when the comparisons of data are being made. One of the fundamental difficulties in following the trends that occur in the diffusivity of oxygen is the inherent multidimensional nature of the data for the materials concerned. For example, the diffusivity is dependent upon temperature, oxygen partial pressure, identity of the A and B cations, the degree of substitution on either the A or B sites, and finally any deviations from the ideal A to B 1:1 occupancy ratio. Most of the trends have not been fully explored, and the data are very sparse. To understand the major trends that have been already recognized, this chapter is divided into several sections covering many aspects of oxygen transport, from the basic concepts and methods of measurement, to a study of the oxygen transport rates in perovskite-related materials.

J.A. Kilner (✉)
Department of Materials, Imperial College, London, London SW7 2AZ, UK
e-mail: j.kilner@imperial.ac.uk

T. Ishihara (ed.), *Perovskite Oxide for Solid Oxide Fuel Cells*,
Fuel Cells and Hydrogen Energy, DOI 10.1007/978-0-387-77708-5_5,

5.1.1 Definitions of Diffusion Coefficients

When a gradient is imposed on the concentration of oxygen, c_o, in a material, an oxygen flux J_o is created, which is described by Fick's first law:

$$\vec{J_o} = -D.\vec{\nabla} c_o \tag{5.1}$$

where D is defined as the self-diffusion coefficient of oxygen. The continuity equation in a small element gives Fick's second law in the presence of a drift velocity v, created by any field (chemical or electrical):

$$\frac{dc_o}{dt} = \nabla \vec{J_o} = -D.\nabla^2 c_o - \bar{v}.c_o \tag{5.2}$$

Philibert [1] has shown that, in a chemical field, Eq. (5.2) can be simplified into the following equation, where D_{Chem} is the chemical diffusion coefficient:

$$\frac{dc_o}{dt} = -D_{Chem}.\nabla^2 c_o \tag{5.3}$$

The chemical diffusion coefficient is related to the self-diffusion coefficient by a term known as the thermodynamic enhancement factor. We will not explore the literature on D_{Chem} as this again is extensive and would only serve to obscure the main theme of this chapter.

5.1.2 The Oxygen Tracer Diffusion Coefficient

The self-diffusion coefficient is usually obtained from measurement of the tracer diffusion coefficient, in this case the oxygen tracer diffusion coefficient $D^* = f\,D$; here, f is called the correlation factor and represents the deviation from randomness of the jumps ($\approx$1). We will follow an analysis given in [2, 3]: for a material that has an oxygen deficiency accommodated by oxygen vacancies which are mobile, D^* can be derived in terms of atomistic parameters from random walk theory.

$$\begin{aligned} D^* &= \frac{z}{6} f(1-c)a_o^2 v_o \exp\left\{-\frac{\Delta G_m}{RT}\right\} \\ &= \frac{z}{6} f(1-c)a_o^2 v_o \exp\left\{-\frac{\Delta H_m}{RT}\right\} \exp\left\{\frac{\Delta S_m}{R}\right\} \end{aligned} \tag{5.4}$$

where z is the number of equivalent near-neighbor sites, $(1 - c)$ represents the fraction of unoccupied equivalent sites and can be replaced by $\left[V_o^{\bullet\bullet}\right]$ (using

Kröger-Vink notation [4]), the mobile oxygen vacancy concentration, expressed as a site fraction, a_0, is the distance between equivalent sites, ν_0 is a characteristic lattice frequency, and $\nu_0 \exp\{\frac{-\Delta H_m}{RT}\}$ is the jump frequency for the migrating ion. ΔH_m and ΔS_m are the enthalpy and entropy of the migrating ion.

Defining the term β as

$$\beta = \frac{z}{6} f \exp\left\{\frac{\Delta S_m}{R}\right\} \tag{5.5}$$

then,

$$D^* = \beta[V_o^{\cdot\cdot}]a_0^2\nu_0 \exp\left\{-\frac{\Delta H_m}{RT}\right\} \tag{5.6}$$

Defining D_v as the vacancy diffusion coefficient, i.e.:

$$D_v = \beta a_0^2 \nu_0 \exp\left\{-\frac{\Delta H_m}{kT}\right\} \tag{5.7}$$

Thus, the oxygen self-diffusion coefficient becomes

$$D^* = [V_o^{\bullet\bullet}]D_v \tag{5.8}$$

At this point, it is instructive to examine both components of this equation and the likely contribution to the tracer diffusion coefficient for these perovskite materials. Here the term $[V_o^{\bullet\bullet}]$ refers to the *mobile* vacancy concentration, which may be different from the *stoichiometric* vacancy concentration (i.e., that determined by the oxidation states of the constituent cations) because of vacancy trapping, as observed in the fluorite oxides [2], or due to vacancy ordering [5]. D_v contains the terms relevant to mobility of the vacancies, i.e., the ease with which the oxygen atoms can jump from an adjacent lattice site into a vacancy. Mizusaki et al. [6] have previously shown that data for D_v, the vacancy diffusivity, show remarkably little variation for a number of perovskite materials. This is a very interesting observation and one to which we return later. It is thus important to understand the changes that occur in the vacancy concentration in these materials and how this affects the oxygen self-diffusion coefficient.

One point to emphasize at this stage concerns the measurement of the self-diffusion coefficients, particularly in the mixed conducting perovskite materials. An extended review and description of diffusion measurement and techniques have been published by Philibert [1]. There are many electrochemical methods whereby various diffusion coefficients can be extracted from ceramic samples. In the main, data obtained using these methods must be treated with some circumspection because there are many sources of possible error, particularly with porous samples when gas diffusion down pores can occur. Sunde et al. [7] have recently pointed out some of these effects when using the potential step method. By far the most direct method of measurement is the use of isotope exchange followed by secondary ion mass spectometry (SIMS) depth profiling, as described in our earlier publications [8,9], often known as the isotope exchange depth profiling technique (IEDP). This technique yields the tracer diffusion coefficient, *D*, and the majority of the data included in this review were obtained using the IEDP/SIMS method.

5.1.3 The Surface Exchange Coefficient

The oxygen surface exchange coefficient, k, is another important kinetic parameter associated with the measurement of oxygen transport rates in these oxide materials. It is a measure of the neutral oxygen exchange flux crossing the surface of the oxide, at equilibrium, as described by the following reaction [10]:

$$\frac{1}{2}O_2 + V_o^{\bullet\bullet} + e' \leftrightarrow O_o^x \tag{5.9}$$

This flux will be dependent upon the surface vacancy concentration, the surface electron concentration, and the dissociation rate of the dioxygen molecule; however, at present, the rate-limiting step has yet to be identified. Kilner et al. [11] have derived a simple relationship for the surface exchange coefficient in terms of the bulk and surface vacancy concentrations, in an attempt to explain the apparent correlation found between the activation enthalpy for the surface exchange coefficient and the diffusion coefficient, in a number of La-based perovskites. Adler et al. [12] have also arrived at a similar relationship for k, by consideration of the AC electrode behavior of symmetrical cells with a "double" cathode structure. As already mentioned, the exact mechanisms of oxygen surface exchange remain elusive; however, the vacancy concentration is clearly a very important parameter.

On a more practical level, the influence of the surface exchange on cathode performance has been recently emphasized by both Adler et al. [12] and Steele [13]. They have both stressed the importance of a characteristic length L_c defined as the ratio of the two parameters D/k. This length is the changeover point at which the permeation flux of oxygen through a thin mixed conducting membrane (such as a cathode) changes from being limited by diffusion in the membrane to being limited by the surface exchange process. For a more detailed discussion of the interpretation of the surface exchange coefficient, the reader is referred to the work of Maier [14] and De Souza [15].

5.1.4 Defect Chemistry and Oxygen Transport

A large number of different materials adopt the perovskite structure, however in terms of the oxygen transport the most studied are materials for applications in solid oxide fuel cells (SOFCs) and oxygen separation membranes. The materials are usually 3,3-perovskites (with the general formula $A^{3+}B^{3+}O_3$) with lanthanum on the A site, a transition metal on the B site, and acceptor doped on the A site, usually with strontium. One of the main features of these mixed conducting perovskites is that they show a large range of stoichiometry, which can vary from hypostoichiometric to hyperstoichiometric; this will alter both the electrical conductivity and the oxygen exchange and transport properties. To understand the oxygen transport properties, we first look at the theoretical defect structure of these mixed conducting perovskites and then compare the predicted behavior with experimental results.

5.1.5 Defect Equilibria

The defect structure of these materials is far from simple, with a number of coincident defect equilibria contributing to the observed behavior. First, we must consider the intrinsic defect processes for a hypothetical mixed conductor $La_{1-x}Sr_xBO_{3\pm\delta}$, which undergoes Schottky disorder together with intrinsic electronic disorder. Following an analysis given in [16]:

$$\text{“0”} \rightarrow V_{La}''' + V_B''' + 3V_O^{\bullet\bullet} \tag{5.10}$$

$$\text{“0”} \rightarrow e' + h^{\bullet} \tag{5.11}$$

Equation (5.11) can also be viewed as the dissociation of the effectively neutral B cation (+3) into two charge states (+2 and +4):

$$2B_B^x \rightarrow B_B' + B_B^{\bullet} \tag{5.12}$$

For perovskites with variable valent B cations, the redox processes that occur in the lattice must also be taken into account. Oxidation of the lattice can lead to oxygen excess material in which cation vacancies form [17].

$$3/2O_2 + 6B_B^x \rightarrow 3O_O^x + V_{La}''' + V_B''' + 6\,B_B^{\bullet} \tag{5.13}$$

In this oxygen excess region, the formation of cation vacancies will have the effect of strongly decreasing the oxygen vacancy concentration through the Schottky equilibrium.

Compensation of the acceptor can occur either by electronic or by a vacancy mechanism, or perhaps more importantly for these materials, by a combination of the two.

$$2SrO + {}^{1}/{}_{2}O_2 + 2B_B^x \rightarrow 2Sr'_{La} + 3O_O^x + 2B_B^{\bullet} \tag{5.14}$$

$$2SrO \rightarrow 2Sr'_{La} + 2O_O^x + V_O^{\bullet\bullet} \tag{5.15}$$

The resulting neutrality condition then contains all the possible defect species

$$3\left[V'''_{La}\right] + 3\left[V'''_{B}\right] + \left[Sr'_{La}\right] + \left[B'_{B}\right] = 2\left[V_O^{\bullet\bullet}\right] + \left[B_B^{\bullet}\right] \tag{5.16}$$

To construct a Brouwer diagram to depict the relative concentration of these defects with P_{O2}, it is necessary to define a series of approximations to the full neutrality condition. The first of these corresponds to a region (R I) where $\delta > 0$ and where the material is hyperstoichiometric.

$$3\left[V'''_{La}\right] + 3\left[V'''_{B}\right] = \left[B_B^{\bullet}\right] \tag{5.17}$$

This region is followed by a stoichiometric region (R II) where the material acts as a controlled valence semiconductor and $\delta = 0$.

$$\left[Sr'_{La}\right] = \left[B_B^{\bullet}\right] \tag{5.18}$$

One of the most important features of these materials is that the compensation of the acceptor does not change abruptly from electronic to vacancy compensation, but there is an extensive region in which the compensation is mixed. Here the material is hypostoichiometric; δ now becomes negative and lies between 0 and –x/2, where x is the concentration of the acceptor (R III).

$$\left[Sr'_{La}\right] = 2\left[V_O^{\bullet\bullet}\right] + \left[B_B^{\bullet}\right] \tag{5.19}$$

Next there is a region (R IV) where the material is vacancy compensated and the value of δ is fixed at $\delta = -x/2$.

$$\left[Sr'_{La}\right] = 2\left[V_O^{\bullet\bullet}\right] \tag{5.20}$$

And finally, the material becomes reduced (R V), and we get the production of oxygen vacancies and electrons:

$$\left[B'_{B}\right] = 2\left[V_O^{\bullet\bullet}\right] \tag{5.21}$$

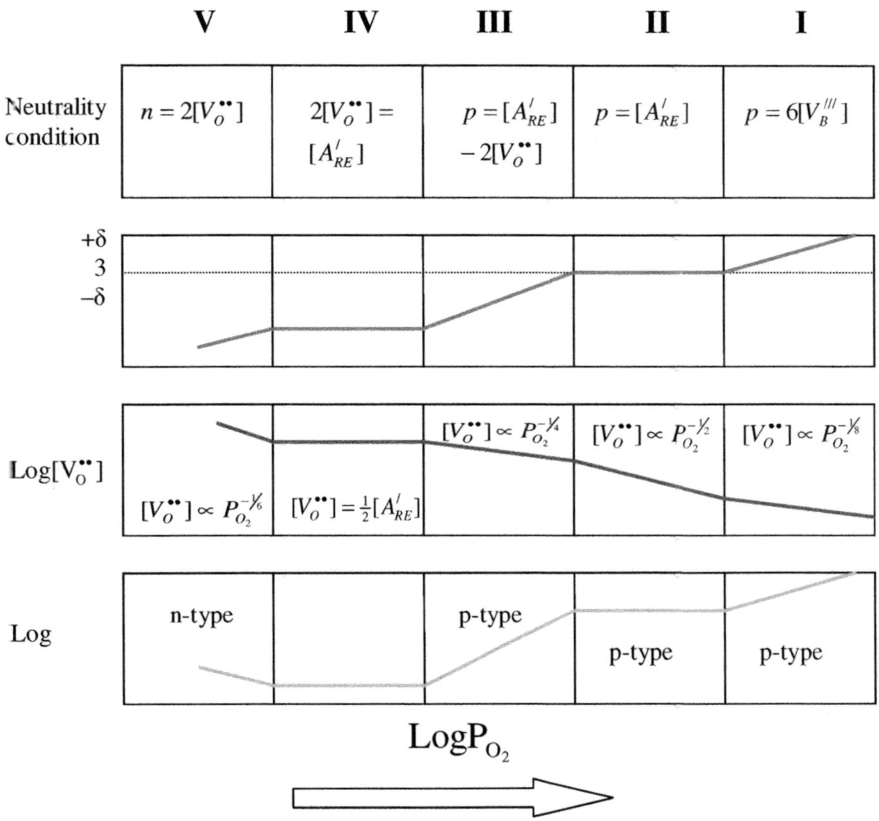

Fig. 5.1 Brouwer diagram for an acceptor-doped $RE_{1-x}A_xBO_3$ oxide showing the oxygen content, δ, oxygen vacancy concentration, $[V_O^{\bullet\bullet}]$, and electrical conductivity, σ, as a function of oxygen partial pressure [59]

The five regions are shown in Fig. 5.1 for a general 3,3 acceptor (A^{2+})-doped perovskite with the rare earth ion (RE) on the A site, $RE_{1-x}A_xBO_3$.

To compare with experimental data, we now need to examine a number of materials. Nonstoichiometry data for three 3,3 acceptor-doped perovskites with the B cations Mn, Fe, and Co are shown in Fig. 5.2 as a function of P_{O2} at 1000°C [18,19]. It is clear that the manganite is hyperstoichiometric (R I, $\delta > 0$) at high P_{O2}s ($P_{O2} \approx 1$), even though it is acceptor doped. As the P_{O2} is lowered, the manganite shows a plateau corresponding to R II behavior, i.e., $\delta = 0$. This behavior implies that the oxygen vacancy concentration will be very low indeed, and this remains true even for quite heavily acceptor doped material. The manganite material shown in Fig. 5.2 does not become hypostoichiometric (R III, $\delta < 0$) until oxygen partial pressures approaching 10^{-10} atm. are reached. Thus, under normal SOFC cathode operating conditions, the manganite materials are expected to have low vacancy concentrations and consequently

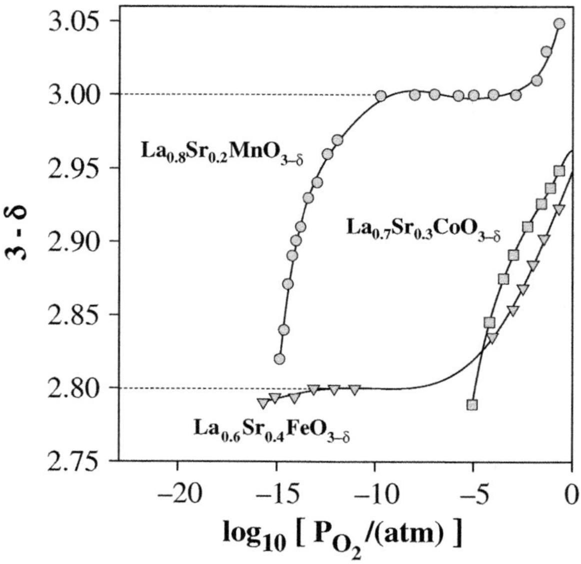

Fig. 5.2 Nonstoichiometry data for the acceptor-doped perovskites $La_{1-x}Sr_xBO_{3-\delta}$ (B = Mn, Fe, and Co) as a function of P_{O2} at 1000°C [18, 19]

low oxygen self-diffusivities. A more extensive set of nonstoichiometery isotherms for the manganite compositions has been published by Tagawa et al. [20].

In comparison to the manganite, the cobaltite and ferrite are seen to be in the mixed compensation region (R III) at high P_{O2}s, where the value of δ lies between 0 and $-x/2$. Thus, fairly high oxygen vacancy concentrations, and consequently oxygen diffusivities, are to be expected under normal cathodic conditions. For the ferrite, a plateau is seen at the lower P_{O2} that corresponds to R IV behavior, indicating that the acceptor is vacancy compensated and the material would become a predominantly ionic conductor. From these data, and the preceding analysis, it is obvious that there are significant changes in the defect populations in these acceptor-doped materials that depend on the temperature, oxygen partial pressure, and perhaps most importantly, the identity of the transition metal in the B-cation site. We now look at the way in which composition changes affect the oxygen transport properties of these oxides.

5.2 Diffusion in Mixed Electronic-Ionic Conducting Oxides (MEICs)

MEICs display some very interesting electrochemical properties and because of this are used as cathodes for the SOFC, both at high and intermediate temperatures, and in permeation membranes for the separation of oxygen.

The materials chosen for the development of commercial application are based on perovskite MEICs; however, most of them have very complex compositions involving both A- and B-site substitution. This situation can lead to problems in any discussion of the literature on these materials because families of materials are often referred to by the use of acronyms. For example, LSM and LSCF are often used to denote cathode materials, but these acronyms cover a range of materials of various levels of Sr substitution, B-site substitution, and often A-site deficiency, leading to sets of materials with very different properties. The exact composition must thus be used to compare the oxygen transport properties within and between each group of materials. As a further note of caution, it is also rare to see detailed analysis of the purity of the materials, which might affect the transport of oxygen, particularly in ceramic samples where the grain boundaries can affect the overall transport rates.

5.2.1 Effect of A-Site Cation on Oxygen Diffusivity

Two types of substitution into the A site in the perovskite structure are possible. Aliovalent doping occurs when the oxidation state of the substituting ion is different from the host ions, thus introducing effective charges for the substitute ion. To maintain electrical neutrality, these charges have to be compensated by the formation of oppositely charged defects. This state can be achieved either by changing the oxidation state of the B cations (electronic compensation) or by the formation of oppositely charged vacancies (ionic compensation), as was discussed earlier for acceptor-doped materials. Isovalent doping occurs when the oxidation state of the substituting and host ions is identical. As a result, no charges are introduced into the A-site sub-lattice and no charge compensation is required; however, there will be elastic strain effects because of the size mismatch of the host and the substituting cation.

Several combinations of elements have been used as A-site host and substitutional ions. Historically, lanthanum has been a host ion of a choice for the past several decades due to its large ionic radius and relative availability. Alkaline earth elements have been used as substitutional ions due to their close size match to the lanthanides and thermodynamic stability at the operational conditions of the SOFC. The effect of alkaline earth doping on the oxygen tracer diffusion coefficient in several families of perovskite compounds is shown in Fig. 5.3a (for $Sm_{1-x}Sr_xCoO_3$ at 793°C [21] and $La_{1-x}Sr_xCoO_3$ at 800°C [22–24]) and Fig. 5.3b (for $La_{1-x}Ca_xCrO_3$ at 900°C [25], $La_{1-x}Sr_xMnO_3$ at 900°C [23, 26], and $La_{1-x}Sr_xFeO_3$ at 1000°C [27]). All these data were obtained from experiments carried out at high oxygen partial pressures ($\sim$ 1 bar), to simulate operation in an SOFC cathode environment. Note from observation of Fig. 5.3a that although all the isothermal diffusivities increase with an increase in the acceptor doping, the changes seen for the cobalt-based materials (by six orders of

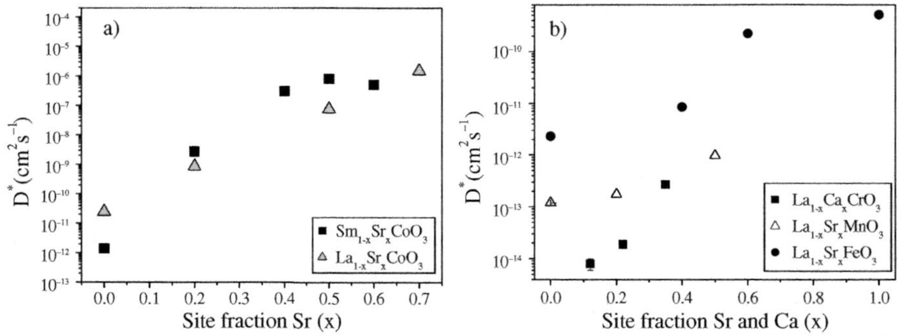

Fig. 5.3 Effect of aliovalent doping on the tracer diffusion coefficient in (**a**) $RE_{1-x}Sr_xCoO_3$ (RE = La, Sm at 800°C [21–24]) and (**b**) $La_{1-x}A_xTmO_3$ (A = Sr, Ca; Tm = Cr, Mn at 900°C [22, 25, 26], and Tm = Fe at 1000°C [27]). Nominal pressure during the experiments was 1 atm except for [22, 27] where a pressure of 34 and 40 torr was used, respectively

magnitude) are much larger that the relatively modest increases seen for the Cr and Mn analogues.

It is generally accepted that oxygen diffusion in these perovskite compounds occurs via oxygen vacancy mechanism. Consequently, the increase in the diffusion coefficient with doping is most likely caused by the increased concentration of oxygen vacancies. However, as we have seen, the identity of the B cation will determine which of the neutrality approximations is valid for each group of materials (i.e., RI, II, III, etc.). Thus, the spectacular increase of the oxygen tracer diffusion coefficient in cobaltites with doping is related to the large values of oxygen hypostoichiometry observed in those compounds (R III) [28]. However, some care is needed in this interpretation at this stage as it must be remembered that the diffusivity arises as a result of the product of the vacancy concentration and the vacancy diffusion coefficient (Eq. 5.8), which we examine in more detail below.

5.2.2 The Effect of B-Site Cation on Oxygen Diffusivity

Only one systematic study has been carried out to investigate the effect of B-site cation on oxygen diffusivity [23]. The effect of doping $La_{0.8}Sr_{0.2}MnO_{3\pm\delta}$ with cobalt is shown in Fig. 5.4. The substitution enhances the diffusivity by five orders of magnitude and the surface exchange coefficient by two orders of magnitude at 1000°C. This is quite an interesting finding, because the level of strontium substitution on the A site remains constant. Clearly, the nature of the B cation is again determining the nature of the neutrality approximation, and it would be apparent from the earlier data given in Fig. 5.2 that we are moving from $La_{0.8}Sr_{0.2}MnO_{3+\delta}$ (R I) to $La_{0.8}Sr_{0.2}CoO_{3-\delta}$ (R III).

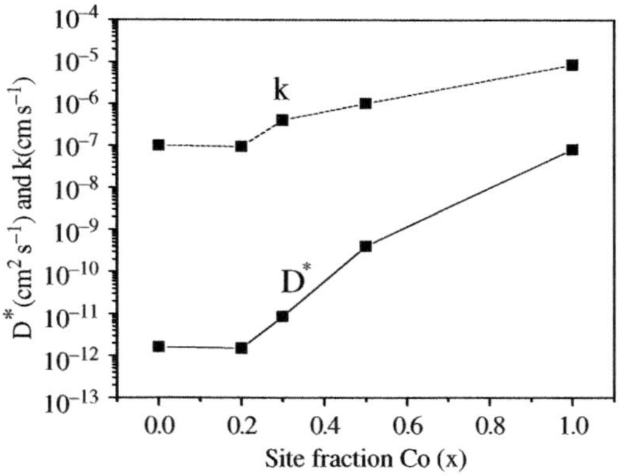

Fig. 5.4 The oxygen self-diffusion coefficient, D, and surface exchange coefficient, k, for $La_{0.8}Sr_{0.2}Mn_{1-x}Co_xO_{3-\delta}$ at 1000°C as a function of Co site y [23]

5.2.3 The Effect of A-Site Cation Vacancies on Oxygen Diffusivity

The effect of A-site vacancies on oxygen diffusion in perovskite materials has been studied in lanthanum-deficient $La_{1-y}MnO_3$ ($y = 0, 0.1$) [26]. It was found that introduction of 10% La vacancies has no effect on the values of oxygen tracer diffusion coefficient. The activation energy for oxygen diffusion increased from the value of 2.49 eV in a stoichiometric sample to 3.05 eV in $La_{0.9}MnO_3$. It was assumed that oxygen vacancies formed by the introduction of La deficiency, are bound in to the defect complexes (e.g., $\{V_{La}''' - V_O^{\bullet\bullet}\}'$). Similarly, no effect of A-site deficiency was observed on the parameters of chemical diffusion of oxygen in $(La_{0.85}Sr_{0.15})_sCoO_{3-\delta}$ (s = 0.98, 1 [29]).

5.2.4 Temperature Dependence of the Oxygen Diffusion Coefficient

In addition to following the isothermal effects, it is also quite important to understand the changes that take place in the observed activation energy for diffusion as a function of the composition. The ease of oxygen ion diffusion in perovskite structure is usually attributed to the value of the apparent activation energy estimated from the Arrhenius plots of diffusion coefficient. This apparent activation energy, E_a, however, consists of several terms, as indicated in the following equation:

$$E_a = \Delta H_m + \Delta H_f + \Delta H_a \tag{5.22}$$

Here ΔH_m is the enthalpy of vacancy migration, ΔH_f is the enthalpy of vacancy formation, and ΔH_a is the enthalpy of vacancy–dopant association, e.g.:

$$Sr'_{La} + V_o^{\bullet\bullet} \Leftrightarrow \{Sr'_{La}V_o^{\bullet\bullet}\}^{\bullet} \tag{5.23}$$

ΔH_f affects the *stoichiometric* vacancy concentration whereas ΔH_a affects the *mobile* vacancy concentration and ΔH_m only enters into the *vacancy* diffusion coefficient. Thus, if we can determine the stoichiometric vacancy concentration (or, more accurately, the mobile vacancy concentration), then we can extract the value of D_v, leading to a value for ΔH_m.

First, let us look at the overall effect of alkaline earth doping on the apparent activation energy of tracer diffusion in $La_{1-x}A_xTmO_3$ (A = Sr, Ca Tm = Mn, Fe, Cr), and $RE_{1-x}Sr_xCoO_3$ (RE = La, Sm), shown in Fig. 5.5(a) and 5.5(b), respectively. Although a large scatter of data is present, presumably due to differences in oxygen partial pressure during diffusional anneals, it is evident that the activation energy increases with doping for Mn- and Cr-based perovskites and decreases with doping for Fe- and Co-based perovskites.

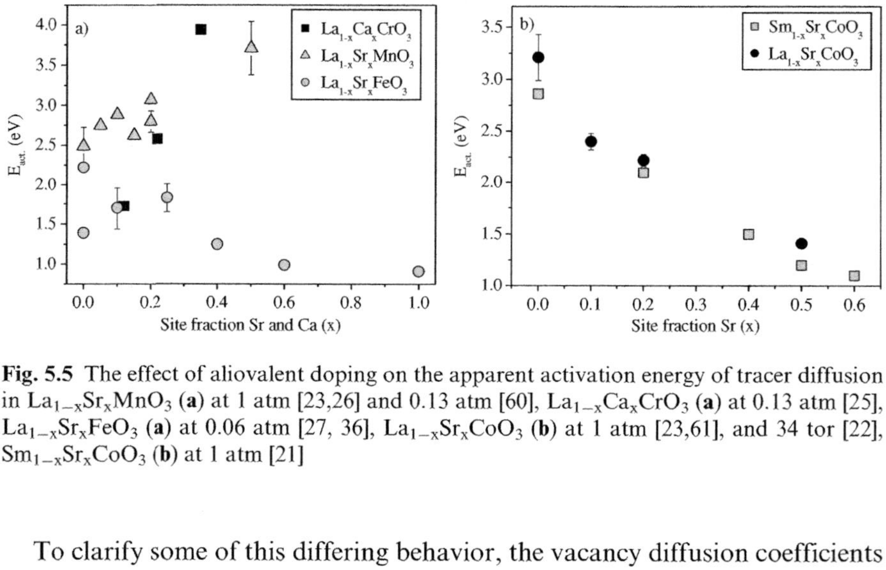

Fig. 5.5 The effect of aliovalent doping on the apparent activation energy of tracer diffusion in $La_{1-x}Sr_xMnO_3$ (**a**) at 1 atm [23,26] and 0.13 atm [60], $La_{1-x}Ca_xCrO_3$ (**a**) at 0.13 atm [25], $La_{1-x}Sr_xFeO_3$ (**a**) at 0.06 atm [27, 36], $La_{1-x}Sr_xCoO_3$ (**b**) at 1 atm [23,61], and 34 tor [22], $Sm_{1-x}Sr_xCoO_3$ (**b**) at 1 atm [21]

To clarify some of this differing behavior, the vacancy diffusion coefficients have been estimated from oxygen tracer diffusion experiments and using thermogravimetric studies to provide the stoichiometric vacancy concentration. A correlation factor, *f*, of 0.69 was used in the calculation. The calculated values of the vacancy diffusion coefficient in several families of perovskite materials are shown in Fig. 5.6. Remarkably, and as mentioned earlier, the values calculated for perovskite oxides, with significantly different oxygen diffusion coefficients and temperature dependencies, appear to have very similar values of vacancy diffusion coefficient. The activation energy of vacancy diffusion in the perovskite structure is ~0.94 ± 0.07 eV *irrespective* of the nature of the constituent ions. This value is close to the migration enthalpy for oxygen vacancies calculated by atomistic simulation techniques for 0.67 eV ($LaMnO_{3-\delta}$) [30] and in the range 0.6–0.8 eV reported for several perovskites by Islam [31].

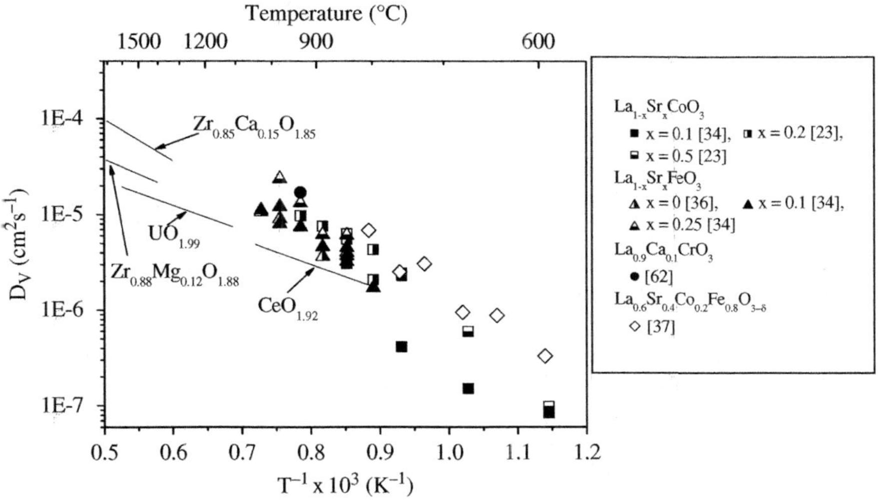

Fig. 5.6 Temperature dependence of the oxygen vacancy diffusion coefficient in perovskites: $La_{1-x}Sr_xCoO_{3-\delta}$ ($x = 0.1$ [34], $x = 0.2$ [23], $x = 0.5$ [23]), $La_{1-x}Sr_xFeO_{3-\delta}$ ($x = 0$ [36], $x = 0.1$ [34], $x = 0.25$ [34]), $La_{0.9}Ca_{0.1}CrO_{3-\delta}$ [62], $La_{0.6}Sr_{0.4}Co_{0.2}Fe_{0.8}O_{3-\delta}$ [37]. Data for fluorite-based oxides ($Zr_{0.85}Ca_{0.15}O_{1.85}$, $Zr_{0.88}Mg_{0.12}O_{1.88}$, $UO_{1.99}$, and $CeO_{1.92}$) have been taken from reference [34]

There are several very interesting implications from this finding. It would suggest that the differences seen between the activation energies measured for the different perovskite materials are the result of changes in the values of either the association or vacancy formation energies. It is difficult to discriminate between the two components; however, some observations are helpful.

1. The majority of the materials under discussion here are composed of perovskites made nonstoichiometric by the substitution of La by Sr. The extent of vacancy trapping, of the form shown in Eq. (5.23), can be estimated from atomistic calculations of the association enthalpy, ΔH_a. This value has been calculated by Islam [32] for the related Sr-doped lanthanum gallate material to be essentially *zero*, implying that trapping would not occur and that the vacancies are essentially free to participate in the migration process. The major part of this trapping energy has been shown to be the elastic contribution due to the size mismatch between the host and the substitutional cation [2], and hence the close size match of the La^{3+} (1.36 Å) and Sr^{2+} (1.44 Å) ions leads to a minimization of this term.
2. The close match of D_v from very different materials was obtained using the *stoichiometric* vacancy concentrations determined by Thermogravimetric Analysis (TGA); this implies that all these vacancies are mobile and hence that trapping is negligible.

If we can discount the effects of vacancy trapping, then the major differences seen in the activation enthalpy are ascribable to differences in the vacancy formation energy. Tagawa et al. [33] have noted that the partial molar enthalpy for oxidation/reduction in the LSM materials is substantial and thus account for the large values of activation energy seen in these materials.

The decrease in activation energy seen for the cobalt- and iron-based perovskites with increasing the substitution of La with Sr is also caused by changes in the vacancy formation component. In early work the energy of vacancy formation was calculated to decrease with alkaline earth doping in $La_{1-x}Sr_xFeO_{3-\delta}$ and $La_{1-x}Sr_xCoO_{3-\delta}$ [34]. Later, Lankhorst and Bouwmeester [35] provided a model for this effect for $La_{1-x}Sr_xCoO_{3-\delta}$ by considering terms due to the gradual filling of a broad conduction band by the electrons produced by vacancy formation.

Interestingly, the vacancy diffusion coefficients in several oxides with the fluorite structure have also been included in Fig. 5.6 and display values close to the ones found in the perovskites but with a slightly lower activation energy of 0.47–0.62 eV [36]. This result implies that the vacancy diffusion coefficient (mobility) of the oxygen vacancies in these different oxide structures is very close, and that the major discriminator for oxygen diffusion is the free vacancy concentrations and the degree of vacancy trapping. The lack of trapping in the Sr-substituted La perovskites is certainly an important factor in achieving the very high oxygen diffusion coefficients measured for these materials.

5.2.5 *The Effect of Oxygen Pressure*

Rather surprisingly little work has been done on the effect of oxygen partial pressure on the diffusivity of oxygen, probably because of the experimental difficulties in obtaining reliable pressure dependence data. One interesting study that adds weight to the previous argument about the vacancy diffusion coefficient is shown in Fig. 5.7 [37]. This work on the mixed conducting, mixed B-site perovskite $La_{0.6}Sr_{0.4}Co_{0.2}Fe_{0.8}O_{3-\delta}$ shows the oxygen diffusivity decreasing with increasing oxygen pressure at 800°C. If the stoichiometric vacancy concentration determined from TGA measurements taken at the same temperature is used to derive the vacancy diffusion coefficient, we see again that the isothermal vacancy diffusion coefficient is constant and corresponds well to the values obtained in Fig. 5.6.

5.3 Oxygen Diffusion in Ionic Conducting Perovskites

There are several La-based perovskites that display predominantly ionic conductivity: these are the materials based on $LaYO_3$ [38], $LaAlO_3$ [39], and $LaGaO_3$ [40]. In the gallate, the oxygen ion conductivity is enhanced by doping both the

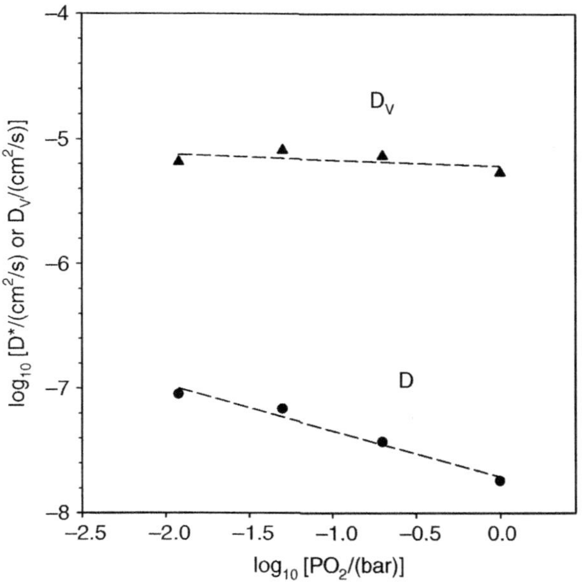

Fig. 5.7 Dependence of the tracer diffusion coefficient D^* and the vacancy diffusion coefficient D_V on the partial pressure in $La_{0.6}Sr_{0.4}Co_{0.2}Fe_{0.8}O_{3-\delta}$ at 800°C [37]

A- and B sites with Sr and Mg, respectively. The oxygen ion diffusivity is easily obtained in these materials from the ionic conductivity, and only a few measurements have been made of the tracer diffusivity. For the main material of the series $La_{1-x}Sr_xGa_{1-y}Mg_yO_{3-(x+y)/2}$ (LSGM 8282) [40], the measured oxygen diffusivity is higher than that in 9.5 mol% yttria-stabilized zirconia (YSZ), as shown in Fig. 5.8. This finding again is most probably due to the finding in the section on MEICs that there is very little trapping of the vacancies in these La-based perovskites, yielding very high mobile vacancy concentrations.

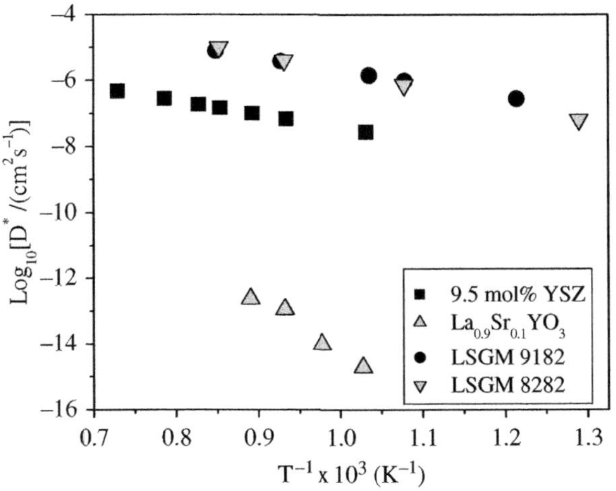

Fig. 5.8 Arrhenius plot of the oxygen ion diffusivity in ionic conductors with fluorite (9.5 mol% YSZ [63]) and perovskite ($La_{0.9}Sr_{0.1}YO_3$ [38], LSGM 9182 [40], and LSGM 8282 [40]) structures

5.4 Oxygen Diffusion in Perovskite-Related Materials

There is a large number of materials with structures related to the parent perovskite structure; these include the Ruddleston-Popper series ($A_{n+2}B_{n+1}O_{3n+4}$), of which the first member has the K_2NiF_4 structure ($n = 1$). Materials of this type have been investigated recently for their oxygen diffusion properties, most notably the rare earth-based $RE_2NiO_{4\pm\delta}$. These materials have an oxygen excess at high oxygen partial pressures and display diffusion by oxygen interstitials. These materials, although interesting, are not reviewed further in this chapter.

Some interesting findings have very recently been published on materials with the double perovskite structure. Doubling the perovskite formula gives the composition $A_2B_2O_6$. By having 50% substitution on the A site by a substitutional (A*) ion gives the formula $AA^*B_2O_6$. If the substitutional ion has a lower valence, then a material such as $AA^*B_2O_{5+\delta}$ can be obtained. It has been recently recognized that ordering of A and A* ions can play an important role in the oxygen diffusion in these perovskite-related materials. High-resolution electron microscopy study of the double perovskites $GdBaCo_2O_{5+\delta}$ ($0 \leq \delta \leq 1$) revealed that Gd and Ba ions were ordered in the alternative (0 0 1) layers with oxygen vacancies located predominantly in the GdO planes [41]. This particular arrangement of oxygen vacancies appeared to facilitate oxygen transport. The high values of oxygen tracer diffusion (around 10^{-9} cm^2 s^{-1} at 575°C) and low values of the activation energy for oxygen tracer diffusion of 0.60 eV were observed in $GdBaCo_2O_{5+\delta}$ [42] and $PrBaCo_2O_{5+\delta}$ [43].

5.5 Correlations Between Oxygen Diffusion Parameters

Many attempts have been made to rationalize the differences seen in the diffusion of oxygen in the perovskites in terms of a simplified parameter such as the tolerance factor, $t \equiv \frac{(r_A+r_O)}{\sqrt{2}(r_B+r_O)}$. Hayashi et al. [44] evaluated a large amount of available data on the ionic conductivity in perovskite materials. They came to the conclusion that the highest ionic conductivity occurred when the perovskite has a tolerance factor around 0.96, large specific free volume (unit cell volume minus volume of constituent ions), and a ratio of dopant to host ion ionic radii of around 1.05. As a result, perovskites with Sr substitution for La exhibit the highest values of ionic conductivity. In addition, computer simulations showed that Sr ion substituted on La site has the lowest solution energy in several series of perovskite compounds [45,46]. Several attempts have been made to correlate parameters of diffusion or ionic conductivity (mostly the activation energy) with the structure and chemical composition of the perovskite material [11, 44, 47–49]. For example, a decrease in the activation energy of oxygen diffusion in perovskites was observed with the decrease of the average metal–oxygen bond [47], increase in the free volume [47], and increase in the critical radius of saddle point, which is traversed by the diffusing oxygen ions and defined by the triangle formed by two

A cations and one B cation [48]. Furthermore, it has been observed that correlations exist between measured transport parameters, e.g., the relationship between the self-diffusion coefficient and the surface exchange coefficient [11].

Recently, the analysis of a large set of available tracer diffusion data in perovskite materials revealed a correlation between the measured values of pre-exponential coefficient and activation energy during oxygen diffusion [49]. In acceptor-doped $LaMnO_3$ $LaCoO_3$, $LaFeO_3$, $LaCrO_3$, and donor-doped $ATiO_3$ (A = Ba, Sr, Ca), a linear correlation was observed between the activation energy of the process and the logarithm of the pre-exponential coefficient (Fig. 5.9). This correlation, called the Meyer–Neldel rule (MNR), or compensation law, is observed in a wide range of thermally activated processes in physics, chemistry, and biology [50]. For example, the parameters of silicon diffusion in silicate

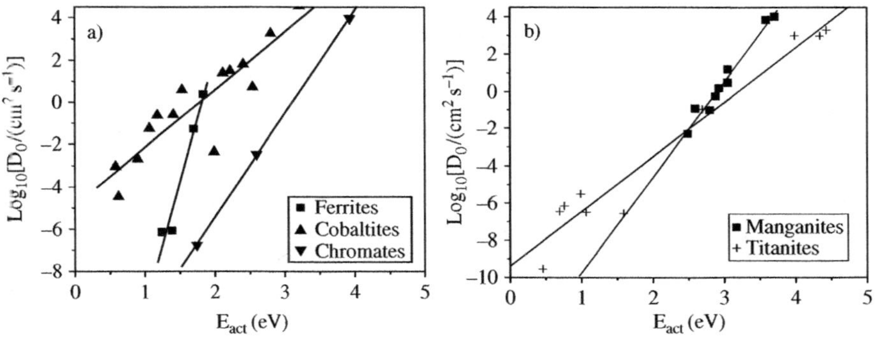

Fig. 5.9 Relationships between the activation energy of oxygen tracer diffusion and the logarithm of pre-exponential coefficient in ferrites (a), chromates (a), cobaltites (a), manganites (b), titanates (b). The lines are a guide to the eye

materials [51], proton diffusion in perovskite-type oxides [52], and Pd self-diffusion [53] were all shown to obey the Meyer–Neldel rule.

The expression for the tracer diffusion coefficient (Eq. (5.4)) can be rewritten as follows:

$$D^* = \frac{z}{6} f(1 - c_0) a_0^2 \nu_0 \times \exp\left\{\frac{\Delta S_m + \Delta S_f + \Delta S_a}{R}\right\} \times \exp\left\{-\frac{\Delta H_m + \Delta H_f + \Delta H_a}{RT}\right\} \qquad (5.24)$$

where all symbols have the meanings defined previously. As discussed earlier, several processes could take place during oxygen diffusion, namely, oxygen ion migration, vacancy formation, and dissociation of the oxygen vacancies from trap sites (e.g., dopant cations). Consequently, the entropies of migration, ΔS_m; vacancy formation, ΔS_f; defect association, ΔS_a; and the enthalpies of migration, ΔH_m; vacancy formation, ΔH_f and defect association, ΔH_a; are included in Eq. (5.24). Although the nonexponential parameters in Eq. (5.24) depend on the chemical composition of perovskite material and the environmental parameters (predominantly temperature and oxygen partial pressure), it is doubtful that their variation can result in the observed values of the pre-exponential coefficient, which differ by several orders of magnitude. Indeed, the correlation factor, f, is estimated to be around 0.67 in perovskite structure [34]; the lattice parameter of a pseudo-cubic perovskite cell, a_0, is around 4 ± 1 Å for a large number of perovskite materials [54]; the characteristic lattice frequency, ν_0, is around 10^{12}–10^{13} Hz [55]; and the number of equivalent near-neighbor sites, z, is 6 in the ideal cubic perovskite. The major variation among the nonexponential terms is expected for the fraction of unoccupied equivalent sites $(1 - c)$. For example, oxygen stoichiometry, δ, varies by three orders of magnitude in Sr-doped cobaltite, $La_{1-x}Sr_xCoO_{3-\delta}$ [18, 56]. At the same time, the variation of D_0 of more than nine orders of magnitude is observed (Fig. 5.9a). Consequently, the MNR is naturally observed only when a linear relationship between any of the entropy and enthalpy terms is present.

There are several models that attempt to relate the entropy and the enthalpy of thermally activated processes in general and diffusion in particular. Zener [57] proposed that the major part of the free energy of the activated state is associated with elastic distortion of the lattice caused by the diffusing ion. An alternative relationship between the entropy, S_a, and the enthalpy of diffusion, E_a, was used by Almond and West [58] to explain the MNR in AgI–Ag_2MoO_4 glasses:

$$S_a = E_a/T_f \tag{5.25}$$

where temperature, T_f, was related either to the melting temperature or temperature of an order–disorder transition in the material. Recently, a model treating multiple excitations, proposed by Yelon et al. [55], showed that the MNR should be naturally expected in some activated processes. They stated that in a system where the activation barrier of a process, δH, is significantly higher than the energy of a single excitation, E_E, the assemblage of several excitations is required to surmount the activation barrier. Consequently, an increase of the activation barrier will result in the increase of the number of excitations required. This, in turn, will increase the configurational entropy of the system.

5.6 Conclusions

This chapter describes the diffusion of oxygen ions in perovskite oxides with particular emphasis on the materials used in design of solid oxide fuel cells. A generic defect model for perovskites with aliovalent substitution on the A site is presented. A limited review of available oxygen tracer diffusion data in several families of perovskite materials (manganite, cobaltites, ferrites, etc.) is given. The oxygen tracer diffusion coefficient is shown to vary by about 10 orders of magnitude and is dependent upon many factors, including temperature, oxygen partial pressure, and the nature of the A- and B-site ions. Oxygen diffusion occurs via a vacancy diffusion mechanism in these materials. It is shown that the concentration of *mobile* oxygen vacancies has a dominant effect on the oxygen diffusivity in perovskite materials. At the same time, the vacancy diffusion does not appear to be dependent on the nature of the A- and/or B-site ions in the perovskite structure and has an activation energy of $\sim$0.94 $\pm$ 0.07 eV. The effect of aliovalent doping on the apparent activation energy of oxygen tracer diffusion is rather complex and depends on the transition metal on the B site. Several experimentally observed correlations between parameters of oxygen diffusion (e.g., Meyer–Neldel rule) are discussed.

References

1. J. Philibert, Atom Movements Diffusion and Mass Transport in Solids. Presses Universitaires de France, Paris (1966)
2. J.A. Kilner, "Fast Oxygen Transport in Acceptor Doped Oxides". Solid State Ionics 129, 13–23 (2000)
3. J.A. Kilner, "Ceramic Electrodes for SOFC's". Bol. De La Soc. Esp. De Ceram. Y Vidr. 37(2–3), 247–255 (1998)
4. F.A. Kröger, The Chemistry of Imperfect Crystals. North-Holland, Amsterdam (1964)
5. K. Vidyasagar, A. Reller, J. Gopalakrishnan, C.N.R. Rao, "Oxygen Vacancy Ordering in Superlatives of the Two Novel Oxides, $La_2Ni_2O_5$ and $La_2Co_2O_5$, Prepared by Low Temperature Reduction of the Parent Perovskites". J. Chem. Soc. Chem. Commun. 1, 7–8 (1985)
6. J. Mizusaki, I. Yasuda, J. Shimayoma, S. Yamauchi, K. Fueki, "Electrical Conductivity, Defect Equilibria and Oxygen Vacancy Diffusion Coefficient of $La_{1-x}Ca_xAlO_{3-\delta}$ Single Crystals". J. Electrochem. Soc. 140, 467–471 (1993)
7. S. Sunde, K. Nisancioglu, T. Gur, "Critical Analysis of Potentiostatic Step Data for Oxygen Transport in Electrically Conducting Perovskites". J. Electrochem. Soc. 143, 3497–3504 (1996)
8. J.A. Kilner, B.C.H. Steele, L. Ilkov, "Oxygen Self-Diffusion Studies Using Negative Secondary Ion Mass Spectrometry". Solid State Ionics 12, 89–97 (1984)
9. Kilner J.A., R.A. de Souza, Measurement of oxygen transport in ceramics by SIMS. In: Poulsen F.W., Bonanos N., Linderoth S., Mogensen M., Zachau-Christiansen B. (eds.) High Temperature Electrochemistry: Ceramics and Metals, Proceedings of the 17th Risø International Symposium on Materials Science, pp. 41–54. Risø National Laboratory, Roskilde (1996)
10. J.A. Kilner, Isotopic exchange in mixed and ionically conducting oxides. In: Ramanarayanan T., Worrell W.L., Tuller H.L. (eds.) Proceedings of 2nd International Symposium on Ionic and Mixed Conducting Ceramics, pp. 174–190, Electrochemical Society, Pennington (1994)
11. J.A. Kilner, R.A. De Souza, I.C. Fullarton, "Surface Exchange of Oxygen in Mixed Conducting Perovskite Oxides". Solid State Ionics 86–88, 703–709 (1996)
12. S.B. Adler, J.A. Lane, B.C.H. Steele, "Electrode Kinetics of Porous Mixed-Conducting Oxygen Electrodes. J. Electrochem. Soc. 143, 3554–3564 (1996)
13. B.C.H. Steele, "Behaviour of Porous Cathodes in High Temperature Fuel Cells". Solid State Ionics 94, 239–248 (1997)
14. J. Maier, "On the Correlation of Macroscopic and Microscopic Rate Constants in Solid State Chemistry". Solid State Ionics 112, 197–228 (1998)
15. R.A. De Souza, "A Universal Empirical Expression for the Isotope Surface Exchange Coefficients (k*) of Acceptor-Doped Perovskite and Fluorite Oxides". Phys. Chem. Chem. Phys. 8, 890–897 (2006)

16. R.A. De Souza, "Ionic Transport in Acceptor Doped Perovskites." University of London, London (1996)
17. J.A.M. van Roosmalen, E.H.P. Cordfunke, R.B. Helmholdt, H.W. Zandberg, "The Defect Chemistry of $LaMnO_{3\pm\delta}$: 2. Structural Aspects of $LaMnO_{3\pm\delta}$". J. Solid State Chem. 110, 100–105 (1994)
18. J. Mizusaki, Y. Minima, S. Yamauchi, K. Fueki, H. Tagawa, "Non Stoichiometry of the Perovskite Type Oxides $La_{1-x}Sr_xCoO_{3-\delta}$". J. Solid State Chem. 80, 102–111 (1989)
19. J.H. Kuo, H.U. Anderson, D.M. Sparlin, "Oxidation-Reduction Behaviour of Undoped and Sr-Doped $LaMnO_3$, Nonstoichiometry and Defect Structure". J. Solid State Chem. 83, 52–60 (1989)
20. H. Tagawa, J. Mizusaki, H. Nambu, C. Nakao, H. Takai, H. Minamiue, Crystal structure, phase relations and oxygen nonstoichiometry in perovskite type oxide $La_{1-x}Sr_xMnO_3$. In: Steele B.C.H. (ed.) Ceramic Oxygen Ion Conductors and Their Technological Applications, Brit. Ceram. Proceedings 56, pp. 113–123. Institute of Materials, London (1996)
21. I.C. Fullarton, J.A. Kilner, B.C.H. Steele, P.H. Middleton, Characterization of oxygen ion transport in selected perovskite structured oxides by O^{18}/O^{16} isotopic exchange and dynamic secondary ion mass spectrometry. In: Ramanarayanan T., Worrell W.L., Tuller H.L. (eds.) Proceedings of 2nd International Symposium on Ionic and Mixed Conducting Ceramics, pp. 9–26. Electrochemical Society, Pennington (1994)
22. T. Ishigaki, S. Yamauchi, J. Mizusaki, K. Fueki, H. Tamura, "Tracer Diffusion Coefficient of Oxide Ions in $LaCoO_3$ Single Crystal". J. Solid State Chem. 54, 100–107 (1984)
23. R.A. De Souza, J.A. Kilner, "Oxygen Transport in $La_{1-x}Sr_xMn_{1-y}Co_yO_{3\pm\delta}$ Perovskites: Part I. Oxygen Tracer Diffusion". Solid State Ionics 106, 175–187 (1998)
24. R.E. van Doorn, I.C. Fullarton, R.A. de Souza, J.A. Kilner, H.J.M. Bouwmeester, A.J. Burggraaf, "Surface Oxygen Exchange of $La_{0.3}Sr_{0.7}CoO_{3-\delta}$". Solid State Ionics 96, 1–7 (1997)
25. I. Yasuda, K. Ogasawara, M. Hishinuma, "Oxygen Tracer Diffusion in Polycrystalline Calcium-Doped Lanthanum Chromates". J. Am. Ceram. Soc. 80(12) 3009–3012 (1997)
26. A.V. Berenov, J.L. MacManus-Driscoll, J.A. Kilner, "Oxygen Tracer Diffusion in Undoped Lanthanum Manganites". Solid State Ionics 122, 41–49 (1999)
27. M.C. Kim, S.J. Park, H. Haneda, J. Tanaka, T. Mitsunashi, S. Shirasaki, "Self-diffusion of Oxygen in $La_{1-x}Sr_xFeO_{3-\delta}$." J. Mater. Sci. Lett. 9, 102–104 (1990)
28. J. Mizusaki, "Nonstoichiometry, Diffusion, and Electrical Properties of Perovskite-Type Oxide Electrode Materials". Solid State Ionics 52, 79–91 (1992)
29. M. Søgaard, P.V. Hendriksen, F.W. Poulsen, M. Mogensen, "A/B-Ratio and Transport Properties of $(La_{0.85}Sr_{0.15})_sCoO_{3-\delta}$ Perovskites". J. Electroceramics 13, 811–816 (2004)
30. S.M. Woodley, D.J. Gale, P.D. Battle, C.R.A. Catlow, "Oxygen Ion Migration in Orthorhombic $LaMnO_{3-\delta}$." J. Chem. Phys. 119, 9737–9744 (2003)
31. S.M. Islam, "Computer Modelling of Defects and Transport in Perovskite Oxides." Solid State Ionics 154–155, 75–85 (2002)
32. M.S. Islam, R.A. Davies, "Atomistic Study of Dopant Site-Selectivity and Defect Association in the Lanthanum Gallate Perovskite". J. Mater. Res. 14, 86–93 (2004)
33. H. Tagawa, N. Mori, H. Takai, Y. Yonemura, H. Minamiue, H. Inaba, J. Mizusaki, T. Hashimoto, Oxygen non-stoichiometry in perovskite-type oxide, undoped and Sr doped $LaMnO_3$. In: Stimming U., Singhal S.C., Tagawa H., Lehnert W. (eds.) Proceedings of the Fifth International Symposium on Solid Oxide Fuel Cells (SOFC-V), pp. 785–794. Electrochemical Society, Pennington (1997)
34. T. Ishigaki, S. Yamauchi, K. Kishio, J. Mizusaki, K. Fueki, "Diffusion of Oxide Ion Vacancies in Perovskite-Type Oxides". J. Solid State Chem. 73, 179–187 (1988)
35. M.H.R. Lankhorst, H.J.M. Bouwmeester, "Determination of Oxygen Nonstoichiometry and Diffusivity in Mixed Conducting Oxides by Oxygen Coulometric Titration". J. Electrochem. Soc. 144, 1268–1273 (1997)

36. T. Ishigaki, S. Yamauchi, J. Mizusaki, K. Fueki, H. Naito, T. Adachi, "Diffusion of Oxide Ions in $LaFeO_3$ Single Crystal". J. Solid State Chem. 55, 50–53 (1984)
37. S.J. Benson, "Oxygen Transport and Degradation Processes in Mixed Conducting Perovskites." University of London, London (1999)
38. E. Ruiz-Trejo, J.A. Kilner, "Oxygen Diffusion and Proton Conduction in $La_{1-x}Sr_xYO_{3-\delta}$." Solid State Ionics 97, 529–534 (1997)
39. D. Lybye, F.W. Poulsen, M. Mogensen, "Conductivity of A- and B-Site Doped $LaAlO_3$, $LaGaO_3$, $LaScO_3$ and $LaInO_3$ Perovskites". Solid State Ionics 128, 91–103 (2000)
40. T. Ishihara, J.A. Kilner, M. Honda, N. Sakai, H. Yokokawa, Y. Takita, "Oxygen Surface Exchange and Diffusion in $LaGaO_3$ Based Perovskite Type Oxides". Solid State Ionics 113–115, 593–600 (1998)
41. A. Maignan, C. Martin, D. Pelloquin, N. Nguyen, B. Raveau, "Structural and Magnetic Studies of Ordered Oxygen-Deficient Perovskites $Ln BaCo_2O_{5+\delta}$, Closely Related to the "122" Structure". J. Solid State Chem. 142, 247–260 (1999)
42. A. Tarancon, S.J. Skinner, R.J. Chater, F. Hernandez-Ramirez, J.A. Kilner, "Layered Perovskites as Promising Cathodes for Intermediate Temperature Solid Oxide Fuel Cells". J. Mater. Chem. 17, 3175–3181 (2007)
43. G. Kim, S. Wang, A.J. Jacobson, L. Reimus, P. Brodersen, C.A. Mims, "Rapid Oxygen Ion Diffusion and Surface Exchange Kinetics in $PrBaCo_2O_{5+x}$ with a Perovskite Related Structure and Ordered A Cations". J. Mater. Chem. 17, 2500–2505 (2007)
44. H. Hayashi, H. Inaba, M. Matsuyama, N.G. Lan, M. Dokiya, H. Tagawa, "Structural Consideration on the Ionic Conductivity of Perovskite-Type Oxides". Solid State Ionics 122, 1–15 (1999)
45. E. Ruiz-Trejo, M.S. Islam, J.A. Kilner, "Atomistic Simulation of Defects and Ion Migration in $LaYO_3$." Solid State Ionics 123, 121–129 (1999)
46. M.S. Islam, "Ionic Transport in ABO_3 Perovskite Oxides: A Computer Modelling Tour". J. Mater. Chem. 10, 1027–1038 (2000)
47. R.L. Cook, A.F. Sammells, "On the Systematic Selection of Perovskite Solid Electrolytes for Intermediate Temperature Fuel-Cells". Solid State Ionics 45, 311 (1991)
48. J.A. Kilner, R.J. Brook, "A Study of Oxygen Ion Conductivity in Doped Non-Stoichiometric Oxides". Solid State Ionics 6, 237 (1982)
49. A.V. Berenov, J.L. MacManus-Driscoll, J.A. Kilner, "Observation of the Compensation Law During Oxygen Diffusion in Perovskite Materials". Int. J. Inorg. Mater. 3, 1109–1111 (2001)
50. E. Peacock-López, H. Suhl, "Compensation Effect in Thermally Activated Processes". Phys. Rev. B 26, 3774–3782 (1982)
51. F. Béjina, O. Jaoul, "Silicon Diffusion in Silicate Minerals". Earth Planet. Sci. Lett. 153, 229–238 (1997)
52. K.D. Kreuer, "Aspects of the Formation and Mobility of Protonic Charge Carriers and the Stability of Perovskite-Type Oxides". Solid State Ionics 125, 285–302 (1999)
53. A. Steltennpohl, N. Memmel, "Self-Diffusion on Pd(111)." Surf. Sci. 454–456, 558–561 (2000)
54. F.S. Galasso, "Perovskites and High T_C Superconductors." Gordon and Breach Science Publishers, New York (1990)
55. A. Yelon, B. Movaghar, H.M. Branz, "Origin and Consequences of the Compensation (Meyer-Neldel) Law". Phys. Rev. B 46, 12244–12250 (1992)
56. A.N. Petrov, O.F. Kononchuk, A.V. Andreev, V.A. Cherepanov, P. Kofstad, "Crystal Structure, Electrical and Magnetic Properties of $La_{1-x}Sr_xCoO_{3-y}$." Solid State Ionics 80, 189–199 (1995)
57. C. Zener, Theory of diffusion. In: Shockley W., Hollomon J.H., Maurer R., Seitz F. (eds.) Imperfections in Nearly Perfect Crystals, pp. 289–314. John Wiley & Sons, New York (1952)

58. D.P. Almond, A.R. West, "The Activation Entropy for Transport in Ionic Conductors". Solid State Ionics 23, 27–35 (1987)
59. H.U. Anderson, Defect chemistry of p-type perovskites. In: Poulsen F.W., Bentzen J.J., Jacobsen T., Skou E., Østergård M.J.L. (eds.) Proceedings of the 14th Risø International Symposium on Materials Science, pp. 1–18. Risø National Laboratory, Roskilde (1993)
60. I. Yasuda, K. Ogasawara, M. Hishinuma, T. Kawada, M. Dokiya, "Oxygen Tracer Diffusion Coefficient of (La,Sr)$MnO_{3\pm\delta}$." Solid State Ionics 86–88, 1197–1201 (1996)
61. J.L. Routbort, R. Doshi, M. Krumpelt, "Oxygen Tracer Diffusion in $La_{1-x}Sr_xCoO_3$." Solid State Ionics 90, 21–27 (1996)
62. N. Sakai, K. Yamaji, T. Horita, H. Yokokawa, T. Kawada, M. Dokiya, "Oxygen Transport Properties of $La_{1-x}Ca_xCrO_{3-\delta}$ as an Interconnect Material of a Solid Oxide Fuel Cell." J. Electrochem. Soc. 147, 3178–3182 (2000)
63. P.S. Manning, J.D. Sirman, J.A. Kilner, "Oxygen Self-Diffusion and Surface Exchange Studies of Oxide Electrolytes Having the Fluorite Structure". Solid State Ionics 93, 125–132 (1996)

Chapter 6
Structural Disorder, Diffusion Pathway of Mobile Oxide Ions, and Crystal Structure in Perovskite-Type Oxides and Related Materials

Masatomo Yashima

6.1 Introduction

Solid oxides with high ionic conductivity have attracted considerable attention for reasons of their many possible applications, including solid oxide fuel cells (SOFCs), sensors, catalysts, and batteries. Oxide ion (O^{2-}) conductors such as zirconia (ZrO_2) solid solutions [1, 2], bismuth oxide (Bi_2O_3)-based materials [3–6], ceria (CeO_2)-based solid solutions [7, 8], and lanthanum gallate-based compounds [9, 10] have been widely investigated. The development of improved electrolyte and electrode materials requires a better understanding of the mechanism of ionic conduction, and crucial to this is comprehension of the crystal structure at high temperatures at which these materials work most efficiently [5, 6, 8, 10–15]. The detailed structural analysis enables the observation of the structural disorders and diffusion paths of mobile ions in ionic and mixed conductors [5, 6, 8, 10–15].

Doped lanthanum gallate materials $(La_{1-x}Sr_x)(Ga_{1-y-z}Mg_yCo_z)O_{3-\delta}$ are used as electrolytes for SOFCs [9, 16]. Doped lanthanum cobaltites $(La_{1-x}Sr_x)(Co_{1-y}Fe_y)O_{3-\delta}$, mixed ionic-electronic conductors, are widely used as electrode materials for SOFCs [17, 18]. Doped lanthanum titanate $La_{0.64}(Ti_{0.92}Nb_{0.08})O_{2.99}$ has an A-site deficient $ABO_{3-\delta}$ double perovskite-type structure at high temperatures [11, 19, 20]. This material is an ionic conductor [21], and it should be interesting to investigate structural disorder and diffusion paths in a double perovskite-type structure and to compare the results with those of cubic perovskite-type oxides. In this chapter, we review the structural disorder and diffusion paths of oxide ions in the perovskite-structured materials $(La_{0.8}Sr_{0.2})(Ga_{0.8}Mg_{0.15}Co_{0.05})O_{3-\delta}$ [10], $La_{0.64}(Ti_{0.92}Nb_{0.08})O_{2.99}$ [11], $(La_{0.6}Sr_{0.4})CoO_{3-\delta}$ [12], and $(La_{0.6}Sr_{0.4})(Co_{0.8}Fe_{0.2})O_{3-\delta}$ [13]. A_2BO_4-based oxides with K_2NiF_4-type structure have extensively been studied as new mixed ionic-electronic conductors [22, 23],

M. Yashima (✉)
Tokyo Institute of Technology, Yokohama 226–8502, Japan
e-mail: yashima@materia.titech.ac.jp

T. Ishihara (ed.), *Perovskite Oxide for Solid Oxide Fuel Cells*,
Fuel Cells and Hydrogen Energy, DOI 10.1007/978-0-387-77708-5_6,

where A and B are larger and smaller cations. The crystal structure of A_2BO_4-based oxides consists of one perovskite-type block and one rock salt-type AO block. Thus, it is interesting to study the structural disorders and diffusion path of oxide ions in the A_2BO_4-based oxides and compare the results with those of perovskite-type materials. Here we also review the structural disorder and diffusion paths of oxide ions in the K_2NiF_4-type $(Pr_{0.9}La_{0.1})_2(Ni_{0.74}Cu_{0.21}Ga_{0.05})O_{4+\delta}$ [15].

6.2 High-Temperature Neutron Powder Diffractometry

Some $ABO_{3-\delta}$ perovskite-structured materials, where A and B represent larger and smaller cations, are ionic conductors, while some other $ABO_{3-\delta}$ perovskite-type compounds are mixed conductors. Heavy elements such as La and Ba occupy the A site, but because the mobile O anion is a light element, conventional X-ray powder diffractometry is not sensitive to positional and occupational disordering of oxide ions. To investigate the diffusion path of mobile oxide ions, and structural disorder and crystal structure in perovskite-structured ionic and mixed conductors [5, 6, 8, 10–14], we applied a high-temperature neutron powder diffraction method. Our reasons for choosing this method were as follows [24]:

1. The coherent scattering length of the O atom is relatively large compared with its X-ray scattering factor. Figure 6.1 illustrates the relative scattering abilities of the oxygen atom in both methods.
2. At high temperatures, the sample surface is often altered by processes such as sintering, grain growth, cracking, evaporation, and thermal expansion; this can lead to shifts in diffraction peak intensity and position. Thus, it is often difficult to perform structural refinement and electron density analysis using conventional high-temperature X-ray diffraction data measured with

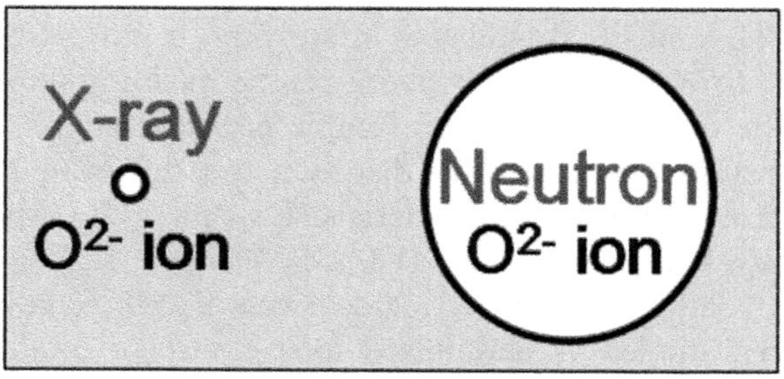

Fig. 6.1 Circles representing the relevant sizes of the square of the X-ray scattering factor (*left*) and the neutron scattering length (*right*) of oxygen atoms in $(La_{0.8}Sr_{0.2})(Ga_{0.8}Mg_{0.15}Co_{0.05})O_{2.8}$ where the size of the square of the X-ray scattering factor of the cation is assumed to be the same as that of the neutron scattering length of the cation

Bragg–Brentano geometry. In contrast, structural analysis based on high-temperature neutron powder diffraction data is less subject to the influence of sintering, grain growth, cracking, evaporation, and thermal expansion.

3. The lack of electronic interference allows for a simple density map. An electron-density map from X-ray diffraction data includes not only the structural disorders but also the electron clouds. In contrast, the nuclear density map from neutron diffraction data does not include the electron clouds.
4. The neutron form factor is independent of the scattering angle, which allows for high precision in the elucidation of atomic displacement parameters and structural disorder.
5. The low absorption of neutron by the furnace itself is less damaging to the data quality.

We devised and fabricated a new furnace for high-temperature neutron diffraction measurements (Fig. 6.2) [24], using molybdenum silicide heaters to heat the sample. The merits of the molybdenum silicide heater are as follows:

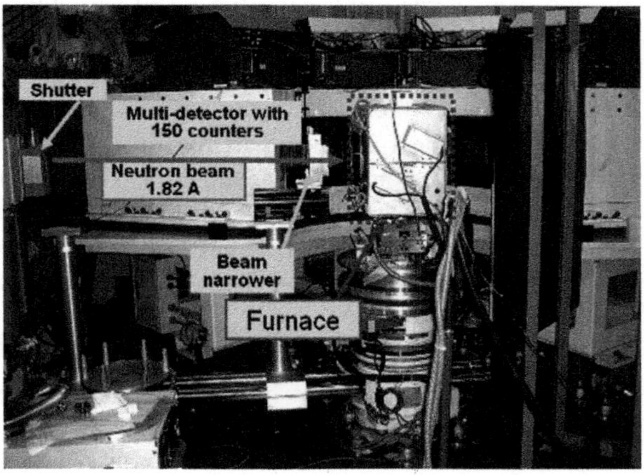

Fig. 6.2 Photograph of the furnace [24] placed on the sample table of the neutron diffractometer HERMES [25]

1. It can be used in air for long periods at temperatures of up to 1900 K without degradation.
2. A furnace based on this heater is superior to a mirror furnace in terms of temperature homogeneity.
3. Low-temperature degradation, which is often seen in $LaCrO_3$ heaters, does not occur.

Using the furnace, neutron powder diffraction measurements were conducted in air from room temperature to 1850 K using a 150-detector system, HERMES [25], installed at the JRR-3 M reactor at the Japan Atomic Energy Agency, Tokai, Japan (Fig. 6.2) [24, 26]. The furnace was placed on a sample table, and neutrons with wavelength 1.82 Å were obtained using the (331) reflection of a Ge monochromator. Although diffraction data where *d* spacing is less than 0.93 Å cannot be measured using HERMES, the diffractometer has sufficient intensity and power to collect data with good counting statistics for nuclear density analysis. Diffraction data were collected in the range $2\theta = 3°–157°$ at step intervals of 0.1°. The sample temperature was kept constant during data collection, and was monitored using a Pt/Pt-13 wt% Rh thermocouple in contact with the sample.

6.3 Data Processing for Elucidation of the Diffusion Paths of Mobile Oxide Ions in Ionic Conductors: Rietveld Analysis, Maximum Entropy Method (MEM), and MEM-Based Pattern Fitting (MPF)

The experimental diffraction data were analyzed by a combined technique involving Rietveld analysis, the maximum entropy method (MEM), and MEM-based pattern fitting (MPF) [10–15]. Rietveld analysis, which is used to refine the crystal structure from the powder diffraction data by a least squares method, was carried out using the RIETAN-2000 program [27], which yields structure factors and their errors after structural refinement. It is known that MEM can be used to obtain a nuclear density distribution map based on neutron structure factors and their errors [5, 6, 8, 10–15, 26–29]; any type of complicated nuclear density distribution is allowed so long as it satisfies the symmetry requirements. MEM calculations were carried out using the PRIMA program [29]. To reduce the bias imposed by the simple structural model in the Rietveld refinement, an iterative procedure known as the REMEDY cycle [29] was applied after MEM analysis (Fig. 6.3). In this procedure, structure factors

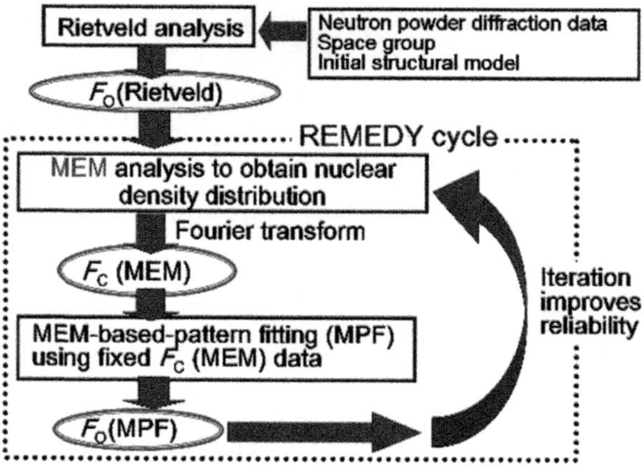

Fig. 6.3 Flow chart of the combined technique involving Rietveld analysis, MEM and MPF. The REMEDY cycle, in which MEM and MPF are performed alternately and repeatedly, improves the reliability of the nuclear density. F_O(Rietveld) is the observed structure factor obtained from the Rietveld analysis. F_C(MEM) is the structure factor calculated from the MEM nuclear density. F_O(MPF) is the observed structure factor, which is obtained from the MPF analysis

F_C(MEM) were calculated by Fourier transform of the nuclear densities obtained by MEM analysis. In the subsequent MEM-based pattern fitting (MPF), the structure factors F_C(MEM) obtained in the previous MEM analysis were fixed, and parameters irrelevant to the structure—e.g., scale factor, profile, unit cell, and background parameters—were refined using RIETAN-2000 [27]. The observed structure factors evaluated after the MPF, F_O(MPF), were then analyzed again by MEM. MPF and MEM analyses were alternated until the reliability indices no longer decreased (REMEDY cycle). The REMEDY cycle allowed us to obtain a reliable nuclear density distribution (Fig. 6.3). When the MEM is successful in obtaining a nuclear density, the reliability factors based on the structure factors (R_F) and on the Bragg intensities (R_I or R_B) in the MPF analysis are lower than those in the Rietveld analysis.

6.4 Diffusion Path of Oxide Ions in the Fast Oxide Ion Conductor $(La_{0.8}Sr_{0.2})(Ga_{0.8}Mg_{0.15}Co_{0.05})O_{2.8}$ [10]

6.4.1 Introduction

Lanthanum gallate-based materials with an $ABO_{3-\delta}$ perovskite-type structure have higher oxide ion conductivity than conventional yttria-stabilized zirconias [9, 30]. The crystal structure of these materials has been the subject of a number of investigations [31–39], and the diffusion path of oxide ions in lanthanum gallates has been studied by computational methods [40, 41] and by diffractometry [36]. Here, we describe the temperature dependence of the diffusion paths and structural disorder of oxide ions in $(La_{0.8}Sr_{0.2})(Ga_{0.8}Mg_{0.15}Co_{0.05})O_{2.8}$ at 1665, 1471, and 1069 K [10]. For comparison, we also describe the nuclear density distribution of $LaGaO_3$ at 1663 K [42]. Comparison of structural disorder in $(La_{0.8}Sr_{0.2})(Ga_{0.8}Mg_{0.15}Co_{0.05})O_{2.8}$ and $LaGaO_3$ is interesting because the oxide ion conductivity of $(La_{0.8}Sr_{0.2})(Ga_{0.8}Mg_{0.15}Co_{0.05})O_{2.8}$ is about 10^3 times higher than that of $LaGaO_3$ (Fig. 6.4 [42]).

6.4.2 Experiments and Data Processing

In this work, we used a material with the chemical composition $(La_{0.8}Sr_{0.2})(Ga_{0.8}Mg_{0.15}Co_{0.05})O_{2.8}$ because doping of Co, Sr, and Mg into lanthanum gallate effectively enhances oxide ion conductivity (Fig. 6.4 [42]) [43]. A high-purity sample was synthesized via solid-state reactions [10]. Chemical analysis of the final product showed a composition of $(La_{0.80(3)}Sr_{0.20(3)})(Ga_{0.80(6)}Mg_{0.15(6)}Co_{0.050(7)})O_{2.8(3)}$, where the number in parentheses is the error in the last digit. Neutron powder diffraction experiments were carried out at 1069.2 ± 1.6, 1470.7 ± 1.3, and 1664.6 ± 1.4 K in air using a furnace with $MoSi_2$ heaters [24], as described above, and the HERMES diffractometer [25]. Neutron

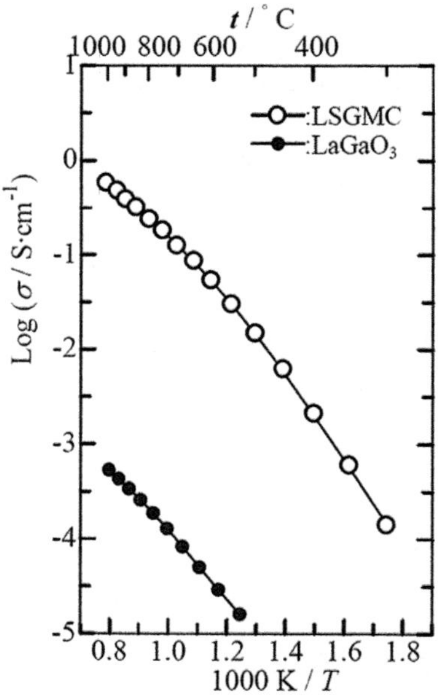

Fig. 6.4 Arrhenius plot of oxide ion conductivity of $LaGaO_3$ (*closed circles*) and $(La_{0.8}Sr_{0.2})(Ga_{0.8}Mg_{0.15}Co_{0.05})O_{2.8}$ (LSGMC, *open circles*) [42]. The structural origin of the difference in ionic conductivity between the two materials can be seen in Fig. 6.6(a,d)

diffraction data for $LaGaO_3$ were obtained in air at 1663 K. The wavelength of the incident neutrons was 1.8207 Å. Powder patterns were obtained in the range $2\theta = 5°–155°$. The experimental diffraction data were analyzed using a combination of the Rietveld method and MPF, with the RIETAN-2000 program [27], and MEM, using the PRIMA program [29].

6.4.3 Results and Discussion

The crystal structure of $(La_{0.8}Sr_{0.2})(Ga_{0.8}Mg_{0.15}Co_{0.05})O_{2.8}$ was successfully refined, assuming an ideal perovskite-type structure with the space group $Pm\bar{3}m$ at 1665 K (Fig. 6.5) and at 1471 K. At 1069 K, the material was analyzed assuming $R\bar{3}c$ symmetry, because the $R\bar{3}c$ reflections forbidden for the $Pm\bar{3}m$ phase exist at this temperature [10]. The refined crystallographic parameters are shown in Table 6.1. The unit-cell volume of the pseudo-fluorite lattice increases with temperature due to thermal expansion. The atomic displacement parameters of the oxygen atom are large and anisotropic (Fig. 6.5(a) and Table 6.1). The isotropic atomic displacement parameters of all cations, and the equivalent isotropic atomic displacement parameters of the oxide ions, increase with increasing temperature (Table 6.1), corresponding to higher oxide ion conductivity at higher temperatures (Fig. 6.4 [42]) [43]. The equivalent isotropic atomic displacement parameters of the oxide ions are higher than those of the cations, suggesting higher diffusivity for the oxide ions.

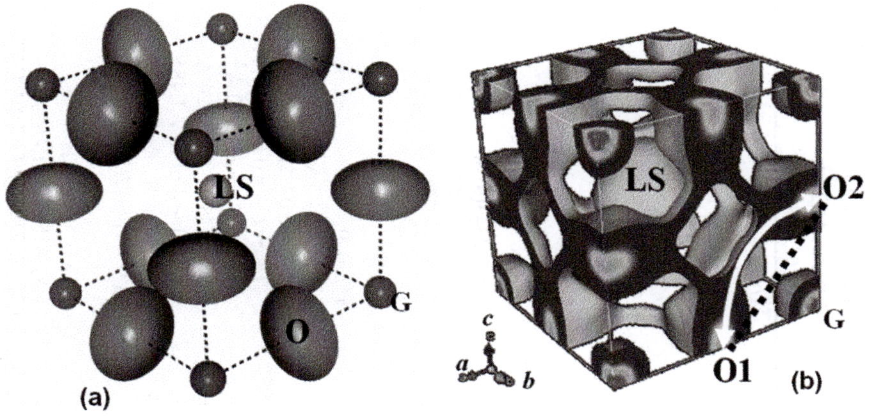

Fig. 6.5 (**a**) Refined crystal structure and (**b**) isosurface of nuclear density at 0.05 fm Å^{-3} of cubic $Pm\bar{3}m$ $(La_{0.8}Sr_{0.2})(Ga_{0.8}Mg_{0.15}Co_{0.05})O_{2.8}$ at 1665 K [10]. LS, $(La_{0.8}Sr_{0.2})$ cation; G, $(Ga_{0.8}Mg_{0.15}Co_{0.05})$ cation

MEM analysis was carried out using the structure factors obtained by Rietveld analysis; 17, 16, and 56 structure factors were obtained for data measured at 1665, 1471, and 1069 K, respectively. We measured the peak intensity of cubic 100 reflection at the lowest 2θ position, because the intensity of the 100 reflection contributes the most information to MEM analysis. MEM calculations were performed using $64 \times 64 \times 64$ and $96 \times 96 \times 235$ pixels for the cubic and trigonal structures, respectively. The R factor based on the Bragg intensities, R_I, was considerably improved by the REMEDY cycle (Table 6.1), indicating the validity of these nuclear density distributions for $(La_{0.8}Sr_{0.2})$-$(Ga_{0.8}Mg_{0.15}Co_{0.05})O_{2.8}$. The isosurface of nuclear density obtained from the REMEDY cycle provided much information on the complexity of structural disorder and diffusion paths of oxide ion in the crystal (Figs. 6.5(b) and 6.6). Simple models consisting of atom spheres were no longer appropriate to describe the positional distribution of the oxide ions.

To visualize the structural disorder and diffusion paths, the MEM nuclear density distribution map in the (100) plane is shown in Fig. 6.6. The oxide ions in the cubic $Pm\bar{3}m$ phase exhibit a large anisotropic distribution, corresponding to large anisotropy in the atomic displacement parameters (Table 6.1). The most striking feature is the diffusion path of the oxide ions. Roughly speaking, the diffusion paths are along the [110], [011], and [101] directions, forming a three-dimensional network of pathways. The diffusion path does not follow the edge of the BO_6 [= $(Ga_{0.8}Mg_{0.15}Co_{0.05})O_{5.6}$] octahedron along the <110> direction (shown as straight dotted line between the ideal O1 and O2 positions in Fig. 6.5(b)), but displays an arc shape (curved solid line with arrows), maintaining a constant distance from the B-site cation (G in Fig. 6.5(b)). This curved feature is consistent with the results obtained by computational methods

Table 6.1 Crystallographic parameters and reliability factors for $(La_{0.8}Sr_{0.2})(Ga_{0.8}Mg_{0.15}Co_{0.05})O_{2.8}$ [10]

Temperature, space group		1069.2 ± 1.6 K, $R\bar{3}c$	1470.7 ± 1.3 K, $Pm\bar{3}m$	1664.6 ± 1.4 K, $Pm\bar{3}m$
Unit-cell parameters		$a = b = 5.5587(9)$ Å; $c = 13.629(3)$ Å	$a = 3.9618(2)$ Å	$a = 3.9744(2)$ Å
Unit-cell volume		$V = 364.7(1)$ Å^3 ($V_p = 60.78(2)$ Å^3)***	$V = 62.184(4)$ Å^3	$V = 62.779(4)$ Å^3
$La_{0.8}Sr_{0.2}$	Wyckoff Position / g	$6a$ / 1.0	$1b$ / 1.0	$1b$ / 1.0
	x, y, z	0, 0, 1/4	1/2, 1/2, 1/2	1/2, 1/2, 1/2
Atomic displacement parameters		$U_{11} = U_{22} = 2U_{12} = 0.025(4)$ Å^2; $U_{33} = 0.033(9)$ Å^2; $U_{13} = U_{23} = 0$ Å^2; $U_{eq} = 0.028$ Å^2	$U = 0.0391(9)$ Å^2	$U = 0.0454(9)$ Å^2
$Ga_{0.8}Mg_{0.15}Co_{0.05}$	Wyckoff Position / g	$6b$ / 1.0	$1a$ / 1.0	$1a$ / 1.0
	x, y, z	0, 0, 0	0, 0, 0	0, 0, 0
Atomic displacement parameters		$U_{11} = U_{22} = 2U_{12} = 0.016(4)$ Å^2; $U_{33} = 0.018(7)$ Å^2; $U_{13} = U_{23} = 0$ Å^2 $U_{eq} = 0.017$ Å^2	$U = 0.0274(9)$ Å^2	$U = 0.0343(9)$ Å^2
O	Wyckoff Position / g	$18e$ / 0.9333	$3d$ / 0.9333	$3d$ / 0.9333
	x, y, z	0.529(4), 0, 1/4	1/2, 0, 0	1/2, 0, 0
Atomic displacement parameters		$U_{eq} = 0.028$ Å^2; $U_{11} = 0.039(10)$ Å^2; $U_{22} = 2U_{12} = 0.018(8)$ Å^2; $U_{33} = 0.087(11)$ Å^2; $U_{13} = 0.5U_{23} = -0.012(3)$ Å^2	$U_{eq} = 0.0712$ Å^2; $U_{11} = 0.0268(12)$ Å^2; $U_{22} = U_{33} = 0.0935(13)$ Å^2; $U_{12} = U_{13} = U_{23} = 0$ Å^2	$U_{eq} = 0.0817$ Å^2; $U_{11} = 0.0292(13)$ Å^2; $U_{22} = U_{33} = 0.0935(13)$ Å^2; $U_{12} = U_{13} = U_{23} = 0$ Å^2
Reliability factors in the Rietveld refinement*		$R_{wp} = 6.53\%$, $R_p = 4.97\%$, Goodness of fit = 1.492, $R_I = 3.16\%$, $R_F = 1.90\%$	$R_{wp} = 6.90\%$, $R_p = 5.20\%$, Goodness of fit = 1.610, $R_I = 3.63\%$, $R_F = 2.23\%$	$R_{wp} = 6.35\%$, $R_p = 4.94\%$, Goodness of fit = 1.485, $R_I = 2.68\%$, $R_F = 2.39\%$
Reliability factors in the final MEM-based whole-pattern fitting*		$R_I = 1.92\%$, $R_F = 1.52\%$	$R_I = 2.03\%$, $R_F = 1.24\%$	$R_I = 1.91\%$, $R_F = 1.34\%$
Reliability factors in the final MEM analysis**		R_F(MEM) = 1.74% wR_F(MEM) = 2.04%	R_F(MEM) = 1.18% wR_F(MEM) = 1.35%	R_F(MEM) = 1.48% wR_F(MEM) = 1.46%

Note: g, occupancy; x, y, z, fractional coordinate.
*Standard Rietveld indices; **reliability factors: MEM analysis; ***V_p, unit-cell volume of the pseudo-perovskite cell.

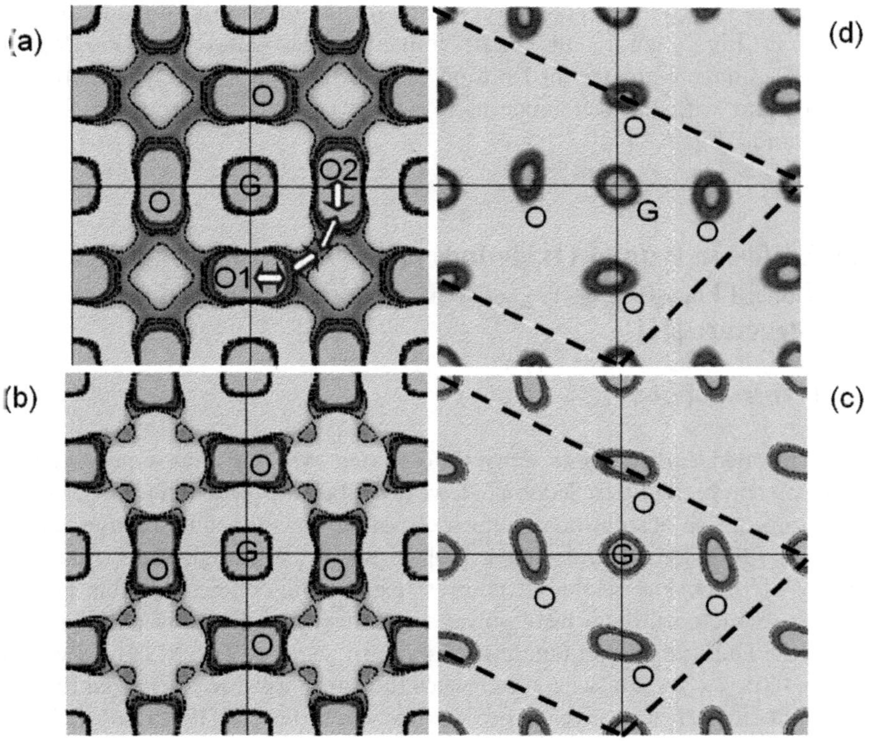

Fig. 6.6 Nuclear density distributions on the (100) plane of cubic $Pm\bar{3}m$ $(La_{0.8}Sr_{0.2})$-$(Ga_{0.8}Mg_{0.15}Co_{0.05})O_{2.8}$ at (**a**) 1665 K and (**b**) 1471 K, and (**c**) on the (012) plane of trigonal $R\bar{3}c$ $(La_{0.8}Sr_{0.2})(Ga_{0.8}Mg_{0.15}Co_{0.05})O_{2.8}$ at 1069 K, with contours in the range from 0.3 to 4.0 fm Å^{-3} (0.3 fm Å^{-3} step) [10]. (**d**) Nuclear density distribution on the (012) plane of trigonal $R\bar{3}c$ $LaGaO_3$ at 1663 K [42]. G and O denote the *B*-site cation $(Ga_{0.8}Mg_{0.15}Co_{0.05})$ and the oxide ion, respectively. The diffusion path of the oxide ion is not along the straight line between the ideal positions, but along the *curved solid line* avoiding the G ion [*white arrows* in (**a**)]. The *thin black straight line and thick black dashed line* in Fig. 6(c, d) show the $Pm\bar{3}m$ and $R\bar{3}c$ unit cells, respectively. In the low-temperature trigonal structure, the oxide ions are localized near the equilibrium position, while in the high-temperature cubic phase the oxide ions are spread over a wide area between the ideal sites

[40, 41] and with the potential map obtained using a probability density function technique [36]. Here, for the first time, we have obtained a diffusion path from the nuclear density distribution and demonstrated its temperature dependence [10]. The nuclear density in the area of the diffusion path is greater at 1665 K (Fig. 6.6(a)) than at 1471 K (Fig. 6.6(b)), which is consistent with an increase in oxide ion conductivity with increasing temperature (Fig. 6.4 [42]) [43]. Notably, the oxide ions in the low-temperature trigonal phase are localized near the equilibrium positions (Fig. 6.6(c)), although they are spread over a wide area between the ideal positions in the high-temperature cubic phase (Fig. 6.6(a, b)). This interesting distribution indicates that the more symmetrical

$Pm\bar{3}m$ phase has a lower activation energy for the migration of oxide ions. As shown in Fig. 6.6(a,d), the oxide ions in cubic $(La_{0.8}Sr_{0.2})(Ga_{0.8}Mg_{0.15}$-$Co_{0.05})O_{2.8}$ have a greater distribution than in trigonal $LaGaO_3$. This finding is consistent with the difference in oxide ion conductivity between the two compounds (Fig. 6.4 [42]).

6.5 Diffusion Path of Oxide Ions in an Oxide Ion Conductor, $La_{0.64}(Ti_{0.92}Nb_{0.08})O_{2.99}$, with a Double Perovskite-Type Structure [11]

6.5.1 Introduction

As mentioned earlier, some perovskite-related $ABO_{3-\delta}$ phases possess high oxide ion conductivity. In Section 6.4, we described the diffusion path of mobile oxide ions in a solid solution of cubic perovskite-type doped lanthanum gallate $(La_{0.8}Sr_{0.2})(Ga_{0.8}Mg_{0.15}Co_{0.05})O_{2.8}$ [10]. However, although there is a wide variety of perovskite-related structures (e.g., A-site-deficient double perovskite-type structure), there were no reports concerning diffusion paths in such materials. The lanthanum titanate solid solution $La_{(2\pm x)/3}(Ti_{1-x}M_x)O_{3-\delta}$ (M = Al or Nb, $0.05 \leq x \leq 0.20$), where δ denotes the concentration of oxygen defects, has an A-site-deficient layered perovskite-type structure [19–21, 44–53], and exhibits high oxide ion conductivity at high temperatures [21, 44]. Yoshioka [21] studied the electrical properties of $La_{(2-x)/3}(Ti_{1-x}Nb_x)O_{3-\delta}$ (x = 0.05–0.15) and reported that a sample with x = 0.10 displayed the highest ionic conductivity ($\sim 10^{-2}$ S cm^{-1} at 973 K). In this section, we describe the crystal structure and pathway of oxide ion diffusion in $La_{0.64}(Ti_{0.92}Nb_{0.08})O_{3-\delta}$ (x = 0.08) [11].

6.5.2 Experiments and Data Processing

The $La_{0.64}(Ti_{0.92}Nb_{0.08})O_{2.99}$ specimen was prepared via solid-state reactions [11, 20]. ICP-OES chemical analysis indicated that the chemical formula of the final product was $La_{0.636(1)}(Ti_{0.921}Nb_{0.079(1)})O_{2.993(1)}$. The value for oxygen (2.993) was calculated based on electrical neutrality and suggests that the amount of oxygen defects is small. Neutron powder diffraction data for $La_{0.64}(Ti_{0.92}Nb_{0.08})O_{2.99}$ were collected at 769 K, 1281 K, and 1631 K using the furnace [24] and HERMES diffractometer [25] as described above. Incident neutrons with a fixed wavelength of 1.8143 Å were used. Powder diffraction data were measured over the range 2θ = 3°–152.62°. The sample temperature was maintained to within ±1.5 K during each measurement. The diffraction data were analyzed by the Rietveld method, followed by application of MEM and MPF, using the RIETAN-2000 [27] and PRIMA [29] programs.

6.5.3 Results and Discussion

Rietveld analysis of the neutron powder diffraction patterns of $La_{0.64}(Ti_{0.92}Nb_{0.08})O_{2.99}$ at 769 K, 1281 K, and 1631 K was performed assuming a tetragonal *P4/mmm* structure ($a = b \approx a_p$, $c \approx 2a_p$; subscript p denotes pseudo-cubic perovskite-type structure). The refined unit-cell and structural parameters and *R* factors are summarized in Table 6.2 [11]. The unit-cell parameters increase with temperature, as does the unit-cell volume due to thermal expansion. Figure 6.7(a) shows the crystal structure of the material

Table 6.2 Refined crystallographic parameters and reliability factors in Rietveld and MPF analyses for $La_{0.64}(Ti_{0.92}Nb_{0.08})O_{2.99}$ [11]

Temperature		769 K ± 1.5 K	1281 K ± 1.5 K	1631 K ± 1.5 K
Atom/Site	Parameter			
	a (Å)	3.8827(2)	3.9019(2)	3.9172(2)
	c (Å)	7.8684(4)	7.9118(4)	7.9249(5)
La1	U (× 10^{-2}Å^2)	1.33(7)	2.83(9)	3.99(12)
La2	U (× 10^{-2}Å^2)	0.8(2)	1.9(3)	1.3(3)
Ti,Nb	z	0.2625(6)	0.2640(7)	0.2621(9)
	U (× 10^{-2}Å^2)	0.54(10)	1.95(12)	2.51(14)
O1	U_{11} (× 10^{-2}Å^2)	2.5(2)	4.5(2)	5.1(3)
	U_{33} (× 10^{-2}Å^2)	1.3(3)	2.3(3)	2.8(4)
	U_{eq} (× 10^{-2}Å^2)	2.12	3.76	4.32
O2	U_{11} (× 10^{-2}Å^2)	3.7(2)	5.5(2)	6.4(3)
	U_{33} (× 10^{-2}Å^2)	0.2(2)	1.1(3)	1.9(4)
	U_{eq} (× 10^{-2}Å^2)	2.58	4.05	4.91
O3	z	0.2340(3)	0.2344(4)	0.2373(5)
	U_{11} (× 10^{-2}Å^2)	2.6(2)	4.4(2)	5.6(3)
	U_{22} (× 10^{-2}Å^2)	0.11(10)	1.04(11)	1.53(14)
	U_{33} (× 10^{-2}Å^2)	3.16(12)	4.81(14)	5.9(2)
	U_{eq} (× 10^{-2}Å^2)	1.97	3.41	4.36
Reliability factors*		R_{wp} = 5.76%, R_p = 4.28% Goodness of fit: 3.16 R_I = 5.38%, R_F = 3.59%	R_{wp} = 5.21%, R_p = 3.83% Goodness of fit: 2.87 R_I = 4.19%, R_F = 4.36%	R_{wp} = 5.00%, R_p = 3.73% Goodness of fit: 2.82 R_I = 4.33%, R_F = 5.06%
Reliability factors**		R_I = 6.03%, R_F = 3.27%	R_I = 4.17%, R_F = 3.13%	R_I = 4.05%, R_F = 3.33%

Note: * Reliability factors in Rietveld analysis. ** Reliability factors in MEM-based pattern fitting. Tetragonal space group *P4/mmm* (No. 123) $Z = 2$. U, Atomic displacement parameter; z, fractional coordinate. Occupancies for La1, La2, O1, O2, and O3 sites are assumed to be 1.0, 0.271, 1.0, 0.9972, and 0.9972, respectively. Occupancies of Ti and Nb atoms at the Ti,Nb site are assumed to be 0.9209 and 0.0791, respectively. In analyses, atom positions were: La1 1*a* (0, 0, 0); La2 1*b* (0, 0, 1/2); Ti,Nb 2*h* (1/2, 1/2, *z*); O1 1*c* (1/2, 1/2, 0); O2 1*d* (1/2, 1/2, 1/2); O3 4*i* (1/2, 0, *z*). Site symmetries give constraints: $U_{12} = U_{13} = U_{23} = 0$ at O1, O2, and O3 sites; $U_{11} = U_{22}$ at O1 and O2 sites. Only independent atomic displacement parameters are given.

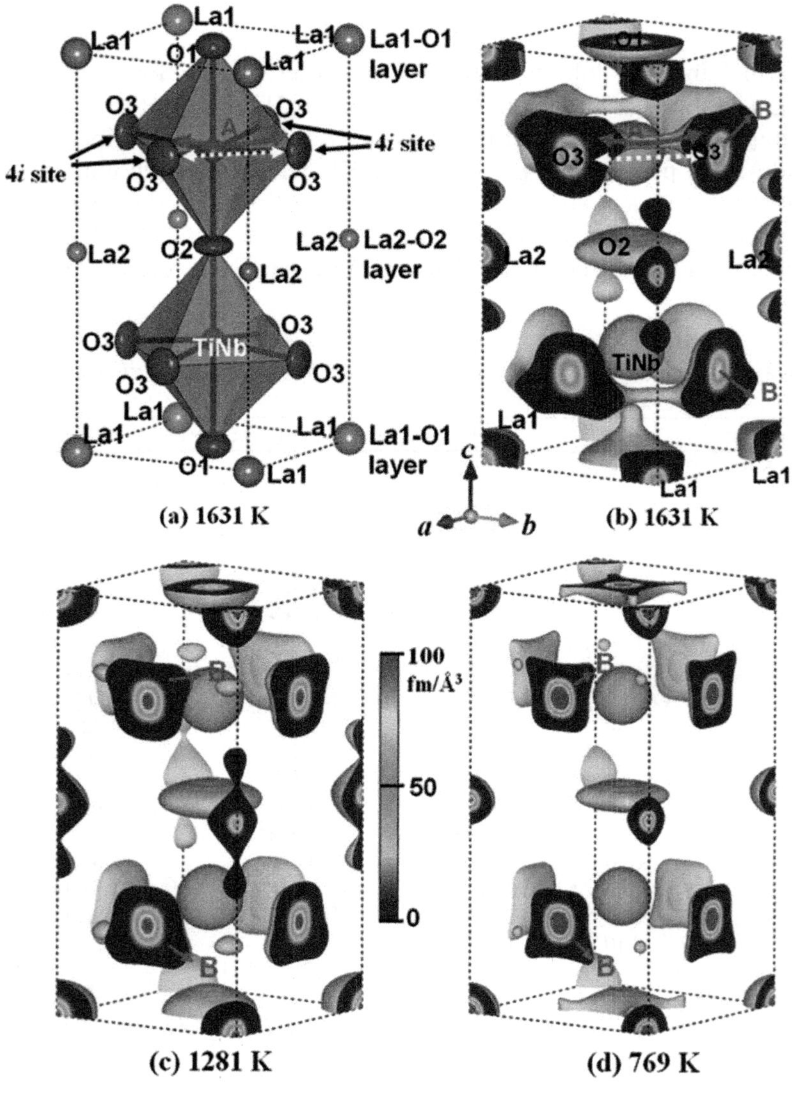

Fig. 6.7 (**a**) Refined crystal structure of double perovskite-type $La_{0.64}(Ti_{0.92}Nb_{0.08})O_{2.99}$ at 1631 K, and (**b, c, d**) isosurfaces of nuclear density distribution at –0.1 fm Å^{-3} (*light blue*) and +0.1 fm Å^{-3} (*yellow*) and nuclear density on the (100) and (001) planes in $La_{0.64}(Ti_{0.92}Nb_{0.08})O_{2.99}$ at 1631 K (**b**), 1281 K (**c**), and 769 K (**d**) [11]. Since the Ti atom has negative scattering length, the Ti,Nb site is drawn with *light blue* equi-density surface at –0.1 fm Å^{-3}. Oxygen atoms at the O3 site have a large spatial distribution to the <101> directions shown by the line with an *arrow* (*B*). The O3 atoms do not move along the straight line shown by the *dotted line with arrows* but along the *curved solid line with arrows* (A in (**a**) and (**b**))

drawn with the refined crystallographic parameters [11]. This is the high-temperature form of $La_{0.64}(Ti_{0.92}Nb_{0.08})O_{2.99}$, which has an A-site-deficient perovskite-type structure with double perovskite $ABO_{3-\delta}$ units along the c axis (number of chemical formula in a unit cell: $Z = 2$), where $A = La_{0.64}$ and $B = (Ti_{0.92}Nb_{0.08})$. The occupancy factors for La at the La1 and La2 sites are $g(\text{La1}) = 1.00$ and $g(\text{La2}) = 0.271$ [11]. The dissimilarity of these values reflects the chemical ordering of La-occupied La1-O1 and La-defective La2-O2 layers [Fig. 6.7(a)]. All the refined atomic displacement parameters increase with temperature (Table 6.2). The equivalent isotropic atomic displacement parameters of the oxygen atoms are larger than those of the cations, suggesting a larger diffusion coefficient for the oxide ions (Table 6.2). The oxygen atoms also display large anisotropy in terms of atomic displacement parameters, suggesting directionality in the movement of oxide ions around the stable positions. Similar large and anisotropic thermal motions of oxide ions were observed for the cubic perovskite-type oxide ion conductor $(La_{0.8}Sr_{0.2})$-$(Ga_{0.8}Mg_{0.15}Co_{0.05})O_{2.8}$ (Fig. 6.5) [10].

MEM analysis was conducted using diffraction data in the range $2\theta = 4.0°–140°$, corresponding to $d > 1.0$ Å (d, spacing of lattice planes), with the structure factors obtained by Rietveld analysis. A total of 59 structure factors were obtained for all data measured at three different temperatures. The 001 reflection appearing at the lowest 2θ position (~13°) was included, as this peak provides information on the disordered arrangement of the oxide ions. MEM calculations were performed with the unit cell divided into $64 \times 64 \times 128$ pixels. Use of the REMEDY cycle resulted in significant improvement in the R factors based on the Bragg intensities (R_I) and structure factors (R_F) (Table 6.2). Figure 6.7(b, c, d) shows the isosurface of nuclear density and the nuclear density distributions on the (100) and (001) planes obtained after the REMEDY cycle. Figure 6.8 shows the temperature dependence of the nuclear density contour map at $z = 0.2$ on the ab plane. Figures 6.7(b, c, d) and 6.8 provide much information on the positional disorder and diffusion paths of mobile oxide ions compared to the simple atomistic model (Fig. 6.7(a)).

At 769 K, the O3 atoms are localized near the stable $4i$ site (1/2, 0, 0.234). The O3 atoms display small bulges in the <101> direction (B in Fig. 6.7(d)), which become larger at 1281 and 1631 K (Fig. 6.7(c,b)). The probability density of each O3 atom is connected with that of its nearest neighbor O3 atoms, indicating diffusion paths (A in Fig. 6.7(a, b)). The diffusion path is along the [100] or [010] direction near the stable O3 positions, and along the [110] or $[1\bar{1}0]$ direction around the center of the paths. The O3 atoms migrate to the nearest neighbor $4i$ site through a triangle formed by adjacent La1, La2, and (Ti,Nb) atoms. The spatial distribution of the O3 atoms becomes larger with increasing temperature (Figs. 6.7 and 6.8). Such an increase in the density of oxide ions with increasing temperature is consistent with the higher conductivity observed at higher temperatures [21]. The O3 atom migrates following a curved route to maintain a relatively constant distance from the (Ti,Nb) atoms (solid curves A in Figs. 6.7(a, b) and 6.8(a)), rather than a direct linear path along the <110>

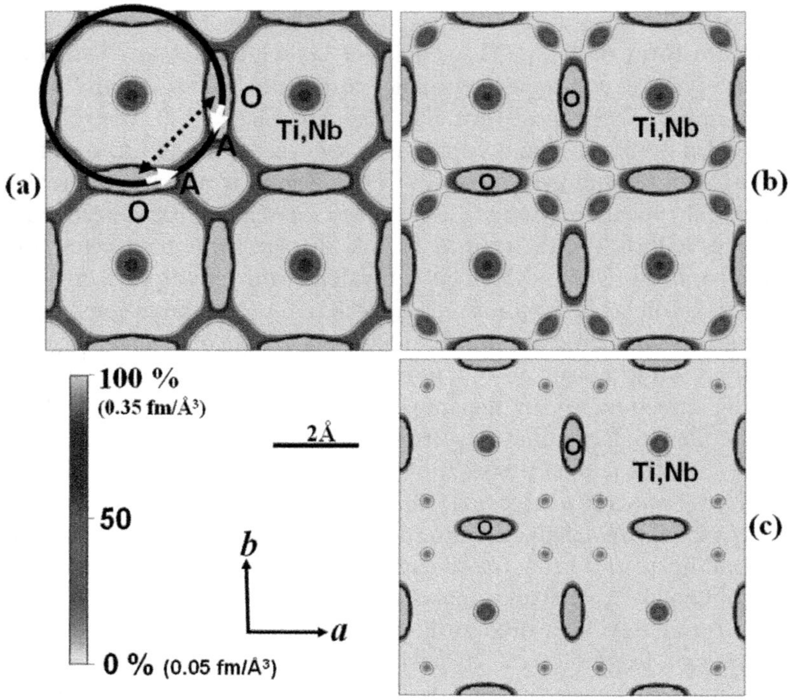

Fig. 6.8 Nuclear density distribution in the *ab* plane at $z = 0.2$ ($0 < x, y < 2$) of double perovskite-type *P*4/*mmm* $La_{0.64}(Ti_{0.92}Nb_{0.08})O_{2.99}$ at (**a**) 1631 K, (**b**) 1281 K, and (**c**) 769 K [11]. Contours are in the range 0.05–0.35 fm $Å^{-3}$ with steps of 0.05 fm $Å^{-3}$. The *solid line* in (**a**) denotes the curved diffusion path of the oxide ions, and the *dotted line* denotes the direct path between ideal positions. At low temperature (769 K), oxide ions are localized near the equilibrium position (see (**c**)); at high temperature (1631 K), the oxide ions are dispersed over a wide area between the regular positions (see (**a**))

direction (straight dotted lines with arrows between regular positions in Figs. 6.7(a, b) and 6.8(a)). Similar curved migration pathways were found in the nuclear density distribution of an ideal cubic perovskite-type compound, $(La_{0.8}Sr_{0.2})(Ga_{0.8}Mg_{0.15}Co_{0.05})O_{2.8}$ [10]. Computer simulations [40, 54] for perovskite-structured LaBO_3 (B = Co, Mn, Ga, Cr, and Fe) compounds also revealed deviations of the migration pathway from the direct path.

The oxide ion conductor $(La_{0.8}Sr_{0.2})(Ga_{0.8}Mg_{0.15}Co_{0.05})O_{2.8}$, which has an ideal perovskite-type structure, exhibits diffusion paths along the [110], $[1\bar{1}0]$, [011], $[01\bar{1}]$, [101] and $[10\bar{1}]$ directions to form a three-dimensional network of equivalent diffusion pathways (Fig. 6.5(b)) [10]. In contrast, in the present double perovskite-type material $La_{0.64}(Ti_{0.92}Nb_{0.08})O_{2.99}$, a two-dimensional diffusion pathway, by which O3 atoms migrate along the [110] and $[1\bar{1}0]$ directions (Fig. 6.7(b)), is present. This two-dimensional feature is attributable to the layered structure of the material, which consists of La-occupied La1-O1,

(Ti,Nb)-O, and La-deficient La2-O2 layers (Fig. 6.7(a)). Two-dimensional lithium cation conduction has also been reported in the orthorhombic layered perovskite-type compound $La_{0.62}Li_{0.16}TiO_3$ [55], in which the Li cation exists and migrates only near the La-deficient La2-O2 layer. This work has thus revealed that oxide ion diffusion in an ionic conductor with a double perovskite structure is two dimensional.

6.6 Crystal Structure and Structural Disorder of Oxide Ions in Cathode Materials, $La_{0.6}Sr_{0.4}CoO_{3-\delta}$ and $La_{0.6}Sr_{0.4}Co_{0.8}Fe_{0.2}O_{3-\delta}$, with a Cubic Perovskite-Type Structure [12, 13]

6.6.1 Introduction

The lanthanum strontium cobaltites $La_{1-x}Sr_xCoO_{3-\delta}$ and $La_{1-x}Sr_xCo_{1-y}Fe_yO_{3-\delta}$, which have a perovskite-type structure, are promising cathode materials for use in conjunction with doped lanthanum gallate electrolyte in SOFCs [56–58]. The crystal structure of trigonal $R\bar{3}c$ $La_{1-x}Sr_xCoO_{3-\delta}$ has been the subject of a number of investigations [59–65]. However, far less attention has been paid to the high-temperature cubic phase of $La_{1-x}Sr_xCoO_{3-\delta}$ and $La_{1-x}Sr_xCo_{1-y}Fe_yO_{3-\delta}$, which is important for application in SOFCs. In this section, we describe the crystal structure and structural disorder of cubic $Pm\bar{3}m$ perovskite-type oxides $La_{1-x}Sr_xCoO_{3-\delta}$ at 1531 K [12] and $La_{0.6}Sr_{0.4}Co_{0.8}Fe_{0.2}O_{3-\delta}$ at 1533 K [13].

6.6.2 Experiments and Data Processing

$La_{0.6}Sr_{0.4}CoO_{3-\delta}$ and $La_{0.6}Sr_{0.4}Co_{0.8}Fe_{0.2}O_{3-\delta}$ specimens were prepared by a solid-state reaction method. Neutron powder diffraction data of $La_{0.6}Sr_{0.4}CoO_{3-\delta}$ were collected in air using the HERMES diffractometer [25] at room temperature and at 1531 K. Neutron diffraction data of $La_{0.6}Sr_{0.4}Co_{0.8}Fe_{0.2}O_{3-\delta}$ were measured in air using the HERMES at 667 K and at 1533 K. The powder patterns were measured in the range $2\theta = 5°–155°$. The wavelength of the incident neutrons was 1.8207 Å. The sample temperature was kept constant during each data collection, using the furnace with $MoSi_2$ heaters [24]. The diffraction data of $La_{0.6}Sr_{0.4}CoO_{3-\delta}$ at 1531 K and of $La_{0.6}Sr_{0.4}Co_{0.8}Fe_{0.2}O_{3-\delta}$ at 667 K and 1533 K were analyzed by the Rietveld method and MPF analysis with the RIETAN-2000 program [27] and the MEM analysis with the PRIMA [29].

6.6.3 Results and Discussion

6.6.3.1 Crystal Structure and Disorder of $La_{0.6}Sr_{0.4}CoO_{3-\delta}$

The neutron diffraction data of $La_{0.6}Sr_{0.4}CoO_{3-\delta}$ at room temperature indicated that the specimen consisted of a single phase of trigonal $R\bar{3}c$ $La_{0.6}Sr_{0.4}CoO_{3-\delta}$. All the peaks in the neutron diffraction pattern of $La_{0.6}Sr_{0.4}CoO_{3-\delta}$ at 1531 K were indexed by a cubic perovskite-type structure with $Pm\bar{3}m$ symmetry (Fig. 6.9(a)), indicating phase transformation from a low-temperature trigonal to a high-temperature cubic phase. Rietveld analysis was performed using the diffraction data in the range $2\theta = 20°–153°$, based on a cubic perovskite-type structure (Fig. 6.9(a)). The La and Sr atoms were placed at the special positions $1b$ 1/2, 1/2, 1/2 in the $Pm\bar{3}m$ symmetry. The Co and O atoms were placed at the $1a$ 0, 0, 0 and $3d$ 1/2, 0, 0 sites, respectively. Isotropic and anisotropic atomic displacement parameters were used for cations and anions, respectively (Table 6.3). The refined crystallographic parameters and reliability factors are shown in Table 6.3. The atomic displacement parameters of the O atom exhibited a large anisotropy (Fig. 6.9(a) and Table 6.3). The occupancy factor of O atoms at the $3d$ site was estimated to be 0.886(6), indicating an oxygen deficiency of $\delta = 0.34(2)$ in the $La_{0.6}Sr_{0.4}CoO_{3-\delta}$ at 1531 K. The averaged valence of the Co cations was estimated to be 2.72 at 1531 K, which is consistent with the calculated bond valence sum of 2.8. Here, the average value of the bond valence parameters, 1.698, was used for the calculation [66].

MPF analysis was conducted using diffraction data in range $2\theta = 20°–153°$, corresponding to $d > 1.07$ Å, based on the structure factors obtained by

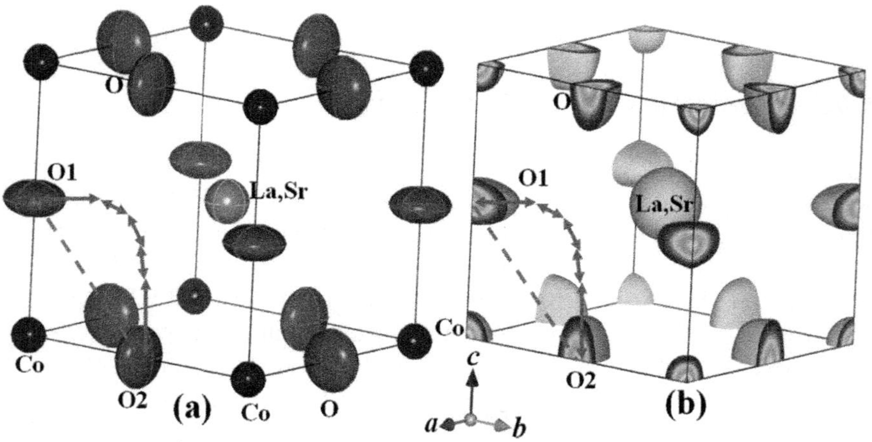

Fig. 6.9 (**a**) Refined crystal structure and (**b**) isosurface of nuclear density at 2 fm Å^{-3} for $La_{0.6}Sr_{0.4}CoO_{3-\delta}$ at 1531 K [12]. The *arrows* denote possible diffusion paths of oxide ions. The *dashed straight line* is the edge of the CoO_6 octahedron

Table 6.3 Refined crystallographic parameters and reliability factors obtained from Rietveld and MPF analysis for $La_{0.6}Sr_{0.4}CoO_{3-\delta}$ at 1531.4 K ($\delta = 0.34(2)$) [12]

Site and atoms	Wyckoff position	g	x	y	z	U (Å^2)
$La_{0.6}Sr_{0.4}$	1b	1.0	1/2	1/2	1/2	0.0425(9)
Co	1a	1.0	0	0	0	0.025(2)
O	3d	0.886(6)	1/2	0	0	0.066*

Note: Cubic space group $Pm\bar{3}m$ (No. 221). Number of formula units of $La_{0.6}Sr_{0.4}CoO_{3-\delta}$ in a unit cell: $Z = 1$. Unit-cell parameters: $a = b = c = 3.9496$ (3) Å, $\alpha = \beta = \gamma = 90°$; unit-cell volume: 61.612(9) Å^3; g, occupancy; x, y, z, fractional coordinates; U, isotropic atomic displacement parameters; *equivalent isotropic atomic displacement parameters; anisotropic atomic displacement parameters of O atom: $U_{11} = 0.027(2)$ Å^2, $U_{22} = U_{33} = 0.085(1)$ Å^2, $U_{12} = U_{23} = U_{31} = 0$ Å^2.
Reliability factors from Rietveld analysis: $R_{wp} = 3.73\%$, $R_p = 2.68\%$, $R_e = 1.86\%$, $R_{wp}/R_e = 2.00$, $R_I = 2.33\%$, $R_F = 1.72\%$. Reliability factors from first MPF analysis: $R_I = 1.71\%$, $R_F = 1.25\%$.

Rietveld analysis. A total of 16 structure factors were obtained. The 100 reflection appearing at the lowest 2θ position (~26.7°) was included, as this peak provides information on the disorder of the oxide ions. MEM calculations were performed with the unit cell divided into 64 × 64 × 64 pixels. The R factor based on the Bragg intensities (R_I) was improved from 2.33% (Rietveld analysis) to 1.71% (MPF), and that based on the structure factors (R_F) was improved from 1.72% to 1.25%. The MEM nuclear density distribution map for the (100) plane is shown in Fig. 6.10. The map reveals that the oxide ions in cubic $Pm\bar{3}m$ $La_{0.6}Sr_{0.4}CoO_{3-\delta}$ exhibit a large thermal motion perpendicular to the Co–O bond, corresponding to the large anisotropy in the atomic displacement parameters (Figs. 6.9 and 6.10). The arrows in Fig. 6.9 and dotted circles in Fig. 6.10 indicate possible diffusion paths of oxide ions in $La_{0.6}Sr_{0.4}CoO_{3-\delta}$.

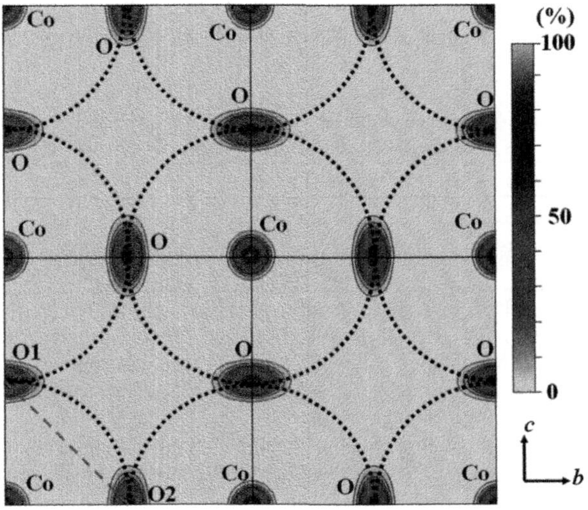

Fig. 6.10 Nuclear density distribution in the (100) plane for $La_{0.6}Sr_{0.4}CoO_{3-\delta}$, measured at 1531 K, with black contours in the range from 2 to 10 fm Å^{-3} (2 fm Å^{-3} steps) [12]. The color scale of 100% corresponds to the maximum density of 46.4 fm Å^{-3}. The *dotted circles* indicate possible oxide-ion diffusion paths. The *dashed straight line* indicates the edge of the CoO_6 octahedron. The *solid straight lines* indicate the unit cell. The figure shows four unit cells

The diffusion path does not follow the edge of the CoO_6 octahedron (shown as straight dashed lines between the ideal O1 and O2 positions in Figs. 6.9 and 6.10), but displays an arc shape (curved solid arrows in Fig. 6.9 and dotted circles in Fig. 6.10), avoiding the Co cation. This possible diffusion path is similar to that observed for $(La_{0.8}Sr_{0.2})(Ga_{0.8}Mg_{0.15}Co_{0.05})O_{2.8}$ [10]. Computer simulations have also indicated a similar curved path for oxide ion migration [40].

6.6.3.2 Crystal Structure and Disorder of $La_{0.6}Sr_{0.4}Co_{0.8}Fe_{0.2}O_{3-\delta}$

Neutron diffraction data for $La_{0.6}Sr_{0.4}Co_{0.8}Fe_{0.2}O_{3-\delta}$ (LSCF6482) at 667 K indicated that the specimen consisted of a single trigonal $R\bar{3}c$ phase. All the peaks in the neutron diffraction pattern of LSCF6482 at 1533 K were indexed by the cubic perovskite-type structure with $Pm\bar{3}m$ symmetry, indicating a phase transformation from the low-temperature trigonal to high-temperature cubic phase between 667 and 1533 K, which is consistent with the literature [67].

Rietveld analysis of LSCF6482 was performed using the neutron diffraction data taken at 667 K in the 2θ range of 20°–153° by a trigonal $R\bar{3}c$ perovskite-type structure. La and Sr atoms were placed at the special position $6a$ 0, 0, 1/4 of the $R\bar{3}c$ symmetry. Co and Fe atoms were put at the $6b$ 0, 0, 0 site. O atom was placed at the $18e$ x, 0, 1/4. In a preliminary analysis, the refined occupancy factor of O atoms at the $18e$ site g(O) was unity within the estimated standard deviation in the Rietveld analysis Thus, the g(O) was fixed to be unity in the final refinement. Isotropic and anisotropic atomic displacement parameters were used for the cations and anions, respectively. The calculated profile agreed well with the observed one [13]. The refined crystal parameters and reliability factors are shown in Table 6.4 [13]. The averaged valence of the Co and Fe cations was estimated to be 3.4 from the occupancy factor at 667 K, which is consistent with the calculated bond valence sum (BVS) value of 3.3. Here the average value of the bond valence parameter of 1.7118 was used for the

Table 6.4 Refined crystal parameters and reliability factors in Rietveld and MPF analyses for $La_{0.6}Sr_{0.4}Co_{0.8}Fe_{0.2}O_{3-\delta}$ at 667 K ($\delta = 0$) [13]

Site and atoms	Wyckoff position	g	x	y	z	U (Å^2)
$La_{0.6}Sr_{0.4}$	$6a$	1.0	0	0	1/4	0.0100(6)
$Co_{0.8}Fe_{0.2}$	$6b$	1.0	0	0	0	0.0061(9)
O	$18e$	1.0	0.5202(3)	0	1/4	0.0228*

Note: Trigonal space group $R\bar{3}c$ (hexagonal setting); number of formula units of $La_{0.6}Sr_{0.4}Co_{0.8}Fe_{0.2}O_{3-\delta}$ in a unit cell: $Z = 6$. Unit-cell parameters: $a = b = 5.46974(18)$ Å, $c = 13.3693$ (5) Å, $\alpha = \beta = 90°$, $\gamma = 120°$; unit-cell volume: 346.40(2) Å^3; g, occupancy; x, y, z, fractional coordinates; U, isotropic atomic displacement parameters; *equivalent isotropic atomic displacement parameter.

Reliability factors in the Rietveld analysis: $R_{wp} = 7.64\%$, $R_p = 6.00\%$, $R_e = 4.75\%$, $R_{wp}/R_e = 1.26$, $R_I = 3.76\%$, $R_F = 2.49\%$. Reliability factors in the first MPF analysis: $R_I = 1.04\%$, $R_F = 0.82\%$.

calculation [66]. Although the BVS is usually applied to the crystal structure at room temperature, we can use it at high temperatures because of small change of unit-cell parameters between different temperatures.

MPF analysis of $R\bar{3}c$ LSCF6482 was conducted using diffraction data in the 2θ range from 20° to 153° with the structure factors obtained from Rietveld analysis. The R factors for the Bragg intensities, R_I, and for the structure factors, R_F, were improved from 3.76% in the Rietveld analysis to 1.04% in the MPF, and from 2.49% to 0.82%, respectively. The resultant nuclear density indicated that the oxide ions are localized near the stable positions (Fig. 6.11(a)).

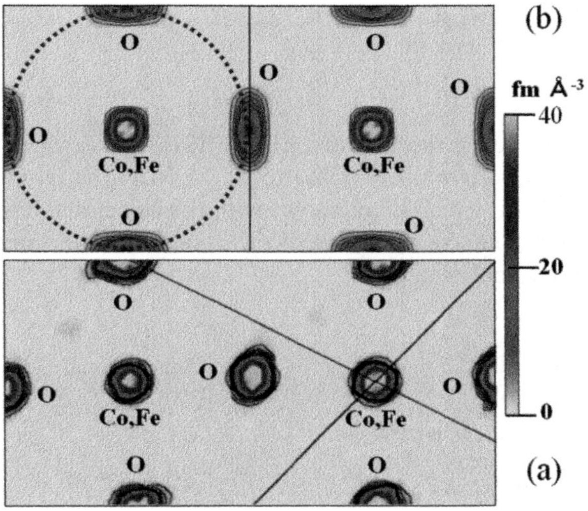

Fig. 6.11 (**a**) Nuclear density distribution on the (012) plane of trigonal $La_{0.6}Sr_{0.4}Co_{0.8}Fe_{0.2}O_{3-\delta}$ measured at 667 K. (**b**) Nuclear density distribution on the (100) plane of cubic $La_{0.6}Sr_{0.4}Co_{0.8}Fe_{0.2}O_{3-\delta}$ measured at 1533 K. *Black contours* in the range from 3 to 13 fm $Å^{-3}$ (3 fm $Å^{-3}$ step). The *dotted circle* indicates the possible oxide-ion diffusion path. The *straight solid line* indicates the unit cell

Rietveld analysis was performed using the diffraction data of LSCF6482 taken at 1533 K in the 2θ range of 20°–153° by a cubic $Pm\bar{3}m$ perovskite-type structure (Table 6.5). La and Sr atoms were placed at the special position 1*b* 1/2, 1/2, 1/2 of the $Pm\bar{3}m$ symmetry. Co and Fe atoms were put at the 1*a* 0, 0, 0 site, whereas the O atom was placed at the 3*d* 1/2, 0, 0 position. The atomic displacement parameters of the O atom exhibited large anisotropy (Fig. 6.11(b) and Table 6.5), which reflects the rotational motion of O atoms in the rigid $(Co,Fe)O_6$ octahedron. Similar anisotropy has been observed in other cubic perovskite-type compounds [10, 12, 68, 69]. The atomic displacement parameters at 1533 K were higher than those at 667 K. The equivalent isotropic displacement parameter of O atom is larger than those of cations (Tables 6.4 and 6.5), suggesting the higher diffusivity of O atoms. The occupancy factor of the O atom at the 3*d* site was estimated to be 0.904(6), indicating an oxygen deficiency of $\delta = 0.288(15)$ in $La_{0.6}Sr_{0.4}Co_{0.8}Fe_{0.2}O_{3-\delta}$ at 1533 K. The change of oxygen deficiency from $\delta = 0$ at 667 K to 0.288 at 1533 K, which was obtained in the Rietveld analyses, is reasonably consistent with the weight loss

Table 6.5 Refined crystal parameters and reliability factors in Rietveld and MPF analyses for $La_{0.6}Sr_{0.4}Co_{0.8}Fe_{0.2}O_{3-\delta}$ at 1533 K ($\delta = 0.288(15)$)

Site and atoms	Wyckoff position	g	x	y	z	U (Å^2)
$La_{0.6}Sr_{0.4}$	1*b*	1.0	1/2	1/2	1/2	0.0416(9)
$Co_{0.8}Fe_{0.2}$	1*a*	1.0	0	0	0	0.0294(11)
O	3*d*	0.904(6)	1/2	0	0	0.071*

Note: Cubic space group $Pm\bar{3}m$, number of formula units of $La_{0.6}Sr_{0.4}Co_{0.8}Fe_{0.2}O_{3-\delta}$ in a unit cell; $Z = 1$. Unit-cell parameters: $a = b = c = 3.9540(3)$ Å, $\alpha = \beta = \gamma = 90°$; unit-cell volume: 61.815(8) Å^3; g, occupancy; x, y, z, fractional coordinates; U, isotropic atomic displacement parameters. *Equivalent isotropic atomic displacement parameters, anisotropic atomic displacement parameters of O atom: $U_{11} = 0.0290(13)$ Å^2; $U_{22} = U_{33} = 0.0921(12)$ Å^2; $U_{12} = U_{23} = U_{31} = 0$ Å^2. Reliability factors in the Rietveld analysis: $R_{wp} = 4.82\%$, $R_p = 3.50\%$, $R_e = 1.70\%$, $R_{wp}/R_e = 2.83$, $R_I = 3.56\%$, $R_F = 3.20\%$. Reliability factor in the first MPF analysis: $R_F = 2.32\%$.

observed in the TG curve (Fig. 6.12). The averaged valence of the *B*-site Co and Fe cations was estimated to be 2.8 from the refined occupancy of O atoms at 1533 K, which is consistent with the calculated bond valence sum value of 2.9. Here, the average value of the bond valence parameter of 1.7118 was used for the calculation [66].

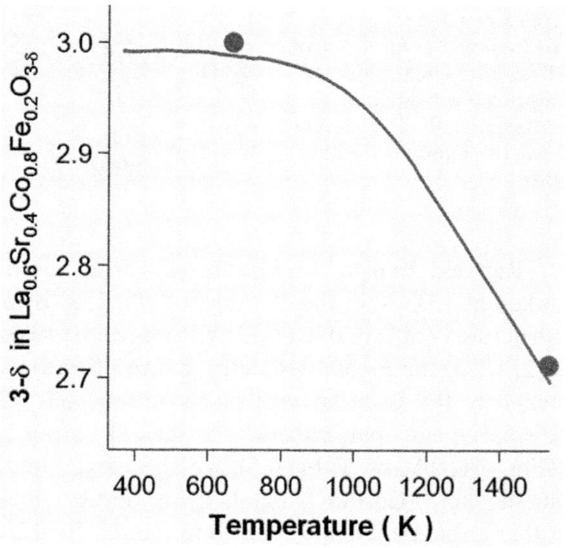

Fig. 6.12 Temperature dependence of $3-\delta$ in $La_{0.6}Sr_{0.4}Co_{0.8}Fe_{0.2}O_{3-\delta}$, where δ is the concentration of oxygen vacancy. *Solid line* was obtained using the weight loss from the TG data of $La_{0.6}Sr_{0.4}Co_{0.8}Fe_{0.2}O_{3-\delta}$ where no oxygen vacancy is assumed at room temperature. *Circle* was calculated from the refined occupancy at the oxygen site in the Rietveld analyses of high-temperature neutron diffraction data

MPF analysis of LSCF6482 was conducted using diffraction data taken at 1533 K in the 2θ range from 20° to 153°, with the structure factors obtained from Rietveld analysis. The *R* factors for the structure factors, R_F was improved from 3.20% in the Rietveld analysis to 2.32% in the MPF. To visualize the structural disorder, the MEM nuclear density distribution map on the (100) plane in LSCF6482 is shown in Fig. 6.11(b). The nuclear density map reveals that the oxide ions in the cubic $Pm\bar{3}m$ LSCF6482 exhibit a large thermal motion perpendicular to the (Co,Fe)-O bond, corresponding to large anisotropy of the atomic displacement parameters (Table 6.5). The dotted circle in Fig. 6.11(b) indicates possible diffusion paths of the oxide ions in LSCF6482. The diffusion path does not follow the edge of the $(Co,Fe)O_6$ octahedron, but displays an arc shape away from the Co,Fe cation. The nuclear density of O atoms in LSCF6482 did not connect with that of nearest neighbor O atoms (Fig. 6.11). On the contrary, the $(La_{0.8}Sr_{0.2})(Ga_{0.8}Mg_{0.15}Co_{0.05})O_{2.8}$ perovskite exhibited connected diffusion paths (Figs. 6.5 and 6.6) [10]. This finding strongly suggests that the diffusivity of oxide ions in LSCF6482 is lower than that in $(La_{0.8}Sr_{0.2})(Ga_{0.8}Mg_{0.15}Co_{0.05})O_{2.8}$. Similarly, the nuclear density around an O site in $La_{0.6}Sr_{0.4}CoO_{3-\delta}$ did not connect with that around the nearest neighbor O site (Figs. 6.9 and 6.10) [12].

It has been commonly assumed that the migrating anion in the $ABO_{3-\delta}$ perovskite-type structure takes a direct linear path along the <110> edge of the BO_6 octahedron. However, based on the results of this work, we suggest a curved diffusion path of the oxide ions in the electrode material LSCF6482 at 1533 K. The oxide ions migrate in the <100> directions near the stable $3d$ position, while they move along the <110> directions around the center of the diffusion path. Similar diffusion paths have been observed in the MEM nuclear density maps of $(La_{0.8}Sr_{0.2})(Ga_{0.8}Mg_{0.15}Co_{0.05})O_{2.8}$ (Figs. 6.5 and 6.6), of $La_{0.64}(Ti_{0.92}Nb_{0.08})O_{2.99}$ (Figs. 6.7 and 6.8), and of $La_{0.6}Sr_{0.4}CoO_{3-\delta}$ (Figs. 6.9 and 6.10) [10–12]. Computer simulations have also indicated a similar curved path for the oxide ion migration in perovskite-type compounds [40]. The diffusion paths of the cubic perovskite-type LSCF6482, $La_{0.6}Sr_{0.4}CoO_{3-\delta}$, and $(La_{0.8}Sr_{0.2})(Ga_{0.8}Mg_{0.15}Co_{0.05})O_{2.8}$ form a three-dimensional network. As described previously, Ali et al. [11] reported a similar but two-dimensional curved diffusion path of oxide ions in an oxide ion conductor, $La_{0.64}(Ti_{0.92}Nb_{0.08})O_{2.99}$, with a double perovskite-type structure. All four $ABO_{3-\delta}$ perovskite-type compounds LSCF6482, $La_{0.6}Sr_{0.4}CoO_{3-\delta}$, $(La_{0.8}Sr_{0.2})(Ga_{0.8}Mg_{0.15}Co_{0.05})O_{2.8}$, and $La_{0.64}(Ti_{0.92}Nb_{0.08})O_{2.99}$ exhibit a curved migration path of the mobile oxide ions, keeping the B–O distance constant to some degree. Thus, this curved feature should be common in perovskite-type ionic and mixed conductors.

6.7 Structural Disorder and Diffusion Path of Oxide Ions in a Doped Pr_2NiO_4-Based Mixed Ionic-Electronic Conductor $(Pr_{0.9}La_{0.1})_2(Ni_{0.74}Cu_{0.21}Ga_{0.05})O_{4+\delta}$ with a K_2NiF_4-Type Structure [15]

6.7.1 Introduction

A_2BO_4-based oxides with K_2NiF_4-type structure have extensively been studied as new mixed ionic-electronic conductors [23, 70–80], where A and B are larger and smaller cations. It has been speculated that the oxide ion conduction in the A_2BO_4-based oxides occurs by diffusion of excess oxide ions along the rock salt-type AO layers [75–78]. However, the diffusion path of the oxide ion has not been determined yet. Pr_2NiO_4-based oxides have high oxygen permeability and high diffusivity of oxide ions [23,78–80]. Here, we describe the diffusion path of oxide ions in a K_2NiF_4-type mixed conductor $(Pr_{0.9}La_{0.1})_2(Ni_{0.74}Cu_{0.21}Ga_{0.05})O_{4+\delta}$ (PLNCG), through a high-temperature neutron powder diffraction study [15].

6.7.2 Experiments and Data Processing

A $(Pr_{0.9}La_{0.1})_2(Ni_{0.74}Cu_{0.21}Ga_{0.05})O_{4+\delta}$ (PLNCG) sample was prepared by a solid-state reaction method. Neutron powder diffraction data of PLNCG were in situ measured at 879.8 and 1288.8 K using a furnace [24] and a 150 detector system HERMES [25] (neutron wavelength of 1.82646 Å). Neutron diffraction profiles at both temperatures indicated a K_2NiF_4-type structure with $I4/mmm$ space group. Neutron diffraction data were analyzed by a combination of Rietveld, MEM, and MPF analyses. A computer program RIETAN-2000 [27] was utilized for the Rietveld and MPF analyses, PRIMA [29] for MEM analysis, and VESTA [81] for visualization of nuclear density (scattering length density) distribution.

6.7.3 Results and Discussion

Rietveld analysis of the neutron diffraction data of PLNCG at 879.8 and 1288.8 K was successfully performed on the basis of the K_2NiF_4-type structure with tetragonal $I4/mmm$ space-group symmetry (Figs. 6.13 and 6.14(a), Table 6.6). The results show that the crystal structure of PLNCG consists of (Ni,Cu,Ga)O_6 octahedron and (Pr,La)-O layers (Fig. 6.14(a)). Refined occupancy factors indicated the excess oxygen of $\delta = 0.0154(3)$ in the PLNCG, $(Pr_{0.9}La_{0.1})_2(Ni_{0.74}Cu_{0.21}Ga_{0.05})O_{4+\delta}$, which is ascribed to the interstitial O3 atom. The O3 atom is located at a 16n site, i.e., $(x, 0, z)$ where $x = 0.666(19)$ and $z = 0.223(9)$ at 1288.8 K (Table 6.6). Figures 6.14(b) and 6.15 show the isosurface and distributions of MEM nuclear density of PLNCG. The oxygen atom at the O2 site (4e; $(0, 0, z)$; $z = 0.1752(4)$ at 1288.8 K) exhibits highly anisotropic thermal motion ($U_{11} = U_{22} = 0.115(3)$ Å^2 and $U_{33} = 0.021(3)$ Å^2) (Table 6.6), which leads to the migration of oxide ions to the nearest neighbor interstitial O3 positions. The striking feature in the nuclear density distribution is the curved O2–O3 diffusion path (Figs. 6.14(b) and 6.15(b)). This curve feature is ascribed to the repulsion between (Pr,La) and O atoms. The distance between the (Pr,La) and O atoms is kept constant to some degree along the diffusion paths. This fact suggests that the large-sized cations such as Pr and La ions at the A site in the A_2BO_4-type conductor are effective in improving

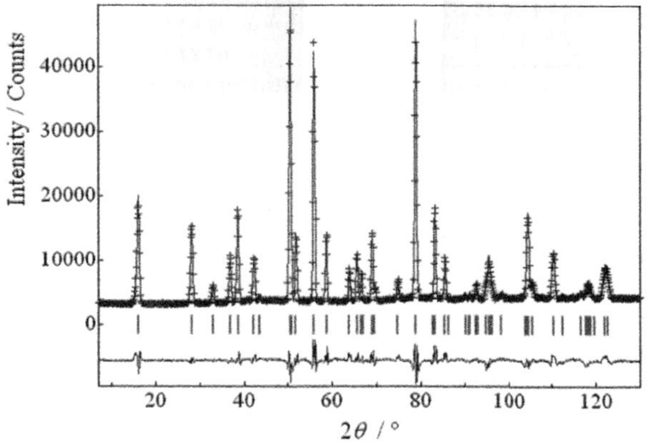

Fig. 6.13 Rietveld fitting results for the neutron diffraction data of $(Pr_{0.9}La_{0.1})_2(Ni_{0.74}Cu_{0.21}$-$Ga_{0.05})O_{4+\delta}$ measured at 1288.8 K. The *red plus symbols* and the *green lines* denote the observed and calculated intensities, respectively. *Short vertical lines* indicate the positions of possible Bragg reflections. The difference between the observed and calculated profiles is plotted at the bottom. The wavelength of the incident neutrons is 1.82646 Å [15]

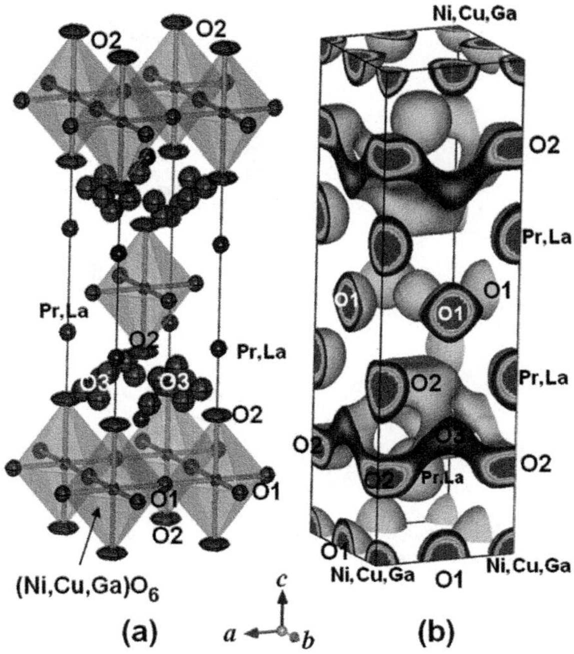

Fig. 6.14 (**a**) Refined crystal structure and isosurface of nuclear density at 0.05 fm $Å^{-3}$ of the mixed oxide-ionic and electronic conductor $(Pr_{0.9}La_{0.1})_2$-$(Ni_{0.74}Cu_{0.21}Ga_{0.05})O_{4+\delta}$ determined in situ at 1288.8 K. Unit cell: tetragonal *I*4/*mmm*, $a = 3.875(3)$ and $c = 12.738(9)$ Å [15]

Table 6.6 Refined crystal parameters and reliability factors in Rietveld and MEM-based-pattern-fitting analyses for $(Pr_{0.9}La_{0.1})_2(Ni_{0.74}Cu_{0.21}Ga_{0.05})O_{4+\delta}$ at 1288.8 K ($\delta = 0.0154(3)$) [15]

Site and atoms	Wyckoff position	Occupancy factor, g	Fractional coordinate x	y	z	Atomic displacement parameter U (Å^2)
$Pr_{0.90}La_{0.10}$	$4e$	1.0	0	0	0.3573 (4)	0.0347 (14)
Ni	$2a$	0.74 (4)	0	0	0	0.0192 (14)
Cu	$2a$	0.21 (4)	0	0	0	= U (Ni)
Ga	$2a$	0.05	0	0	0	= U (Ni)
O1	$4c$	1.0	1/2	0	0	0.0315 (14)
O2	$4e$	1.0	0	0	0.1752 (4)	0.0840*
O3	$16n$	0.019 (3)	0.666 (19)	0	0.223 (9)	0.0630 (28)

Note: Tetragonal space group $I4/mmm$, number of formula units of $(Pr_{0.9}La_{0.1})_2$-$(Ni_{0.74}Cu_{0.21}Ga_{0.05})O_{4+\delta}$ in a unit cell: $Z = 2$. Unit-cell parameters: $a = b = 3.875(3)$ Å, $c = 12.738(9)$ Å, $\alpha = \beta = \gamma = 90°$; unit-cell volume, 191.2(2) Å^3; g, occupancy; x, y, z, fractional coordinates; U, isotropic atomic displacement parameters. *Equivalent isotropic atomic displacement parameters, anisotropic atomic displacement parameters of O2 atom: $U_{11} = U_{22} = 0.115(3)$ Å^2, $U_{33} = 0.021(3)$ Å^2, $U_{12} = U_{23} = U_{31} = 0$ Å^2. Reliability factors in the Rietveld analysis: $R_{wp} = 6.48\%$, $R_p = 4.59\%$, $R_e = 1.35\%$, $R_{wp}/R_e = 4.78$, $R_I = 2.18\%$, $R_F = 1.21\%$. Reliability factors in the first MPF analysis: $R_I = 2.15\%$, $R_F = 0.89\%$.

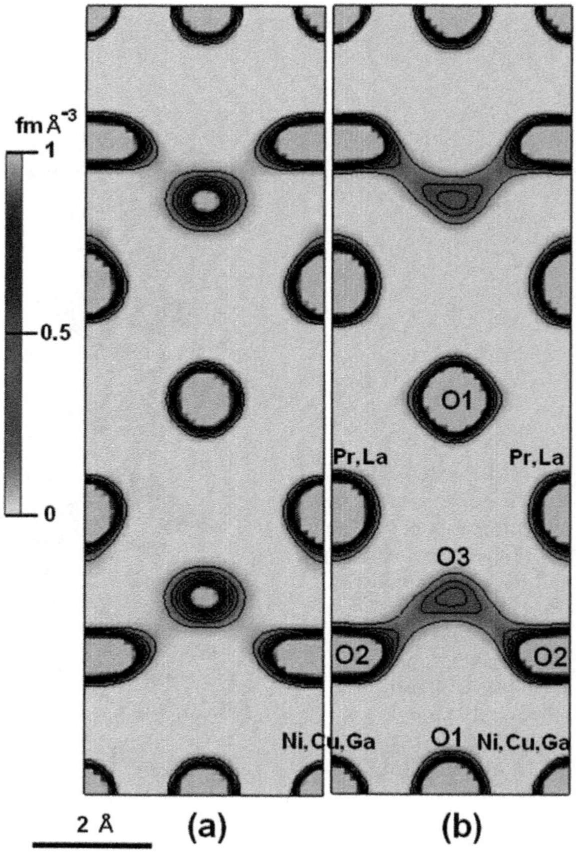

Fig. 6.15 Nuclear density distribution on the (100) plane of the mixed conductor $(Pr_{0.9}La_{0.1})_2$-$(Ni_{0.74}Cu_{0.21}Ga_{0.05})O_{4+\delta}$ at (**a**) 879.8 K and (**b**) 1288.8 K. Contour lines from 0.1 to 1.0 by the step of 0.1 fm Å^{-3} [15]

the oxide ionic conduction on the *A*–O layer. The conduction path is along the <100> directions near the O2 site and roughly along the <301> directions around the center of the paths (Fig. 6.15(b)). The nuclear density distribution also shows the two-dimensional (2D) network of the O2–O3–O2 diffusion paths of oxide ions. The 2D feature is consistent with the anisotropic transport of oxide ions in $La_2NiO_{4+\delta}$ [75]. The nuclear density on the diffusion path at 1238.8 K (Fig. 6.15(b)) is larger than that at 879.8 K (Fig. 6.15(a)), which is consistent with the improved oxygen permeability at higher temperatures [23, 79, 80].

6.8 Conclusions

We analyzed the nuclear density distributions of $(La_{0.8}Sr_{0.2})(Ga_{0.8}Mg_{0.15}Co_{0.05})O_{2.8}$ [10], $La_{0.64}(Ti_{0.92}Nb_{0.08})O_{2.99}$ [11], $La_{0.6}Sr_{0.4}CoO_{3-\delta}$ [12], $(La_{0.6}Sr_{0.4})(Co_{0.8}Fe_{0.2})O_{3-\delta}$ [13], and $(Pr_{0.9}La_{0.1})_2(Ni_{0.74}Cu_{0.21}Ga_{0.05})O_{4+\delta}$ [15] to investigate the diffusion paths and structural disorder of oxide ions at high temperatures.

In the fast oxide ion conductor cubic $(La_{0.8}Sr_{0.2})(Ga_{0.8}Mg_{0.15}Co_{0.05})O_{2.8}$, the oxide ions at the 3*d* site (1/2, 0, 0) are largely distributed along the <100> directions near the equilibrium position and along the <110> directions around the center of the diffusion paths at temperatures of 1665 and 1471 K. The diffusion path of oxide ions in $(La_{0.8}Sr_{0.2})(Ga_{0.8}Mg_{0.15}Co_{0.05})O_{2.8}$ is not a straight line between the ideal positions along the <110> direction, but is in the form of an arc, maintaining a constant distance from the $Ga_{0.8}Mg_{0.15}Co_{0.05}$ cation site (Figs. 6.5(b), 6.6(a, b)). In the trigonal $R\bar{3}c$ $(La_{0.8}Sr_{0.2})(Ga_{0.8}Mg_{0.15}Co_{0.05})O_{2.8}$ at 1069 K and $LaGaO_3$ at 1663 K (Fig. 6.6(c, d)), the oxide ions are localized near the equilibrium positions, whereas in the high-temperature cubic structure, they are spread over a wide area between the ideal positions (Fig. 6.6(a, b)).

In the oxide ion conductor $La_{0.64}(Ti_{0.92}Nb_{0.08})O_{2.99}$, the oxide ions (O3) at the 4*i* site (1/2, 0, *z*) are largely distributed along the <100> directions near the equilibrium position and along the <110> directions around the center of the diffusion paths at temperatures of 769, 1281, and 1631 K (Figs. 6.7 and 6.8). The spatial distribution of these oxide ions becomes greater with increasing temperature (Figs. 6.7 and 6.8). The oxide ions in $La_{0.64}(Ti_{0.92}Nb_{0.08})O_{2.99}$ migrate to nearest neighbor 4*i* sites along the [100] and [010] directions near the equilibrium positions in the vicinity of the (004) plane at 1281 and 1631 K. Around the center of the curved diffusion paths, the oxide ions migrate along the [110] and $[1\bar{1}0]$ directions. The two-dimensionality of the diffusion paths may be attributed to the layered structure of double perovskite-type $La_{0.64}(Ti_{0.92}Nb_{0.08})O_{2.99}$.

For cubic $La_{0.6}Sr_{0.4}CoO_{3-\delta}$ at 1531 K and $La_{0.6}Sr_{0.4}Co_{0.8}Fe_{0.2}O_{3-\delta}$ at 1533 K, the refined anisotropic atomic displacement parameters and the nuclear density maps reveal that the oxide ions exhibit a large thermal motion perpendicular to the Co–O bond (Figs. 6.9, 6.10, and 6.11). Again, oxide ion migration between adjacent anion sites appears to follow a curved path along the <100> directions near the equilibrium position and along the <110> directions around the center of the diffusion paths. High-temperature cubic $La_{0.6}Sr_{0.4}Co_{0.8}Fe_{0.2}O_{3-\delta}$ at 1533 K has larger spatial distribution of oxide ions, compared with low-temperature trigonal $La_{0.6}Sr_{0.4}Co_{0.8}Fe_{0.2}O_{3-\delta}$ at 667 K.

We have presented also the visualization of structural disorder and diffusion path of oxide ions in a K_2NiF_4-type mixed conductor, $(Pr_{0.9}La_{0.1})_2$-$(Ni_{0.74}Cu_{0.21}Ga_{0.05})O_{4+\delta}$ [15]. We have experimentally confirmed that the anisotropic thermal motions of the O2 atom and the interstitial O3 atom are essential for the high oxygen permeability of the K_2NiF_4-type mixed conductor. To design improved K_2NiF_4-type mixed conductors with higher oxide ion diffusivity, it might be useful to adopt larger *A* cations and *A* and *B* cations with higher valences, which yield higher concentration of interstitial O3 atoms.

In the cubic perovskite oxides ($La_{0.6}Sr_{0.4}CoO_{3-\delta}$, $La_{0.6}Sr_{0.4}Co_{0.8}Fe_{0.2}O_{3-\delta}$, $(La_{0.8}Sr_{0.2})(Ga_{0.8}Mg_{0.15}Co_{0.05})O_{2.8}$), the diffusion path forms a three-dimensional network, while in the double-perovskite-type $La_{0.64}(Ti_{0.92}Nb_{0.08})O_{2.99}$ and K_2NiF_4-type $(Pr_{0.9}La_{0.1})_2(Ni_{0.74}Cu_{0.21}Ga_{0.05})O_{4+\delta}$ exhibit a two-dimensional network of oxide ion diffusion paths. All three $ABO_{3-\delta}$ perovskite-type compounds ($La_{0.6}Sr_{0.4}CoO_{3-\delta}$, $La_{0.6}Sr_{0.4}Co_{0.8}Fe_{0.2}O_{3-\delta}$, $(La_{0.8}Sr_{0.2})(Ga_{0.8}Mg_{0.15}Co_{0.05})O_{2.8}$) and the double-perovskite-type $La_{0.64}(Ti_{0.92}Nb_{0.08})O_{2.99}$ exhibit curved oxide ion migration paths, maintaining to some degree a constant *B*–O distance. Around the centers of the diffusion paths, the oxide ions migrate along the <110> directions. It appears that this curved feature is expected to be common in other perovskite-type ionic and mixed conductors. The curve feature is also observed in the K_2NiF_4-type compound $(Pr_{0.9}La_{0.1})_2(Ni_{0.74}Cu_{0.21}Ga_{0.05})$-$O_{4+\delta}$, maintaining to some degree a constant *A*–O distance. Curved migration paths have previously been observed in the nuclear and electron density distribution maps of various materials [14], including the lithium cation conductor $La_{0.62}Li_{0.16}TiO_3$ [55], the fluorite-type structured anion conductors $Bi_{1.4}Yb_{0.6}O_3$ [6] and $Ce_{0.93}Y_{0.07}O_{1.96}$ [8], and the Cu cation conductor CuI, which has a fluorite-type structure [82]. Thus, the curved migration paths are also common among various ionic and mixed conductors.

Acknowledgments The author acknowledges all the authors and collaborators of the joint papers mentioned in the references. In particular, the author expresses special thanks to Dr. K. Nomura for useful discussion. We also thank Dr. K. Ohoyama and Mr. K. Nemoto for use of the HERMES diffractometer. Figures 6.5, 6.6, 6.7, 6.8, 6.9, 6.10, 6.11, 6.14, and 6.15 were drawn using the VENUS [29] and VESTA [81] programs developed by Dr. R. Dilanian, Dr. K. Momma, and Dr. F. Izumi. This research was supported in part by the Ministry of Education, Culture, Sports, Science and Technology of Japan (Monbu-Kagaku-sho).

References

1. E.C. Subbarao, Advances in Ceramics Vol. 3, Science and Technology of Zirconia I, Heuer A.H. and Hobbs L.W. (eds.), p. 1. American Ceramic Society, Columbus, OH (1981)
2. M. Yashima, M. Kakihana, M. Yoshimura, Solid State Ionics 86–88, 1131 (1996)
3. T. Takahashi, H. Iwahara, J. Appl. Electrochem., 3, 65 (1972)
4. N.M. Sammes, G.A. Tompsett, H. Näfe, F. Aldinger, J. Eur. Ceram. Soc. 19, 1801 (1999)
5. M. Yashima, D. Ishimura, Chem. Phys. Lett. 378, 395 (2003)
6. M. Yashima, D. Ishimura, Appl. Phys. Lett. 87, 221909 (2005)
7. H. Inaba, H. Tagawa, Solid State Ionics 83, 1 (1996)
8. M. Yashima, S. Kobayashi, T. Yasui, Faraday Discussions 134, 369 (2007)
9. T. Ishihara, H. Matsuda, Y. Takita, J. Am. Chem. Soc. 116, 3801 (1994)
10. M. Yashima, K. Nomura, H. Kageyama, Y. Miyazaki, N. Chitose, K. Adachi, Chem. Phys. Lett. 380, 391 (2003)
11. R. Ali, M. Yashima, F. Izumi, Chem. Mater. 19, 3260 (2007)
12. M. Yashima, T. Tsuji, J. Appl. Crystallogr. 40, 1166 (2007)
13. M. Yashima, T. Kamioika, Solid State Ionics 178, 1939 (2008)
14. M. Yashima, Solid State Ionics 179, 797 (2008)
15. M. Yashima, M. Enoki, T. Wakita, R. Ali, Y. Matsushita, F. Izumi, T. Ishihara, J. Am. Chem. Soc. 130, 2762 (2008)
16. M. Feng, J.B. Goodenough, K. Huang, C. Milliken, J. Power Sources, 63, 47 (1996)
17. Y. Teraoka, H.M. Zhang, K. Okamoto, H. Yamazoe, Mater. Res. Bull. 23, 51 (1998)
18. T. Horita, K. Yamaji, N. Sakai, H. Yokokawa, A. Weber, E. Ivers-Tiffee, Solid State Ionics 138, 143 (2000)
19. R. Ali, M. Yashima, M. Tanaka, H. Yoshioka, T. Mori, S. Sasaki, J. Solid State Chem. 164, 51 (2002)
20. M. Yashima, M. Mori, T. Kamiyama, K.I. Oikawa, A. Hoshikawa, S. Torii, K. Saitoh, K. Tsuda, Chem. Phys. Lett. 375, 240 (2003)
21. H. Yoshioka, J. Am. Ceram. Soc. 85, 1339 (2002)
22. M.F. Hundley, R.S. Kwok, S.W. Cheong, J.D. Thompson, Z. Fisk, Physica C 172, 455 (1991)
23. S. Miyoshi, T. Furuno, H. Matsumoto, T. Ishihara, Solid State Ionics 177, 2269 (2006)
24. M. Yashima, J. Am. Ceram. Soc. 85, 2925 (2002)
25. K. Ohoyama, T. Kanouchi, K. Nemoto, M. Ohashi, T. Kajitani, Y. Yamaguchi, Jpn. J. Appl. Phys. Part 1. 37, 3319 (1998)
26. M. Yashima, In: Proceedings of an International Conference on Solid → Solid Phase Transformations in Inorganic Materials 2005, Vol. 2, Howe J.M., Laughlin D.E., Lee J.K., Dahmen U., Soffa W.A. (eds.), p. 493. TMS: The Minerals, Metals & Materials Society, Warrendale, Pennsylvania (2005)
27. F. Izumi, T. Ikeda, Mater. Sci. Forum 321–324, 198 (2000)
28. M. Sakata, T. Uno, M. Takata, C.H. Howard, J. Appl. Crystallogr. 26, 159 (1993)
29. F. Izumi, R.A. Dilanian, In: Recent Research Developments in Physics, Vol. 3, Part II, p. 699. Transworld Research Network, Trivandrum (2002)
30. K. Nomura, S. Tanase, Solid State Ionics 98, 229 (1997)
31. W. Marti, P. Fischer, F. Altorfer, H.J. Scheel, M.J. Tadin, J. Phys. Condens. Matter 6, 127 (1994)
32. W. Marti, P. Fischer, J. Schefer, F. Kubel, Z. Kristallogr. 211, 891 (1996)
33. P.R. Slater, J.T.S. Irvine, T. Ishihara, Y. Takita, J. Solid State Chem. 139, 135 (1998)
34. C.J. Howard, B.J. Kennedy, J. Phys. Condens. Matter 11, 3229 (1999)
35. L. Vasylechko, D. Savytskiia, A. Matkovskia, M. Berkowskic, M. Knappd, U. Bismayer, J. Alloy Comp. 328, 264 (2001)
36. M. Lerch, H. Boysen, T. Hansen, J. Phys. Chem. Solids 62, 445 (2001)

37. N.P. Vyshatko, V. Kharton, A.L. Shaula, E.N. Naumovich, F.M.B. Marques, Mater. Res. Bull. 38, 185 (2003)
38. M. Kajitani, M. Matsuda, A. Hoshikawa, K. Oikawa, S. Torii, T. Kamiyama, F. Izumi, M. Miyake, Chem. Mater. 15, 3468 (2005)
39. M. Kajitani, M. Matsuda, A. Hoshikawa, S. Harjo, T. Kamiyama, T. Ishigaki, F. Izumi, M. Miyake, Chem. Mater. 17, 4235 (2007)
40. M.S. Islam, J. Mater. Chem. 10, 1027 (2000)
41. M.S. Khan, M.S. Islam, D.R. Bates, J. Phys. Chem. B 102, 3099 (1998)
42. K. Nomura, M. Yashima, In: Proceedings of the Symposium on Powder Diffraction III, KEK Proceedings 2005-19, Ida T., Kamiyama T. (eds.), p. 45. High Energy Accelerator Research Organization, Tsukuba, Japan (2006)
43. T. Ishihara, T. Akbay, H. Furutani, Y. Takita, Solid State Ionics 113–115, 585 (1998)
44. H. Yoshioka, S. Kikkawa, J. Mater. Chem. 8, 1821 (1998)
45. D. Suvorov, M. Valant, S. Skapin, D. Dolar, J. Mater. Sci. 33, 85 (1998)
46. M. Yashima, R. Ali, H. Yoshioka, Solid State Ionics 128, 105 (2000)
47. R. Ali, M. Yashima, M. Yoshimura, H. Yoshioka, J. Am. Ceram. Soc. 84, 468 (2001)
48. M. Yashima, R. Ali, M. Tanaka, T. Mori, Chem. Phys. Lett. 363, 129 (2002)
49. M. Yashima, M. Mori, R. Ali, M. Tanaka, T. Mori, Chem. Phys. Lett. 371, 582 (2003)
50. R. Ali, M. Yashima, J. Synchrotron Radiat. 10, 28 (2003)
51. C.J. Howard, Z. Zhang, Acta Crystallogr. B 60, 249 (2004)
52. R. Ali, F. Izumi, M. Yashima, J. Am. Ceram. Soc. 89, 3805 (2006)
53. V. Vashook, L. Vasylechko, N. Trofimenko, M. Kuznecov, P. Otchik, J. Zosel, U. Guth, J. Alloys Compd. 419, 271 (2006)
54. M. Cherry, M.S. Islam, C.R.A. Catlow, J. Solid State Chem. 118, 118 (1995)
55. M. Yashima, M. Itoh, Y. Inaguma, Y. Morii, J. Am. Chem. Soc. 127, 3491 (2005)
56. O. Yamamoto, Y. Takeda, R. Kanno, M. Noda, Solid State Ionics 22, 241 (1987)
57. J. Mizusaki, Y. Mima, S. Yamauchi, K. Fukui, H. Tagawa, J. Solid State Chem. 80, 102 (1989)
58. S.B. Adler, J.A. Lane, B.C.H. Steele, J. Electrochem. Soc. 143, 3554 (1996)
59. J. Kirchnerova, D.B. Hibbert, J. Mater. Sci. 28, 5800 (1993)
60. V.G. Sathe, A.V. Pimpale, V. Siriguri, S.K. Paranjpe, J. Phys. Condens. Matt. 8, 3889 (1996)
61. N.L.N.P. Closset, R.H.E. van Doorn, H. Kruidhof, Powd. Diff. 11, 31 (1996)
62. R. Sonntag, S. Neov, V. Kozhukharov, D. Neov, J.E. ten Elshof, Physica B 241–243, 393 (1998)
63. R.H.E. Van Doorn, A.J. Burggraaf, Solid State Ionics 128, 65 (2000)
64. V.V. Sikolenko, E.V. Pomjakushina, S.Y. Istomin, J. Mag. Mag. Mater. 258–259, 300 (2003)
65. T. Hanashima, S. Azuhata, K. Yamawaki, N. Shimizu, T. Mori, M. Tanaka, S. Sasaki, Jpn. J. Appl. Phys. Part A 43, 4171 (2004)
66. R.E. Brese, M. O'Keeffe, Acta Crystallogr. B 47, 192 (1991)
67. S. Wang, M. Katsuki, M. Dokiya, T. Hashimoto, Solid State Ionics 159, 71 (2003)
68. M. Yashima, M. Tanaka, J. Appl. Crystallogr. 37, 786 (2004)
69. J. Hutton, R.J. Nelmes, J. Phys. C: Solid State Phys. 14, 1713 (1981)
70. V.V. Kharton, A.P. Viskup, A.V. Kovalevsky, E.N. Naumovich, F.M.B. Marques, Solid State Ionics 143, 337 (2001)
71. S. Kato, M. Ogasawara, M. Sugai, S. Nakata, Solid State Ionics 149, 53 (2002)
72. A. Manthiram, F. Prado, T. Armstrong, Solid State Ionics 152–153, 647 (2002)
73. C.T. Li, H. Hu, H. Zhang, Y. Chen, J. Jin, N.R. Yang, J. Membr. Sci. 226, 1 (2003)
74. V.V. Kharton, E.V. Tsipis, A.A. Yaremchenko, J.R. Frade, Solid State Ionics 166, 327 (2004)
75. S.J. Skinner, J.A. Kilner, Solid State Ionics 135, 709 (2000)
76. L. Minervini, R.W. Grimes, J.A. Kilner, K.E. Sickafus, J. Mater. Chem. 10, 2349 (2000)

77. J.M. Bassat, P. Odier, A. Villesuzanne, C. Marin, M. Pouchard, Solid State Ionics 167, 341 (2004)
78. E. Boem, J.-M. Bassat, P. Dordor, F. Mauvy, J.-C. Grenier, P. Stevens, Solid State Ionics 176, 2717 (2005)
79. S. Miyoshi, T. Furuno, O. Sangoanruang, H. Matsumoto, T. Ishihara, J. Electrochem. Soc. 154, B57 (2007)
80. T. Ishihara, S. Miyoshi, T. Furuno, O. Sangoanruang, H. Matsumoto, Solid State Ionics 177, 3087 (2006)
81. K. Momma, F. Izumi, J. Appl. Crystallogr. 41, 653 (2008)
82. M. Yashima, Q. Xu, A. Yoshiasa, S. Wada, J. Mater. Chem. 16, 4393 (2006)

Chapter 7
Perovskite Oxide for Cathode of SOFCs

Tatsuya Kawada

7.1 Introduction

Solid oxide fuel cells (SOFC) can achieve high efficiency without using costly precious metal catalysts, which is regarded as a great advantage of SOFC compared to polymer electrolyte fuel cells (PEFC) and phosphoric acid fuel cells (PAFC). It does not mean, however, there are no technical issues concerning the electrode materials for SOFCs. Extensive studies are still needed to look for a better material that shows high performance at lower temperatures and high stability at higher temperatures. For a cathode material, various oxides have been proposed so far [1–4]. Most of them have the perovskite-type structure, ABO_3, or related structures.

The transport properties of perovskite-type oxides are dependent mainly on the B-site cations. Among them, Mn-based perovskites and Co/Fe-based perovskites are most frequently used for high-temperature and intermediate-temperature SOFCs, respectively. Recently, Ni-based K_2NiF_4-type oxides are also being investigated [5]. Their composition and microstructure are still to be optimized based on the defect chemistry, electrochemistry, and thermodynamics.

The scientific bases for the electrode reaction kinetics are also to be established. In a porous cathode, oxygen adsorbed from the gas phase on the cathode particles is dissociated and transported via diffusion on the surface or through the bulk. The oxygen potential profile inside the cathode layer is established according to the rates of those processes under current flow. Thus, the properties of the electrode particles under operation are not correctly understood without the knowledge of electrode kinetics.

In this chapter, general features of perovskite-type oxides are summarized from the point of view of the required properties as a cathode material. Then,

T. Kawada (✉)
Graduate School of Environmental Studies, Tohoku University, 1-1 Aoba, Aramaki, Aoba-ku, Sendai 980-8579, Japan
e-mail: kawada@ee.mech.tohoku.ac.jp

T. Ishihara (ed.), *Perovskite Oxide for Solid Oxide Fuel Cells*,
Fuel Cells and Hydrogen Energy, DOI 10.1007/978-0-387-77708-5_7,

discussion is presented on the properties and electrochemistry of some typical materials for high-temperature and intermediate-temperature SOFC cathodes.

7.2 Properties Required for a Cathode Material

A cathode material for SOFC should meet various requirements in catalytic activity, thermodynamic stability, and compatibility. The following are the requirements for the cathode materials and the proposed approach to find and design a suitable material for high stability and a high-performance cathode.

7.2.1 Catalytic Activity

Oxygen reduction proceeds on the electrode surface or at the electrode–electrolyte interface. The electrode material catalyzes the oxygen molecules to be dissociated into atoms, charged, and incorporated into the electrolyte (Fig. 7.1). In selection of the cathode material, the electrocatalytic activity is an important parameter to be considered. The surface reaction rate constant in oxygen isotope exchange is a good measure for the catalytic activity. Kilner et al. [4] compared various oxides in terms of isotope diffusion coefficient and found a positive correlation between those parameters. A high mixed electronic and ionic conductor may be a promising candidate in terms of electrode performance.

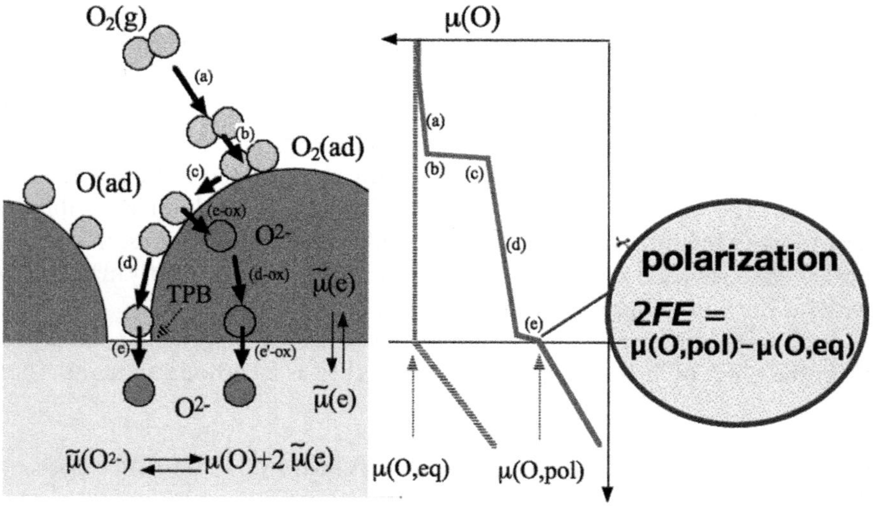

Fig. 7.1 Schematic view of cathode reaction in SOFC

For further improvement, the reaction mechanism should be clarified. Although many reports have been published so far on the reaction kinetics, information obtained from the experiments is limited, and because of the variety of reaction models the reaction mechanism still remains unclear. The development of in situ observation technique will be necessary. Recently, some efforts were reported on in situ techniques. Lu et al. [6] applied infrared emission spectroscopy to observe the adsorbed species on a (Sm,Sr)CoO_3 cathode under operation. They suggested O_2^- is the most probable adsorbate (Fig. 7.2). Murai et al. [7] employed polarization-modulated IR reflection absorption spectroscopy and found response in a similar frequency region. Quantum mechanical calculations are also made by several researchers [8].

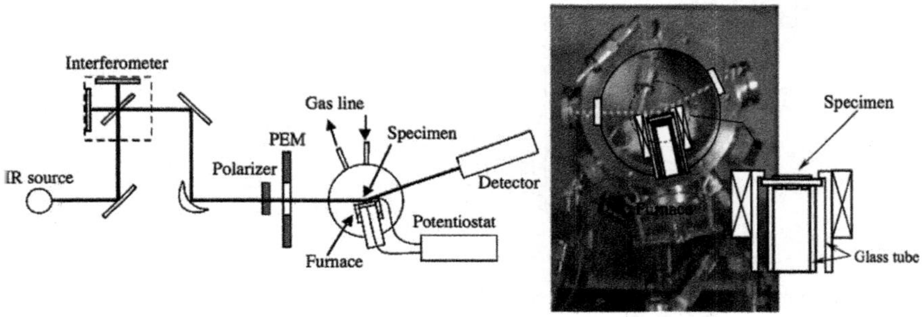

Fig. 7.2 Schematic view of in-situ electrochemical PMIRRAS system

Recently, several oxides were reported to show an extremely high surface exchange rate. Baumann et al. [9] compared several Co- and Fe-based perovskites in a controlled shape and found $Ba_{0.5}Sr_{0.5}Co_{0.8}Fe_{0.2}$ shows 100 times smaller electrochemical resistance than the (La,Sr)(Co,Fe)O_3 family, which is often used for the intermediate-temperature cathode. Sase et al. [10] reported that existence of the $(La,Sr)_2CoO_4$ phase on an (La,Sr) CoO_3 electrode enhances the oxygen exchange reaction rate (Fig. 7.3). Although stability should be carefully examined for these materials, further improvement of catalytic activity may be possible by research on those materials.

7.2.2 Electronic Conductivity

An electrode transfers electrons from the current collectors to the reaction sites. The importance of electronic conductivity depends on the structure of the cell stack. For a porous electrode that is fabricated on electrolyte as a thin layer, the lateral current transport often becomes a serious problem.

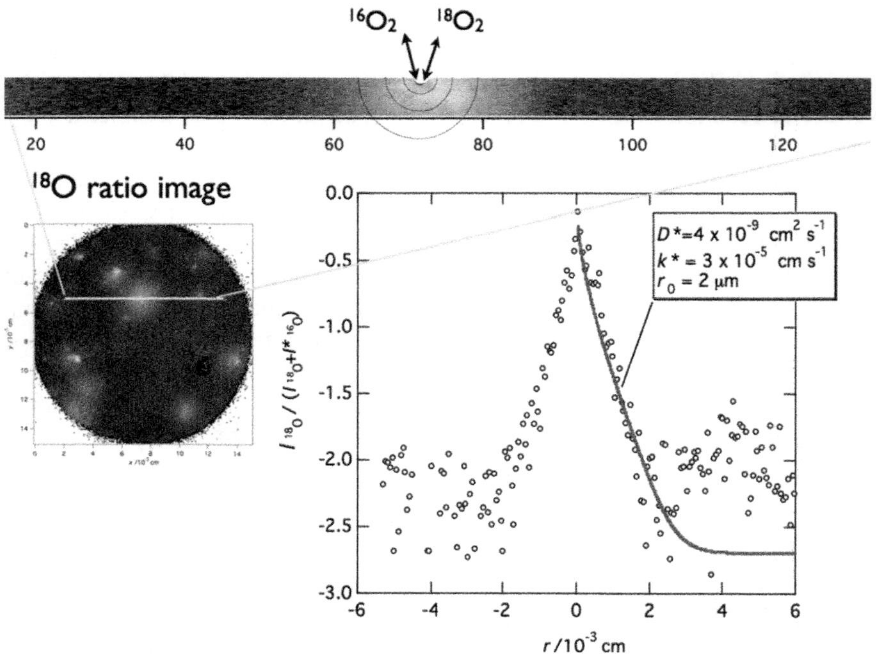

Fig. 7.3 Enhancement of surface oxygen exchange rate on (La, Sr)CoO_3 around the deposited second phase of (La, Sr)$_2CoO_4$

Especially, segment-in-series type stacks will have large current collection loss if the electrode has low conductivity. Even for planar stacks, the electrons may not be supplied sufficiently to the place if the current collection points are separated. Generally, electronic conductivity higher than 100 S cm^{-1} is preferred for a SOFC electrode. If the electronic conductivity is 10 S cm^{-1} and the electrode thickness is 50 μm, the resistance to transport electrons to the distance of 1 mm is as high as 2 Ω cm^{-1}. Because area-specific resistance of a practical cell is below 1 Ω cm^{-2}, it will cause constriction of the current into the vicinity of the current collector.

The electronic conductivity of (Ln, RE)MO_3 (RE, rare earth ions; M, transition metal ions) perovskite is mainly dependent on the B-site cation. Among the third period transition metals, Mn, Co, and Fe are investigated for cathodes. Especially, Co-based perovskite shows high conductivity, which shows metallic behavior when doped with RE higher than 0.5. (La, Ca)CrO_3 is used for an interconnect, and the electronic conductivity in air is 61 S cm^{-1} [11] at 1273 K, which may be tolerable for a cathode. However, electrocatalytic activity is reported to be low. $LnNiO_3$ is known to show high conductivity, but it is not stable in air. Instead, the K_2NiF_4-type structure is stable and is investigated as a cathode material.

7.2.3 Oxygen Transport (Bulk or Surface)

In a porous electrode, reaction takes place most easily at the gas–electrode–electrolyte boundaries (triple-phase boundary, TPB). Oxygen adsorbed on the electrode must be transported to the electrolyte through the surface or bulk diffusion. Bulk diffusivity has been studied for various perovskites and related oxides.

For electrodes with low oxygen diffusivity, the surface is the major diffusion path for the adsorbed oxygen. Because it is difficult to distinguish surface and bulk oxygen, knowledge of surface diffusion is limited. Contribution of surface diffusion and effective diffusion length are estimated by modeling and parameter fitting of the ac and dc polarization results. Comprehensive studies are necessary to design an electrode with fast surface diffusion.

Kawada et al. [12] attempted to obtain the surface diffusion coefficient on an oxide electrode by measuring the surface oxygen potential gradient under current flow. An oxygen potential microprobe was fabricated by coating a porous yttrium-stabilized zirconia (YSZ) layer on a tip of a thin Pt-Rh wire probe (TA Instruments; thermal probe) (Fig. 7.4). They estimated the surface diffusion coefficient on $La_{0.8}Sr_{0.2}MnO_3$ to be around 10^{-5} at ~700°C. Due to the difficulties in experimental setup, uncertainties remained in the obtained values. Further studies are necessary to clarify the contribution of surface diffusion.

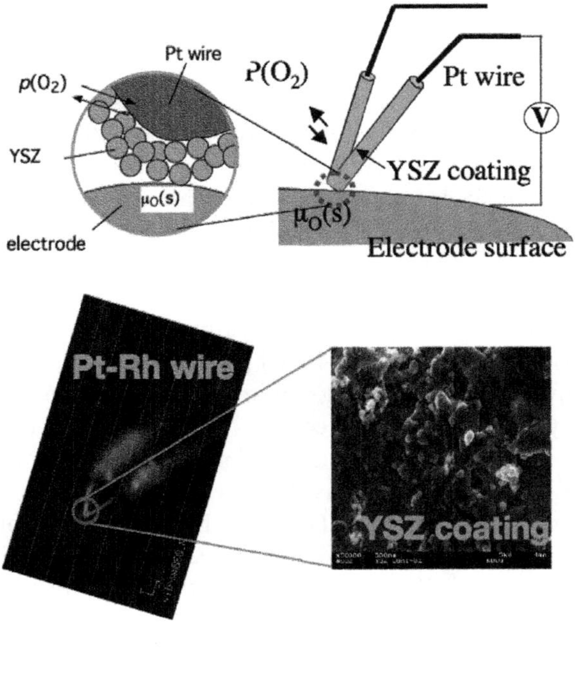

Fig. 7.4 Microprobe for local oxygen potential measurement

7.2.4 Chemical Stability and Compatibility

Chemical stability of the material is essential. Not only the thermal stability in air but also the compatibility with the electrolyte and interconnect materials must be considered. The chemical stability of the perovskite-type oxides can be estimated from the valence stability of the cations in constituent binary oxides and the stabilization energy of forming the perovskite lattice from the separated oxides. Yokokawa et al. [13] summarized the stabilization energy for perovskite-type oxides and found that they have strong correlation with the tolerance factor of perovskite lattice. The stabilization energy enables existence of 3+ or 4+ ions for Cr, Mn, Fe, Co, etc., even though they are not stable as binary oxides. When the (Ln, RE)MO_3 perovskites are in contact with YSZ electrolyte, they may be decomposed to $Ln_2Zr_2O_7$ or $REZrO_3$ at the interface. Those compounds have low electrical conductivity and cause degradation of cell performance. Among (La, Sr)MO_3 (M = Mn, Co, Fe), only Mn-based oxides have a stability region with YSZ.

7.2.5 Morphological Stability

Electrode microstructure must be maintained during long-term operation. The morphological stability of the cathode is important, especially when the cell stack has the cathode-support configuration. Microstructure change may be caused by sintering of the electrode particles or cation drift under an oxygen potential gradient. When current flows through the electrode–electrolyte interface, the oxygen potential at the interface becomes out of equilibrium with the gas phase because of oxygen flux on the electrode particles or across the interfaces. This flux makes the electrode–electrolyte interface more "reducing" than the gas phase, and an oxygen potential gradient is developed inside the electrode layer. According to the Gibbs–Dühem equation, this becomes a driving force for cations to move from the electrolyte side to the gas-phase side.

The morphological stability is attributed to cation diffusivity. It must be low enough to keep the microstructure in the fabrication state for a long-term operation. Among the candidate perovskites, $LaMnO_3$ is known to have high cation diffusivity in oxidizing atmospheres, as is discussed later. Other perovskites, Co-based or Fe-based perovskites, may also have sufficient cation diffusivity to cause change in morphology during a long-term operation [14]. Systematic study is necessary for predicting the durability.

Lattice expansion is also an important factor in the selection of cathode material. $LaMnO_3$-based perovskites are known to have a tolerable thermal expansion coefficient (11×10^{-6} K^{-1}) when used on most electrolyte materials [15]. $LaCoO_3$-based oxide, however, shows a higher expansion coefficient ($\sim20\times10^{-6}$ K^{-1}). The cause of the lattice expansion is not only "thermal" expansion but also "chemical" expansion due to the formation of oxygen vacancy [16]. The chemical expansion linearly depends on the vacancy concentration. Information on oxygen nonstoichiometry is important for the analysis of chemical expansion.

7.3 General Description of Cathode Reaction and Polarization

7.3.1 Oxygen Electrode Process

Oxygen incorporation reaction at a cathode can be broken down into several processes that are connected in series and in parallel, i.e., gas-phase diffusion, adsorption, dissociation, surface or bulk diffusion, and incorporation into the electrolyte (Fig. 7.1). At every step, a driving force is required to promote the reaction or the mass transport, and this causes an energy loss, which is called "overpotential" or "polarization." Although a large number of studies have been published so far, complete understanding has not been attained on the cathode reaction mechanism [3]. Most work is based on dc/ac electrochemical measurements, which provide only macroscopic and averaged information on the whole electrode process. In actuality, however, a reaction site is not uniform and distributes three dimensionally around the triple-phase boundaries of electrode, electrolyte, and gas phases.

In modeling and analyzing the electrode process, it is often assumed that the local equilibrium is held at every point inside the electrode and electrolyte, where local chemical potential of oxygen is determined using electrochemical potentials of oxide ion and electron as follows:

$$\mu_{\mathrm{O}} = \eta_{\mathrm{O}^{2-}} - 2\eta_{\mathrm{e}^-} \tag{7.1}$$

Additionally, in most cases, quasi-equilibrium is assumed between the electrons in the electrode and in the electrolyte at the interface [17–19]. Possible high tunneling current at the interface of ionic conductors might rationalize this assumption [20]. Based on these assumptions, overpotential is attributed to the variation of oxygen potential at the interface from the equilibrium with the surrounding atmospheres:

$$E = \frac{1}{2F}(\mu_{\mathrm{O,i}} - \mu_{\mathrm{O,e}}) = \frac{RT}{2F}\ln\frac{a_{\mathrm{O,i}}}{a_{\mathrm{O,e}}} \tag{7.2}$$

where $a_{\mathrm{O,i}}$ and $a_{\mathrm{O,e}}$ are the oxygen activity in the electrolyte close to TPB under current flow and that in equilibrium with gaseous oxygen molecules, respectively. The origin of the oxygen potential shift is described as the chemical reaction and mass transport, as shown in Fig. 7.1. Mass transport limitation, if it is dominant in the whole electrode reaction, will give oxygen potential gradient inside or at the surface of the electrode. If electrode surface processes such as oxygen adsorption, dissociation, and incorporation are slow, an oxygen potential gap will appear between the gas phase and the electrode interface. The electrochemical reaction is discussed in terms of purely chemical processes, and the oxygen potential profile is determined by the relative rates of those processes [19]. Recently, Fleig [21] proposed a concept of implying electron transfer overpotential for the surface reaction of the electrode. In his discussion, the electron transfer overpotential is connected to the local oxygen potential of the electrode via the concentration variation of the surface dipoles. If the reaction rate equation is not concerned, the discussion on the oxygen potential profile is same in any case.

Knowledge of oxygen potential profile in the electrode is important in designing the composition and morphology of a practical cathode. As was discussed in the previous section, the oxygen potential gradient in the cathode layer acts as a driving force for cation diffusion. The morphological change, or in some cases, the compositional change, might take place under an oxygen potential gradient [22].

7.3.2 Equivalent Circuit for a Cathode–Electrolyte Interface

Oxygen potential profile inside a cathode, although it is important, is not easy to be determined. Since transient response of current or potential reflects the process of forming the potential gradient, the ac impedance signal contains useful information. For understanding the ac response of the cathode–electrolyte system, equivalent circuit analysis based on the mass transport equation is useful [23].

Transport of oxide ion and electron in an oxide material is given by the following:

$$J_{O^{2-}} = -\frac{\sigma_{O^{2-}}}{(-2F)}\frac{d\eta_{O^{2-}}}{dx} \quad \text{and} \quad J_{e^-} = -\frac{\sigma_{e^-}}{(-F)}\frac{d\eta_{e^-}}{dx} \tag{7.3}$$

if cross terms are negligible. An equivalent circuit for one-dimensional bulk transport is represented as Fig. 7.5. The ionic path and the electronic path are hypothetically shown as two separate vertical lines with the resistances for a small element, dx, in the position x. Electrochemical potentials, $\eta_O{}^{2-}$ and η_e, are defined for the ionic and electronic paths, respectively. Current flows between the two lines when charge carrier changes, i.e., local defect concentration changes. The local equilibrium (7.1) leads that the potential difference across the horizontal line, $\eta_O{}^{2-} - \eta_e$, gives local oxygen potential, μ_O. Thus, the circuit element in the horizontal line must represent a capacity to change oxygen nonstoichiometry when oxygen potential changes. This type of capacitance that comes from the chemical change is called chemical capacitance [24]. For a porous electrode, the surface reaction takes place. If it is roughly regarded as one-dimensional path, the Faraday current is written as the impedance that

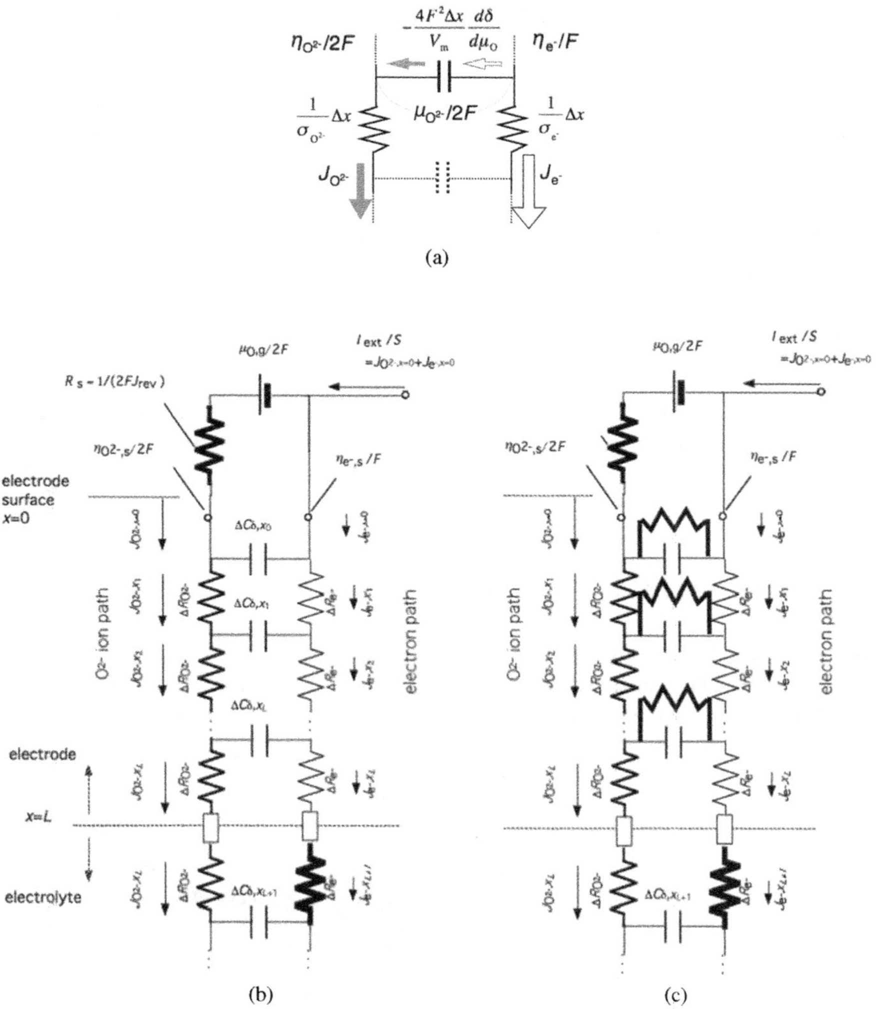

Fig. 7.5 Equivalent circuit for **(a)** a local mixed conductor, **(b)** dense electrode/electrolyte system, and **(c)** porous electrode–electrolyte system

connects the electron and ion paths. Thus, the electrode–electrolyte system can be represented as shown in Fig. 7.5(c).

With the equivalent circuit, the rate-determining step and the oxygen potential distribution can be discussed in terms of the relative value of the circuit elements. When the surface reaction is the rate-determining step, the resistance in the horizontal lines is much larger than that in the vertical lines. In that case, the local oxygen potential, i.e. the potential difference between the vertical lines, is uniform throughout the electrode layer. The chemical capacitances in the horizontal lines are equally charged with the applied potential. Then, observed impedance will be represented by a R-C parallel circuit with the capacitance that comes from the oxygen nonstoichiometry in the whole electrode layer. In an actual mixed conductor electrode, the transport and surface reaction are colimiting the total reaction rate [25].

In contrast, if the observed capacitance is smaller than that expected from the nonstoichiometry, the resistance at the electrode–electrolyte interface is possibly the rate-determining step. In such a case, improvement of the electrode will not be made without improving the ionic contact at the interface.

The equivalent circuit analysis may be similarly applied to a more realistic electrode by using further numerical calculation in a three-dimensional model.

7.4 Cathode for High-Temperature SOFC: (La, Sr)MnO_3

Manganese-based perovskites are widely recognized as the materials best suited for the cathode of a high-temperature SOFC that uses a zirconia-based electrolyte and operates at temperatures higher than 800°C. In this section, (La, Sr)MnO_3 (LSM) is chosen for further discussion.

7.4.1 Transport Properties and Electrochemical Reaction

Temperature dependence of electrical conductivity of (La, Sr)MnO_3 is shown in Fig. 7.6 [26]. The conductivity in air is high enough for a planar SOFC. For a tubular or segment-in-series type stack, however, the current path is longer

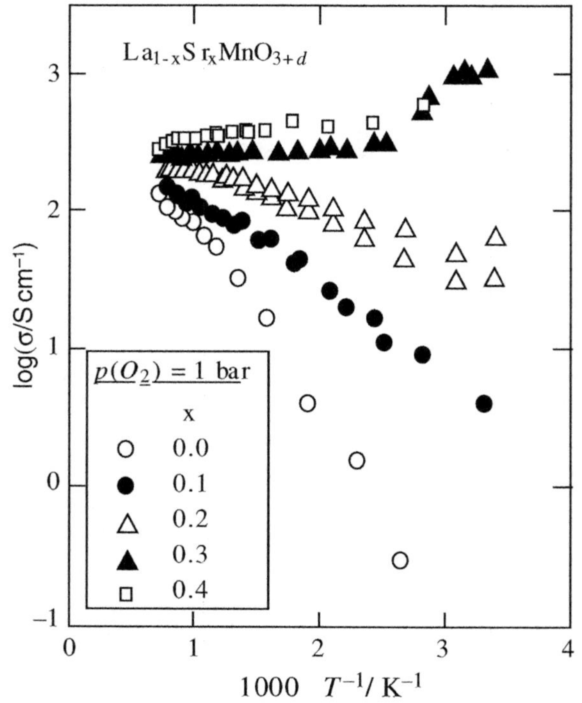

Fig. 7.6 Electronic conductivity of $La_{1-x}Sr_xMnO_3$ in air

through the electrode layer, and ohmic loss in the cathode can be a problem. In such a case, thicker current collection layers are formed on the active layer of (La, Sr)MnO_3. The composition and the morphology are optimized to keep gas flow as well as high conductivity.

A distinct feature of this material is the existence of large "oxygen excess" nonstoichiometry under oxidizing atmospheres [27]. As it is attributed to the formation of cation vacancies, it does not enhance oxygen diffusivity. Reported oxygen tracer diffusion coefficient is smaller than 10^{-12} cm^2 s^{-1} at 1173 K [28]. It corresponds to the oxide ion conductivity of around 4×10^{-7} S cm^{-1}, which gives the specific resistance of higher than 200 Ωcm^2 even though the electrode is as thin as 1 μm. Thus, oxygen bulk transport cannot play a major role in cathode reaction mechanism. Bulk transport becomes significant only when large overpotential is applied to the electrode.

Typical current-potential curves for (La,Sr)MnO_3 reported by Tsuneyoshi et al. [29] are shown in Fig. 7.7. The data are taken for both the cathodic and anodic polarization. The empirical equation is derived from the reaction order analysis as follows:

$$j = k\left(a_{\mathrm{O}} - P_{\mathrm{O}_2} a_{\mathrm{O}}^{-1}\right) \tag{7.4}$$

$$\text{or } j = k a_{\mathrm{O,rev}}\left(\exp\left(\frac{2F(E - E_{\mathrm{rev}})}{RT}\right) - P_{\mathrm{O}_2}\exp\left(\frac{-2F(E - E_{\mathrm{rev}})}{RT}\right)\right) \tag{7.4'}$$

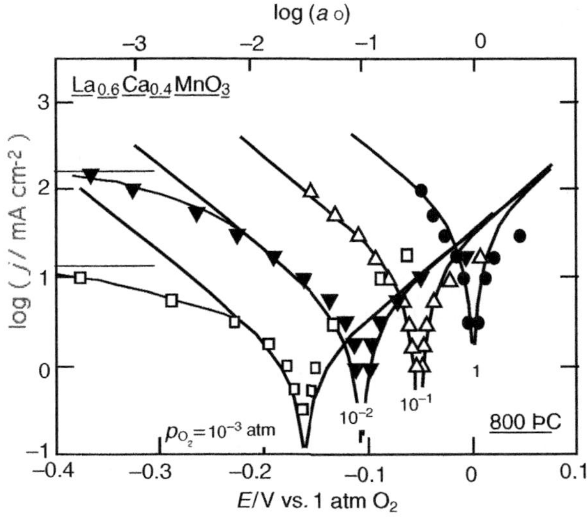

Fig. 7.7 Typical I-V curves observed with (La, Ca)MnO_3 electrode[28]

although it overestimates the anodic current under high oxygen partial pressures. Further analyses on the electrode reaction kinetics were made by ac impedance analysis. Kamata et al. [30]. assumed that the surface diffusion process controls the electrode reaction rate. They found the electrode conductivity, i.e., the inverse specific resistance, depends on $P_{O2}^{1/2}$ in diluted oxygen, and explained the dependence with the Langmuir adsorption model. Yasumoto et al. [31] introduced the effect of oxygen excess nonstoichiometry to explain the deviation. With their surface diffusion model, however, they do not specify the diffusion length or the width of the active electrode reaction site. The kinetics and the oxygen reaction pathway are still to be clarified.

A relatively large transient behavior is also a characteristic feature of the LSM electrode. Many authors reported that the performance of an LSM electrode is improved in minutes or hours just after the current load is applied; this may consists of both reversible and irreversible factors. When a large overpotential is applied, LSM acts as a mixed electronic/ionic conductor, and the bulk diffusion of oxygen begins to play an important role in the kinetics; this gives an expression for the reversible change of the performance. On the other hand, the irreversible change may come from the morphology or the composition change of the electrode. As was discussed in a previous section, an oxygen potential gradient is applied inside the electrode layer under operation. The cations drift from the interface to the outside and may modify the microstructure around the active area. It may increase the number of triple-phase boundaries [32] and improve the performance. It can also affect the relative stabilization energy of (La, Sr)MnO_3 and $SrZrO_3$, which may cause the disappearance of the resistive layer at the interface. These behaviors make the electrode kinetics of LSM complicated.

In a practical application, LSM is often used as a composite with YSZ particles [33] to increase the electrochemical reaction site. As YSZ can make a separate ionic path, the reaction site is made three dimensionally inside the electrode layer. The width of the active reaction area is determined by the resistance ratio of the diffusion and the interface reaction. Because electronic conductivity is decreased by mixing YSZ, a current collection layer of LSM or other material is necessary.

7.4.2 Chemical and Morphological Stability of LSM

The advantage of (La, Sr)MnO_3 over the other transition metal perovskites is the compatibility with a YSZ-based electrolyte. The thermal expansion coefficient matches well, and moreover it can make a stable interface with YSZ. However, for long-term stability, the interface stability may become a problem. According to thermodynamic calculation by Yokokawa et al. [34], (La, Sr)MnO_3 may react with YSZ to form $SrZrO_3$ or $La_2Zr_2O_7$ if the activity of La or Sr become high even though they are in their stability region. As (La, Sr)MnO_3 allows A-site-deficient composition, it is effective to incorporate excess Mn to decrease the activity of La and Sr. In a long-term operation, or at high-temperature processing, Mn may diffuse into the YSZ layer, causing the remaining La and Sr to have higher activity. Thus, the use of Mn excess composition is a safer choice to have a stable interface. Even though Mn diffuses into YSZ, it is not harmful for the electrochemical reaction [35].

Another possible problem of LSM cathode is morphological instability. As discussed before, LSM generates cation vacancies in oxidizing atmospheres. Thus, oxidation and reduction cycles vary the number of cation vacancies, which causes the creation and annihilation of the crystal lattices. Miyoshi et al. [27] reported a drastic change of the surface morphology of $LaMnO_3$ in oxidation and reduction runs (Fig. 7.8). Irreversible change in the sample dimension is also reported by Mori et al. [36] with thermal cycle experiments. If LSM is used in a cathode-support cell stack, the morphological change will be harmful for the mechanical stability. Existence of cation vacancies in LSM also causes the diffusion of cations from the interface to outside under current load. In a short time, it may increase the number of TPB and improve the performance. However, if the cations continue to move in a long-term operation, the interface resistance will tend to increase. Also, the mechanical strength of the interface will deteriorate because of the weakening of the adhesion. Cation vacancy concentration decreases by increasing the concentration of Sr dopant in the La site (Fig. 7.9) [37]. Thus, composition with higher Sr concentration is

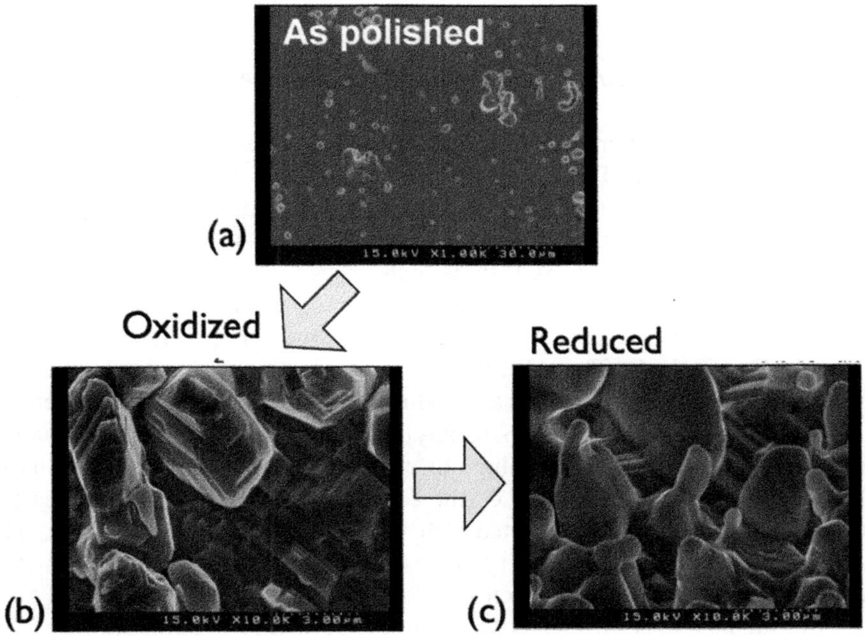

Fig. 7.8 Morphological instability of $LaMnO_3$. The sample was sintered at 1673 K in air for 4 h and polished (**a**). It was oxidized in 1 bar O_2 at 1273 K for 300 h (**b**), and then reduced in 10^{-2} bar O_2 at 1273 K for 300 h (**c**)

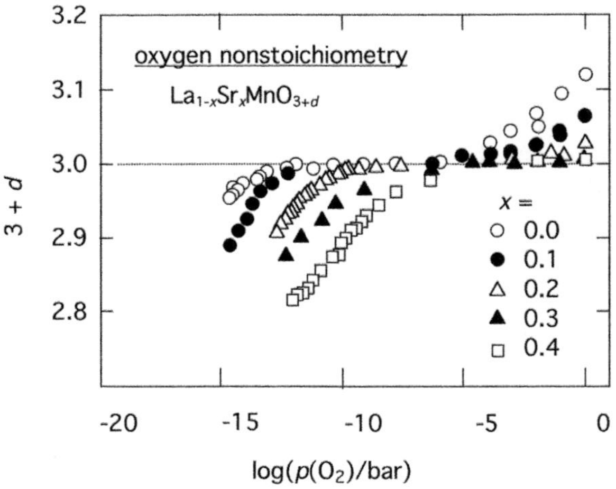

Fig. 7.9 Oxygen nonstoichiometry of $La_{1-x}Sr_xMnO_{3+\delta}$

preferable in terms of morphological stability. Addition of too much Sr, however, increases the risk of formation of $SrZrO_3$ at the interface. Usually, the composition around $La_{0.7}Sr_{0.3}MnO_3$ is used in practical cells.

7.5 Cathode for Intermediate-Temperature SOFC: (La, Sr)CoO_3, (La, Sr)(Co, Fe)O_3

7.5.1 General Features of Co-Based Perovskite Cathode

Reducing operation temperature leads to a severe increase in the overpotential of an LSM cathode. For operation at temperatures below 700°C, a high-performance cathode is required. Lanthanum cobaltite is a typical material for an intermediate-temperature cathode. Strontium-substituted lanthanum cobaltites, (La, Sr)CoO_3 (LSC), show high oxygen diffusivity due to the large number of oxygen vacancies formed even under oxidizing atmospheres [38]. In contrast to LSM, the bulk diffusion path shown in Fig. 7.1 has the major contribution in LSC. The effective electrode reaction site spreads over the surface of the electrode particles and reduces the overall cathode overpotential.

Another merit of using (La, Sr)CoO_3 (LSC) is its high electronic conductivity. At high temperatures, with high Sr content, the electronic conductivity of LSC shows metallic behavior; i.e., the conductivity decreases with increasing temperature [38]. It is represented well with a itinerant electron model. The absolute value of the conductivity ranges from 10^3 to 3×10^3 S cm^{-1}, which is useful for effective current collection.

Although LSC is an attractive material for a SOFC cathode, its use is restricted because of the instability of LSC on zirconia-based electrolytes. It is well known that LSC reacts with YSZ and forms $SrZrO_3$ at the interface. When YSZ or ScSZ is used as the electrolyte, a protective interlayer is indispensable.

Ceria doped with gadolinia (GDC) or samaria (SDC) is used as the interlayer in some intermediate-temperature SOFCs. Even with the interlayer, however, issues of long-term stability may still remain [39]. If $LaGaO_3$-based oxide is used as the electrolyte, the interface stability problem is less significant. LSC-related cathodes are widely used on $LaGaO_3$. For long-term stability, however, the interdiffusion of Co and Ga should be avoided.

Another problem is the large thermal expansion coefficient of (La, Sr)CoO_3, as discussed in the previous section. To avoid the problem of thermal expansion, (La, Sr)(Co, Fe)O_3 is chosen in practical systems.

7.5.2 *Electrochemical Reaction of a Model Electrode: A (La,Sr)CoO_3 Dense Film*

A mixed conductor electrode has complicated reaction pathways (as shown in Fig. 7.1). In kinetic studies, dense film electrodes are often employed to simplify the reaction route. Pulsed laser deposition (PLD) [40] or magnetron sputtering are used to deposit a dense film on an electrolyte substrate.

The rate-determining step of a dense $La_{0.6}Sr_{0.4}CoO_3$ electrode deposited on a doped ceria electrolyte was investigated using an isotope exchange technique [41]. After being exposed to $^{18}O_2$-enriched gas, the sample was quenched and analyzed with a secondary ion mass spectrometer (SIMS). Figure 7.10 shows

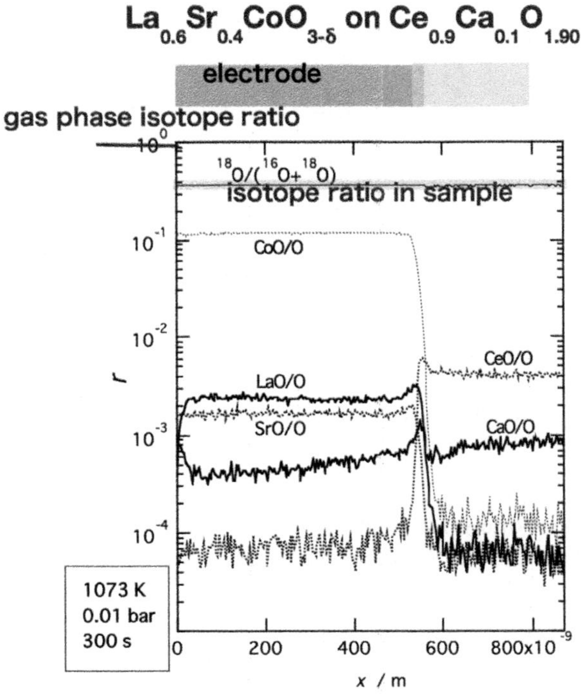

Fig. 7.10 Isotope diffusion profile in $La_{0.6}Sr_{0.4}CoO_{3-\delta}$ dense film electrode on $Ce_{0.9}Ca_{0.1}O_{1.90}$ electrolyte

typical results. A transport barrier, i.e., isotope concentration gap or gradient, was not observed at the electrode–electrolyte boundary nor inside the electrode but only at the gas–electrode boundary. This result clearly shows that the rate-determining step is the surface process. Figure 7.11 shows a typical impedance response [23]. The main part of the impedance is well fitted by a simple R-C semicircle for which the capacitance was as large as 0.01–1 F cm^{-2}. The equivalent circuit in Fig. 7.5(b) suggests that the chemical capacitance due to the oxygen nonstoichiometry in the electrode will be detected if the surface process is the rate-determining step. Observation of the large capacitance was thus consistent with the above results of isotope exchange. (Further quantitative analysis of the capacitance revealed that the observed capacitance was slightly smaller than expected. Oxygen vacancy formation energy in the film might be larger than that in the bulk. Further studies are ongoing to clarify the properties of the film electrode.)

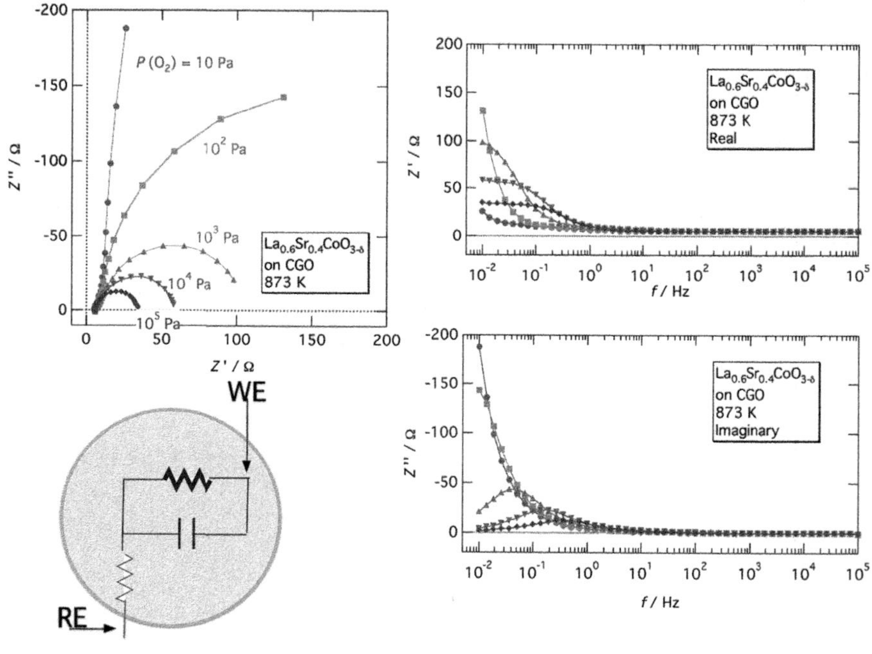

Fig. 7.11 A typical impedance response of $La_{0.6}Sr_{0.4}CoO_{3-\delta}$ film electrode on $Ce_{0.9}Gd_{0.1}O_{1.95}$ electrolyte

Figure 7.12 shows typical dc polarization curves of a dense $La_{0.6}Sr_{0.4}CoO_{3-d}$ electrode on a GDC electrolyte under various oxygen partial pressures [42]. Because RDS is the surface process, the polarization curve gives the rate of the overall reaction:

$$1/2\ O_2\ (gas) <=> O^{2-} + 2\ H\ (inside\ the\ electrode) \qquad (7.5)$$

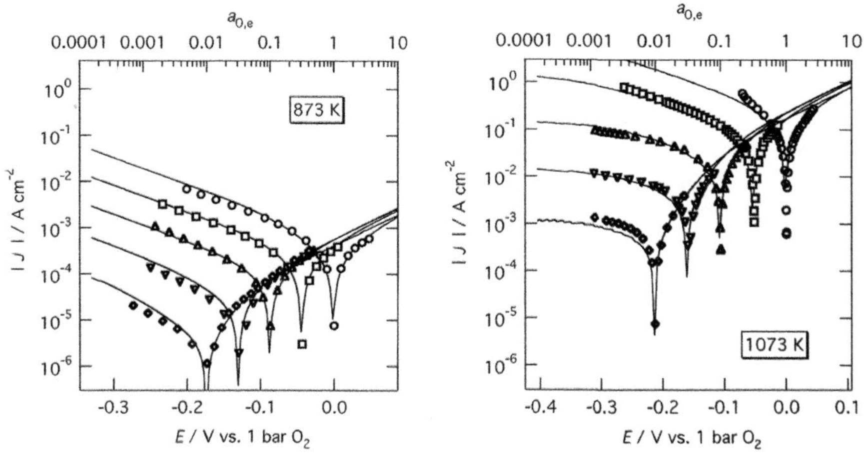

Fig. 7.12 Typical I-V curves for a dense film $La_{0.6}Sr_{0.4}CoO_{3-\delta}$ electrode on $Ce_{0.9}Gd_{0.1}O_{1.95}$ electrolyte

According to Eq. (7.1), reaction order analysis was attempted for the polarization results. Apparent reaction orders, however, were not simple constants for the present system, which appeared to depend on temperature. To quantify the polarization curves, two series reactions and the respective empirical equations were assumed: one for the adsorption/desorption reaction:

$$J = \alpha k_S \left(a_{O,S}{}^2 P_{O_2}{}^{-1/2} - a_{O,S}{}^{-1} P_{O_2} \right) \tag{7.6}$$

and the other for the oxygen transfer between the surface and the bulk [43]:

$$J = \alpha k_{SS} \left(a_{O,e} a_{O,S}{}^{-1} - a_{O,e}{}^{-1} a_{O,S} \right) \tag{7.7}$$

where $a_{O,S}$ is the oxygen activity on the electrode surface and $a_{O,e}$ is that inside the electrode. With the combination of the temperature-dependent parameters, k_S and k_{SS}, the I–V curves were well represented at various temperatures. This analysis, however, is not based on physical meaning of the polarization. Further consideration is necessary to obtain a more realistic rate equation.

7.5.3 Electrochemical Response of (La, Sr)CoO$_3$ on Zirconia with and Without Ceria Interlayer

Although an early paper reported that the reaction between (La, Sr)CoO_3 and YSZ electrolyte is suppressed at reduced temperatures, a long-term operation with a model electrode clearly showed that $La_{0.6}Sr_{0.4}CoO_3$ reacts with YSZ

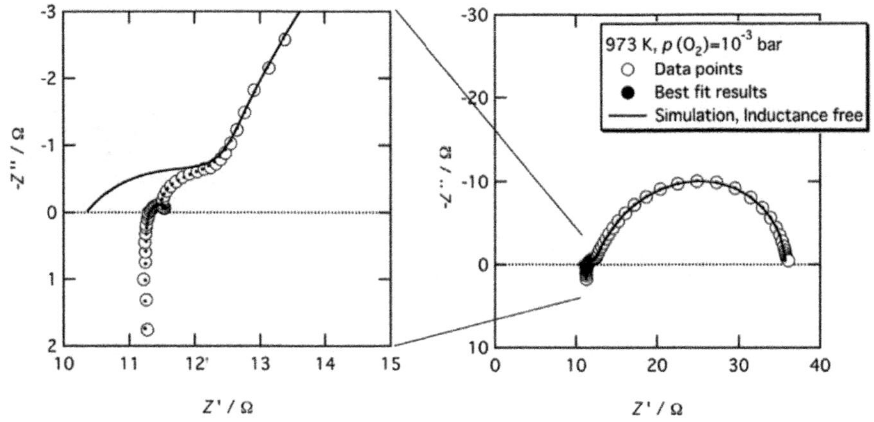

Fig. 7.13 Complex impedance of a dense film $La_{0.6}Sr_{0.4}CoO_{3-\delta}$ electrode on YSZ electrolyte with $Ce_{0.9}Gd_{0.1}O_{1.95}$ interlayer (cited from Ref. 45)

even at 700°C [44]. In the impedance response, the interface resistance between $La_{0.6}Sr_{0.4}CoO_3$ and YSZ was clearly separated from the surface reaction by the difference of their chemical capacitance. The interface resistance increased according to the parabolic rate law when the sample was kept at 700°C. The parabolic rate constant k was estimated to be 10^{-17} to 10^{-18} $cm^2\ s^{-1}$ at 973 K.

The reaction is suppressed by applying a protective interlayer between the cathode and the electrolyte. Ceria-based oxides are used as the interlayer. To test the validity of GDC interlayer, a model (La, Sr)CoO_3 electrode was deposited by PLD after depositing a thin layer of GDC on a YSZ single crystal. With the interlayer, the impedance and current versus potential curves were quite similar to that observed for (La, Sr)CoO_3 electrode on GDC electrolyte [45]. The GDC interlayer was found to act effectively as a protective layer. Careful examination, however, revealed that a small resistance element exists in high-frequency region besides the main arc in impedance response as shown in Fig. 7.13. This high-frequency element and ohmic resistance gradually increased with time. Interdiffusion of YSZ and GDC or LSC and GDC might cause increase of the interface resistance.

Recently, Sakai et al. [46] reported that significant cation diffusion or second-phase segregation were observed at the interface between (La, Sr)(Co, Fe)O_3 and ceria-based electrolytes. Although effects on electrochemical performance are not yet clear, further details should be studied for long-term durability of intermediate-temperature electrodes.

7.6 Summary

Perovskite-type oxides based on Mn, Co, Fe, or K_2NiF_4-type oxide with Ni are studied as cathode materials for SOFCs. For high-temperature SOFCs, $LaMnO_3$-based materials are mainly used because of the high compatibility with zirconia-based electrolytes. Although many studies have been carried out for modeling electrochemical reaction kinetics, common understanding among the researchers has not been reached. Morphological and chemical instabilities caused by cation vacancy formation may be the reasons for complicated behavior of the LSM electrodes, and it is a problem in a long-term operation. For intermediate-temperature SOFCs, Co- and Fe-based perovskites are considered. High oxide ion conductivity makes the reaction site wider, so that high performance is obtained. Recently, several cathode candidates with even higher performance have been proposed. Although high performance is obtained with Co- and Fe-based perovskites, chemical stability should be carefully examined.

References

1. O. Yamamoto, Y. Takeda, R. Kanno, M. Noda, Solid State Ionics 22, 241 (1987)
2. H.U. Anderson, Solid State Ionics 52, 33 (1992)
3. S.B. Adler, Chem. Rev. 104, 4791 (2004)
4. J.A. Kilner, R.A. De Souza, I.C. Fullarton, Solid State Ionics 86–88, 703 (1996)
5. E. Boehm, J.M. Bassat, M.C. Steil, P. Dordor, F. Mauvy, J.C. Grenier, Solid State Sci. 5, 973 (2003)
6. X. Lu, P.W. Faguy, M. Liu, J. Electrochem. Soc., 149, A1293 (2002)
7. T. Murai, K. Yashiro, A. Kaimai, T. Otake, H. Matsumoto, T. Kawada, J. Mizusaki, Solid State Ionics, 31–34, 2399 (2005)
8. Y.-M. Choi, D.S. Mebane, M.C. Lin, M. Liu, Chem. Mater. 19, 1690 (2007)
9. F.S. Baumann, J. Fleig, G. Cristiani, B. Stuhlhofer, H.U. Habermeier, J. Maier, J. Electrochem. Soc. 154, B931 (2007)
10. M. Sase, F. Hermes, K. Yashiro, T. Otake, A. Kaimail, T. Kawada, J. Mizusaki, N. Sakai, K. Yamaji, T. Horita, H. Yokokawa, Proceedings of the 7th European SOFC Forum, Luzern, 2006, Bossel U. (ed.). (CD-ROM) B-66 (2006)
11. N. Sakai, T. Horita, K. Yamaji, Y.P. Xiong, H. Kishimoto, M.E. Brito, H. Yokokawa, Solid State Ionics 177, 1933 (2006)
12. T. Kawada, M. Kudo, A. Kaimai, Y. Nigara, J. Mizusaki, In: S. Singhal, M. Dokiya (eds.), SOFC VIII. The Electrochemical Society, Inc., New Jersey, p. 470 (2003)
13. H. Yokokawa, N. Sakai, T. Kawada, M. Dokiya, J. Solid State Chem. 94, 106 (1991)
14. M. Palcut, K. Wiik, T. Grande, J. Phys. Chem. B, 111, 2299 (2007)
15. A. Fossdal, M. Menon, I. Wærnhus, K. Wiik, M.-A. Einarsrud, T. Grande, J. Am. Ceram. Soc. 87, 1952 (2004)
16. X. Chen, J. Yu, S. B. Adler, Chem. Mater. 17, 4537 (2005)
17. J. Mizusaki, K. Amano, S. Yamauchi, K. Fueki, Solid State Ionics, 22, 313 (1987)
18. M. Kleits, F. Petibon, Solid State Ionics, 92, 65 (1996)
19. S. B. Adler, Solid State Ionics 135, 603 (2000)
20. I. Riess, M. Gödickemeier, L.J. Gauckler, Solid State Ionics, 90, 91 (1996)
21. J. Fleig, Phys. Chem. Chem. Phys. 7, 2027 (2005)
22. H. Schmalzried, J. Chem. Soc. Faraday Trans. 85, 1273 (1990)
23. T. Kawada, J. Suzuki, M. Sase, A. Kaimai, K. Yashiro, Y. Nigara, J. Mizusaki, J. Electrochem. Soc. 149, E252 (2002)
24. J. Jamnik, J. Maier, J. Electrochem. Soc. 146, 4183 (1999)
25. S.B. Adler, Solid State Ionics 111, 125 (1998)
26. J. Mizusaki, Y. Yonemura, H. Kamata, K. Ohyama, N. Mori, H. Takai, H. Tagawa, M. Dokiya, K. Naraya, T. Sasamoto,. H. Inaba, T. Hashimoto, Solid State Ionics 132, 167 (2000)

27. S. Miyoshi, J.-O. Hong, K. Yashiro, A. Kaimai, Y. Nigara, K. Kawamura, T. Kawada, J. Mizusaki, Solid State Ionics, 154–155, 257 (2002)
28. I. Yasuda, K. Ogasawara, M. Hishinuma, T. Kawada, M. Dokiya, Solid State Ionics 86–88, 1197 (1996)
29. K. Tsuneyoshi, K. Mori, A. Sawata, J. Mizusaki, H. Tagawa, Solid State Ionics 25, 263 (1989)
30. H. Kamata, A. Hosaka, J. Mizusaki, H. Tagawa, Solid State Ionics 106, 237 (1998)
31. K. Yasumoto, M. Shiono, H. Tagawa, M. Dokiya, K. Hirano, J. Mizusaki, J. Electrochem. Soc. 149, A531 (2002)
32. S.P. Jian, J. G. Love, Solid State Ionics 158, 45 (2003)
33. E. P. Murray, T. Tsai, S. A. Barnett, Solid State Ionics, 110, 235 (1998)
34. H. Yokokawa, N. Sakai, T. Kawada, M. Dokiya, Solid State Ionics 40/41, 398 (1990)
35. T. Kawada, N. Sakai, H. Yokokawa, M. Dokiya, Solid State Ionics 50, 189 (1992)
36. M. Mori, J. Electrochem. Soc. 152, A732 (2005)
37. J. Mizusaki, Y. Yonemura, H. Kamata, K. Ohyama, N. Mori, H. Takai, H. Tagawa, M. Dokiya, K. Naraya, T. Sasamoto,. H. Inaba, T. Hashimoto, Solid State Ionics 132, 167 (2000)
38. J. Mizusaki, Y. Mima, S. Yamauchi, K. Fueki, H. Tagawa, J. Solid State Chem. 80, 102 (1989)
39. T. Kawada, D. Ueno, M. Sase, K. Yashiro, T. Otake, A. Kaimai, J. Mizusaki, Solid Oxide Fuel Cells IX, (Electrochemical Society), PV 2005–07, pp. 1695 (2005)
40. K. Masuda, A. Kaimai, K. Kawamura, Y. Nigara, T. Kawada, J. Mizusaki, H. Yugami, H. Arashi. In: Solid Oxide Fuel Cell V, U. Stimming, S.C. Singhal, H. Tagawa, W. Lehnert (eds.), PV 97–40, p. 473. The Electrochemical Society Proceedings Series, Pennington, NJ (1997)
41. T. Kawada, K. Masuda, J. Suzuki, A. Kaimai, K. Kawamura, Y. Nigara, J. Mizusaki, H. Yugami, H. Arashi, N. Sakai, H. Yokokawa, Solid State Ionics, 121, 271 (1999)
42. T. Kawada, M. Sase, K. Yashiro, T. Otake, A. Kaimai, J. Mizusaki, Proc. 26th Risø International Symposium on Materials Science: Solid State Electrochemistry, S. Linderoth et al. (eds.). Risø National Laboratory, Roskilde, Denmark, 2005, pp. 23–38 (2005)
43. T. Kawada, M. Kudo, A. Kaimai, Y. Nigara, J. Mizusaki. In: "SOFC VIII", S. Singhal, M. Dokiya (eds.). The Electrochemical Society, Inc., Pennington, NJ, pp. 470–477 (2003)
44. M. Sase, D. Ueno, K. Yashiro, A. Kaimai, T. Kawada, J. Mizusaki, J. Phys. Chem. Solids, 66, 343 (2005)
45. T. Kawada, D. Ueno, M. Sase, K. Yashiro, T. Otake, A. Kaimai, J. Mizusaki. In: SOFC IX, S.C. Singhal, J. Mizusaki (eds.), PV2005-07. The Electrochemical Society, Pennington, NJ, p. 1695 (2005)
46. N. Sakai, H. Kishimoto, K. Yamaji, T. Horita, M.E. Brito, H. Yokokawa, J. Electrochem. Soc., 154, B1331 (2007)

Chapter 8
Perovskite Oxide Anodes for SOFCs

J.T.S. Irvine

8.1 Introduction

The solid oxide fuel cell (SOFC) is one of the most exciting systems for future power generation because of its fuel flexibility and very high potential efficiency. Up to now, most SOFC development has been based upon the yttria-stabilized zirconia (YSZ) electrolyte due to its high oxygen ion conductivity, good stability under SOFC operating conditions, and high mechanical strength; however, as its ionic conductivity is not very high at lower temperatures, it has limited applicability below 750°C. Alternative electrolyte materials, such as $Ce_{0.9}Gd_{0.1}O_{2-\delta}$ (CGO) [1] and $La_{0.85}Sr_{0.15}Ga_{0.9}Mg_{0.1}O_{3-\delta}$ (LSGM) [2], have been proposed, although there are also some limitations for these materials. For example, CGO exhibits significant n-type electronic conduction at low P_{O2} above 600°C, which limits its application temperature range. The limitations and advantages of LSGM are discussed in detail elsewhere in this volume; however, the most important limitation with respect to anode chemistry is its reactivity with NiO [3, 4], which means that Ni-based anodes are quite difficult to manufacture. One possible solution is to develop perovskite-based anodes, which leads to the attractive concept of the all-perovskite solid oxide fuel cell [5–7].

Perovskite oxide materials offer excellent thermal and mechanical stability, physical compatibility with electrolyte materials, and relatively low cost and have attracted interest in their application as fuel electrodes in SOFC designs. The present fuel electrode of choice is a nickel–YSZ cermet; the nickel acts as a fuel oxidation catalyst and provides electronic conductivity, while the YSZ provides oxygen ion conductivity. The challenge is to develop a single-phase oxide material that provides all these benefits and at the same time reduces the problems of coking, nickel sintering, and sulfur poisoning sometimes encountered with the nickel–YSZ cermet. Although all ceramic cathodes, $La_{1-x}Sr_xMnO_3$ or $La_{1-x}Sr_xCo_{1-y}Fe_yO_3$ with or without YSZ addition, currently

J.T.S.Irvine (✉)
School of Chemistry, University of St-Andrews, Fife, Scotland KY16 9ST, UK
e-mail: jtsi@st-and.ac.uk

T. Ishihara (ed.), *Perovskite Oxide for Solid Oxide Fuel Cells*,
Fuel Cells and Hydrogen Energy, DOI 10.1007/978-0-387-77708-5_8,

function well as SOFC cathodes [8, 9], an ideal all-ceramic anode material is not available. The requirements for SOFC anode materials are good chemical and mechanical stability under SOFC operating conditions, high ionic (O^{2-}/H^{+}) and electronic conductivity over a wide range of P_{O2}, and good chemical and thermal compatibility with electrolyte and interconnect materials, high surface oxygen exchange kinetics, and good catalytic properties for the anode reactions.

In this chapter, we review the status of current research on perovskite-based SOFC anodes, taking particular note of all perovskite concept developments. A rationale for the application of perovskites as fuel electrodes is given, and the defect chemistry of perovskites is then reviewed. Attention is given to reducible transition metal ions, doping, nonstoichiometry, and the effects these parameters have on chemical stability and both the magnitude and mechanism of conduction. Findings are summarized, and the need for an optimal doping strategy is identified.

8.2 Anode Materials for SOFCs

The most commonly used anode materials for zirconia-based SOFCs are Ni–ZrO_2 cermets, which display excellent catalytic properties for fuel oxidation and good current collection. Unfortunately, these also exhibit some disadvantages, such as low tolerance to sulfur [10] and carbon deposition [11] when using hydrocarbon fuels, and poor redox cycling, causing volume instability if the microstructure is not carefully optimized. The nickel metal in the cermet tends to agglomerate after prolonged operation, leading to a reduced three-phase-boundary and increasing cell resistance. As nickel is such a good catalyst for hydrocarbon cracking, these cermets can only be utilized in hydrocarbon fuels if excess steam is present to ensure complete fuel reforming, thus diluting fuel and adding to system cost. For example, nickel is a very good catalyst for methane cracking (8.1), which causes carbon deposition when methane is used as the fuel without sufficient steam provision.

$$CH_4 = C + 2H_2 \tag{8.1}$$

The deposited carbon not only may block the porosity of the anode but also disrupts the cermet microstructure, breaking Ni–ZSZ linkages, and finally causes degradation of cell performance.

Mixed metal oxides have been targeted as possible alternatives; they are potentially less likely to promote carbon buildup, due to the higher availability of oxygen throughout the anode (Fig. 8.1), and are less likely to suffer from sulfur poisoning [12].

The challenge is to develop a single-phase oxide material that catalyzes fuel oxidation, reduces significantly the problems described above, and possesses a level of mixed electronic–ionic conductivity comparable with the nickel–YSZ cermet.

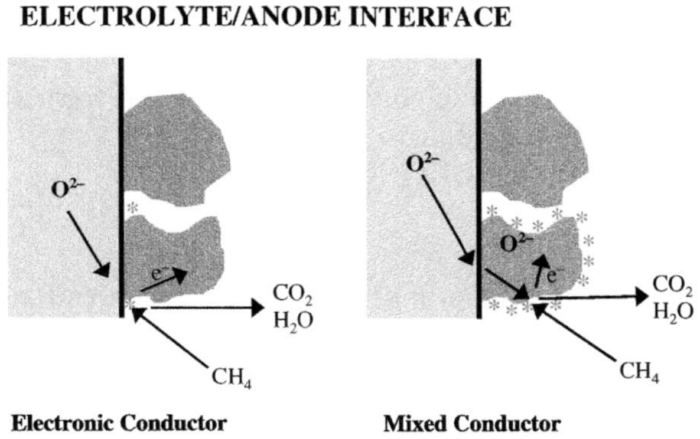

Fig. 8.1 Enhancement of electrode reactive surface area for fuel oxidation in an SOFC anode by mixed conducting materials

8.3 Perovskite Chemistry

Perovskite oxide materials possess the general stoichiometry ABO_3. Conventionally, the A cation is larger than the B cation. In the archetype, the A cation has an oxidation state of +2 and the B cation has the oxidation state +4. These materials comprise three different ionic species, each with its own equilibrium defect concentration due to three different activation energies for defect formation, which, combined with the constraint of electroneutrality, make for diverse and potentially useful defect chemistry, particularly when considering electronic, hole, and ionic conduction under atmospheres of different oxygen partial pressures [13].

Perovskite oxides generally exhibit excellent thermal and mechanical stability. They remain stable well above 1000°C, and therefore the temperatures of SOFC operation do not present a problem. This condition is in contrast with the Ni-cermet, where nickel sintering and agglomeration are potential hazards. It is also important that the anode exhibits good physical and mechanical compatibility with the dense electrolyte layer onto which it is deposited. The electrolyte choice is usually YSZ, although there is interest in alternative electrolyte materials such as lanthanum gallate. Examination of unit-cell parameters and thermal expansion coefficients reveals that perovskites generally show good compatibility with the electrolyte material. Several perovskites have also been shown to be stable in the kind of reducing environment encountered at the fuel electrode, and specific perovskites have exhibited the levels of ionic and electronic conductivity necessary to be realistically viable as an SOFC anode.

Perovskites containing a transition metal are of particular interest because of the availability of multiple oxidation states, which facilitate electrocatalytic processes and provide mechanisms for electronic conductivity. For example, under reducing atmospheres the transition metal ions change to lower oxidation states, effectively freeing up electrons to pass current. Typical examples of such species are titanium, niobium, and vanadium. $SrNbO_3$ has an electronic conductivity of 10^4 Scm^{-1} under reducing conditions [14], and Petric reported a conductivity of 10^3 cm^{-1} for $SrVO_3$ under similar conditions [15]. Unfortunately, it was also found that these compounds could not be fabricated in air, and their stability under fuel electrode conditions was questionable. As well as the increased electronic conductivity, there may also be an increase in the oxide ion contribution to conductivity due to the formation of oxygen vacancies on reduction, according to Eq. (8.2):

$$O_O^X \leftrightarrow 2e' + {}^1\!/_2 O_2 + V_O^{\bullet\bullet} \tag{8.2}$$

By raising the intrinsic oxygen vacancy concentration in this manner, there is an increase in available hopping sites. The increase in vacant sites facilitates oxygen transport through the crystal and hence raises the potential for oxide ion conductivity. The coordination of these oxide vacancies to the B-site ion is particularly important in determining mobility. Generally the B-site ion is six coordinate, octahedral. If fivefold or fourfold coordinates are also well known for a particular transition metal ion, then an obvious mechanism for oxide ion conduction exists; however, this would not be the case for B-site ions such as Cr(III), which strongly prefer sixfold coordination; this is illustrated in the chart shown in Fig. 8.2.

	Sc	Ti	V	Cr	Mn	Fe	Co	Ni	Cu	Zn
MO_5			V^V		✓	✓	✓	✓	Cu^{II}	
MO_4*tet*			V^V		?	✓	?	?		
MO_4*Sq*								?	Cu^{II}	
Fuel	✓	✓	✓	✓	✓	?				

Fig. 8.2 Comparison between different perovskite B-site ions comparing stability under fuel conditions and ability to reduce coordination number to allow vacancy oxide ion conductivity of transition metals in perovskite oxides

8.4 Doping, Nonstoichiometry, and Conductivity

Defect concentrations and hence defect chemistry of perovskites can be controlled and tailored significantly by doping. Figure 8.3 shows schematically the various possibilities.

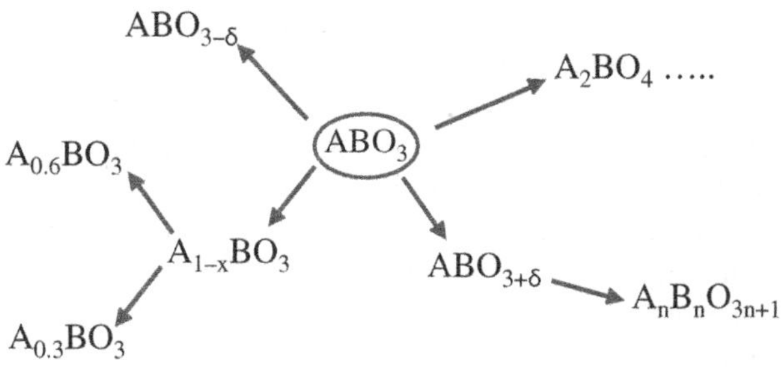

Fig. 8.3 Possible domains of perovskite nonstoichiometry

By substitution of parent cations with similar-sized cations of different valence, defects can be introduced into the structure. Oxygen ion vacancies or interstitials can be generated by substitution of B-site ions with cations of lower or higher valence, respectively, producing compounds of stoichiometry $AB_{(1-x)}B'_xO_{(3\,-/+\,\delta)}$. A-site vacancies can be introduced by substitution of A-site ions with cations of higher valence, giving compounds of stoichiometry of $A_{(1-x-)}A'_xBO_3$. Substitution of A-site ions with lower-valence cations results in oxygen vacancy formation giving compounds of stoichiometry $A_{(1-x)}A'_xBO_{(3-\delta)}$. The effects of such extrinsic defect concentrations on both carrier concentration and conductivity mechanism are discussed next.

Oxygen ion vacancy concentrations can be increased by partial substitution of the tetravalent B-site ions with lower-valency cations, as shown in Eq. (8.3):

$$O_O^X + B_B^X + MO \leftrightarrow M''_B + V_O^{\bullet\bullet} + BO_2 \tag{8.3}$$

M_B'' is a divalent cation, and $V_O^{\bullet\bullet}$ is an induced oxygen ion vacancy. It is expected that these additional vacant sites facilitate oxygen transport through the crystal by increasing the number of potential carriers.

The most important example of a perovskite that exhibits high oxide ion conductivity when doped is lanthanum gallate, $LaGaO_3$. Slater et al. [16] performed neutron diffraction and conductivity studies on $La_{0.9}Sr_{0.1}Ga_{0.8}Mg_{0.2}O_{2.85}$ and found it to exhibit significantly different structure and properties when compared with the undoped compound. In this case, both the A-site and the B-site have been doped to create oxygen vacancies, the resulting material possessing an oxide ion conductivity of 6.6×10^{-2} at 1000 K. Figure 8.4 shows schematically structural changes that may account for the dependence of activation energy on temperature involving the tilting of GaO_6 octahedra. The oxide ion conductivity attained is

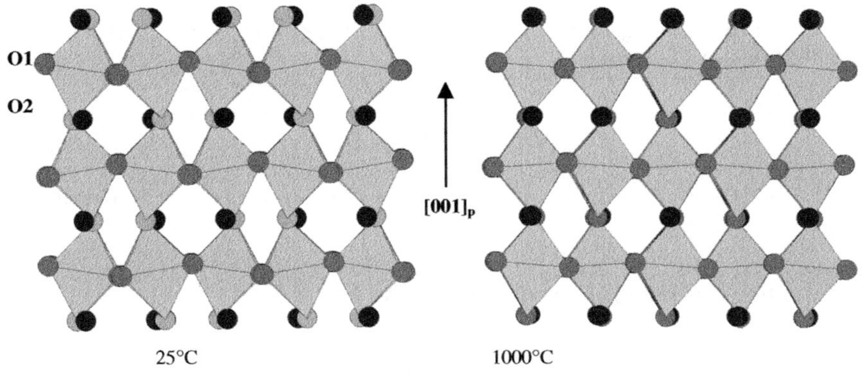

Fig. 8.4 Comparison between structure of $La_{0.9}M_{0.1}Ga_{0.8}Mg_{0.2}O_{2.85}$ at ambient temperature and 1000°C illustrating significant changes in octahedral tilting with temperature: views down $[110]_p$, La atoms in *black*, O atoms, and MO_6 octahedra shown [16]

higher than YSZ at the same temperature and has sparked considerable ongoing research into its application as a SOFC electrolyte [16–18].

Strontium titanate is an archetypal example of perovskite and exhibits a wide range of defect chemistry that aptly illustrate the different factors that may influence electronic conductivity. The effects of changing oxygen partial pressure upon undoped and differently doped strontium titanates are shown in Fig. 8.5. Important aspects are the extended p-type behaviou of the B-site-doped sample at higher P_{O2}, the p-type behavior of the undoped sample at higher P_{O2}, and the very high n-type conductivity of the A-site-deficient sample at lower P_{O2} values. Equation 8.4 shows how such a doped material is likely to exhibit p-type conductivity at the expense of ionic conductivity in high oxygen partial pressures. The ambient oxygen atoms fill the positively charged vacancies, generating a pair of holes in accordance with electroneutrality.

$$1/2O_2 + V_O^{\bullet\bullet} \leftrightarrow O_O^X + 2\,h^{\bullet} \tag{8.4}$$

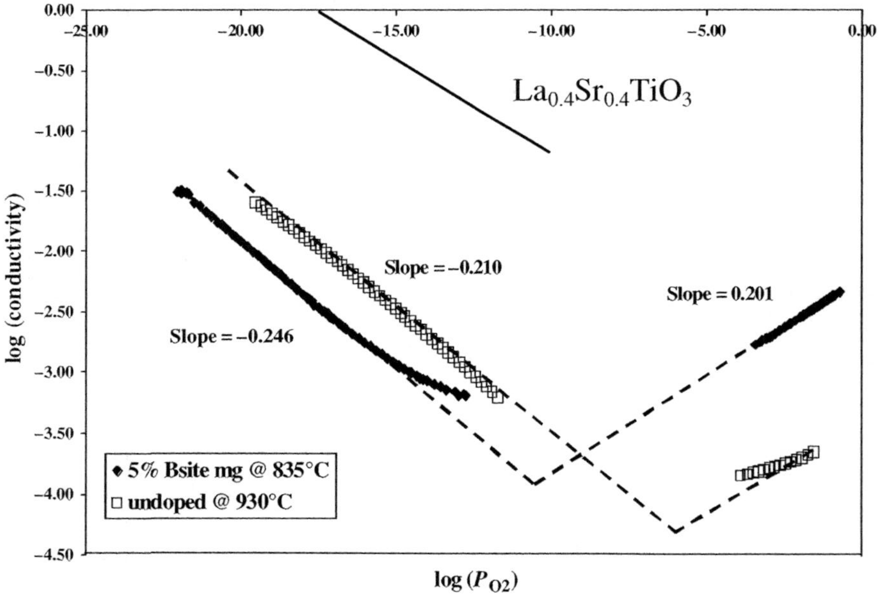

Fig. 8.5 $SrTiO_3$, conductivity variation for different doping scenarios with oxygen partial pressure at 930°C [13]

Equations (8.3) and (8.4) combine to give Eq. (8.5):

$$B_B^X \leftrightarrow M_B'' + 2\,h^{\bullet} \tag{8.5}$$

The p-type behavior of the undoped sample at high P_{O2} has various possible explanations, the most likely being simply that equilibrium intrinsic oxygen vacancies are filled by ambient oxygen atoms generating holes. The high n-type conductivity at low P_{O2} seen for the A-site-deficient sample can be explained with reference to the equilibria described by Eqs. (8.6) and (8.7).

$$2e' + 1/2O_2 + V_O^{\bullet\bullet} \leftrightarrow O_O^X \quad (8.6)$$

$$O_O^X + A_A^X \leftrightarrow AO + V_O^{\bullet\bullet} + V_A'' \quad (8.7)$$

The large value of V_A'' achieved by doping reduces the number of intrinsic Schottky defects pushing Eq. (8.7) to the left and hence reducing $V_O^{\bullet\bullet}$. For a given P_{O2}, Eq. (8.6) shifts left to oppose the change, facilitating the removal of lattice oxygen by hydrogen and the associated generation of free electrons. Figure 8.6 shows the temperature dependence of the n-type conduction in a strontium titanate, which indicates metallic behavior [12].

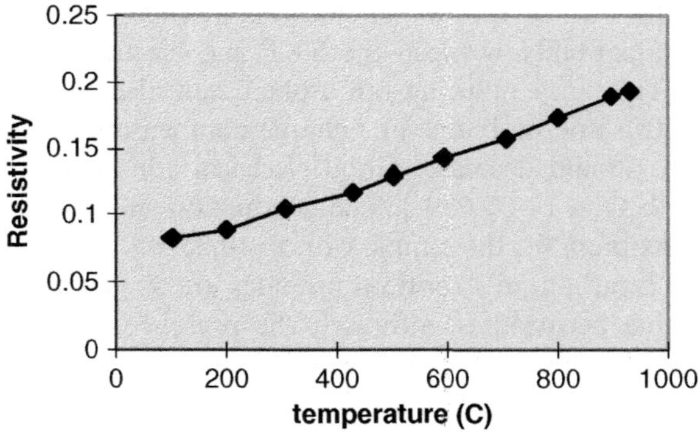

Fig. 8.6 Resistivity (in ohm-cm) vs. temperature for A-site-deficient $Sr_{0.875}Ti_{0.75}Nb_{0.25}O_{(3-\delta)}$ exhibiting metallic conductivity [12]

8.5 Perovskite Anode Materials

Perovskite oxides can accommodate a large content of oxygen vacancies; hence, some perovskites are good oxygen ionic conductors. The small B site in the perovskite allows first-row transition elements to be introduced in the lattice. These elements exhibit multivalency under different conditions, which may be a source of high electronic conductivity. Good ionic and mixed conductivity is thus found in several perovskite oxides. As already mentioned, such mixed conductivity is beneficial to electrode performance. P-type perovskite materials are widely considered for SOFC and other applications [19]. Mixed conducting perovskites, such as $La_{1-x}Sr_xMnO_3$ with modest oxide ionic conductivity or $La_{1-x}Sr_xCo_{1-x}Fe_xO_3$ with quite high oxide ionic conductivity, have been used as SOFC cathode materials [8, 20]. $La_{1-x}M_xCrO_3$ (M = Ca, Sr), a purely electronic conductor, has also been widely used as the interconnector for SOFCs [8].

Perovskites have also been widely investigated as potential SOFC anode materials. Among these materials, chromites and titanates are promising [21, 22]. Interesting results have been obtained with lanthanum strontium titanates [23] and especially cerium-doped lanthanum strontium titanate [24]; however, it is now thought that the cerium-doped anodes are in fact two phases consisting of a ceria–perovskite assemblage [24].

It was also reported that Y-doped $SrTiO_3$ exhibits high electrical conduction under SOFC anodic conditions [25–27]. For example, the optimized composition of $Sr_{0.86}Y_{0.08}TiO_{3-\delta}$ exhibits a conductivity of 82 S/cm at a P_{O2} of 10^{-19} atm at 800°C. However, the sample was pre-reduced in pure argon or 7% H_2/Ar at 1400°C before conductivity measurements. It is supposed that the conductivity of the materials would be significantly lower if the sample were only reduced below 1000°C in this case less Ti^{4+} was reduced to Ti^{3+}, which is the source of the high electronic conductivity. The high-temperature pre-reduction process for such titanates makes it difficult to co-fire the anode and cathode. The conductivity of $Sr_{0.86}Y_{0.08}Ti_{0.9}Sc_{0.1}O_3$ is only about 1–2 S/cm when reduced in situ in 5% H_2 at 900°C [28]. No phase changes were found for a mixture of Y-doped $SrTiO_3$ (SYT) with YSZ or LSGM on calcining at 1400°C for 10 h, indicating good chemical compatibility between the SYT and electrolyte materials. The conductivity of $SrTiO_3$ in a reducing atmosphere can also be improved by replacing titanium with some niobium. For charge compensation, the strontium content at the A site should decrease. Good electrical conductivity was observed for $Sr_{1-x}Ti_{1-x/2}Nb_xO_{3-\delta}$ ($x \leq 0.4$) [29] on reduction in low oxygen partial pressure, with a maximum for the sample with $x = 0.25$, $\sigma = 5.6$ S/cm at 930°C ($P_{O2} = 10^{-18}$ atm).

Lanthanum strontium titanates are usually treated in the literature as simple cubic perovskites, although the presence of extra oxygen beyond the ABO_3 stoichiometry plays a critical role in both the structure and the electrochemical properties, as summarized in Fig. 8.1. The lower members of the $La_4Sr_{n-4}Ti_nO_{3n+2}$ series, $n < 7$, are layered phases, having oxygen-rich planes in the form of crystallographic shears joining consecutive blocks. These planes become more sporadic with increasing n (i.e., decreasing the oxygen content) until they are not a crystallographic feature, rendering local oxygen-rich defects randomly distributed within a perovskite framework, $n > 11$ [30, 31]. The presence of such disordered defects appears to strongly affect the redox characteristics of the oxide, as indicated by marked effects on conductivity induced by mild reduction (Fig. 8.7). Unfortunately, although the materials from this lanthanum strontium titanate oxygen excess series are much easier to reduce, and hence exhibit much higher electronic conductivity than their oxygen stoichiometric analogues, they do not exhibit very good electrochemical performance [32]. This detriment is attributed to the inflexibility of the coordination demands of titanium, which strongly prefers octahedral coordination in the perovskite environment.

To make the B-site coordination more flexible and hence to improve electrocatalytic performance, Mn and Ga were introduced to replace Ti in $La_4Sr_8Ti_{12}O_{38-z}$-based fuel electrodes. Mn supports p-type conduction in oxidizing conditions and has been previously shown to promote electroreduction under

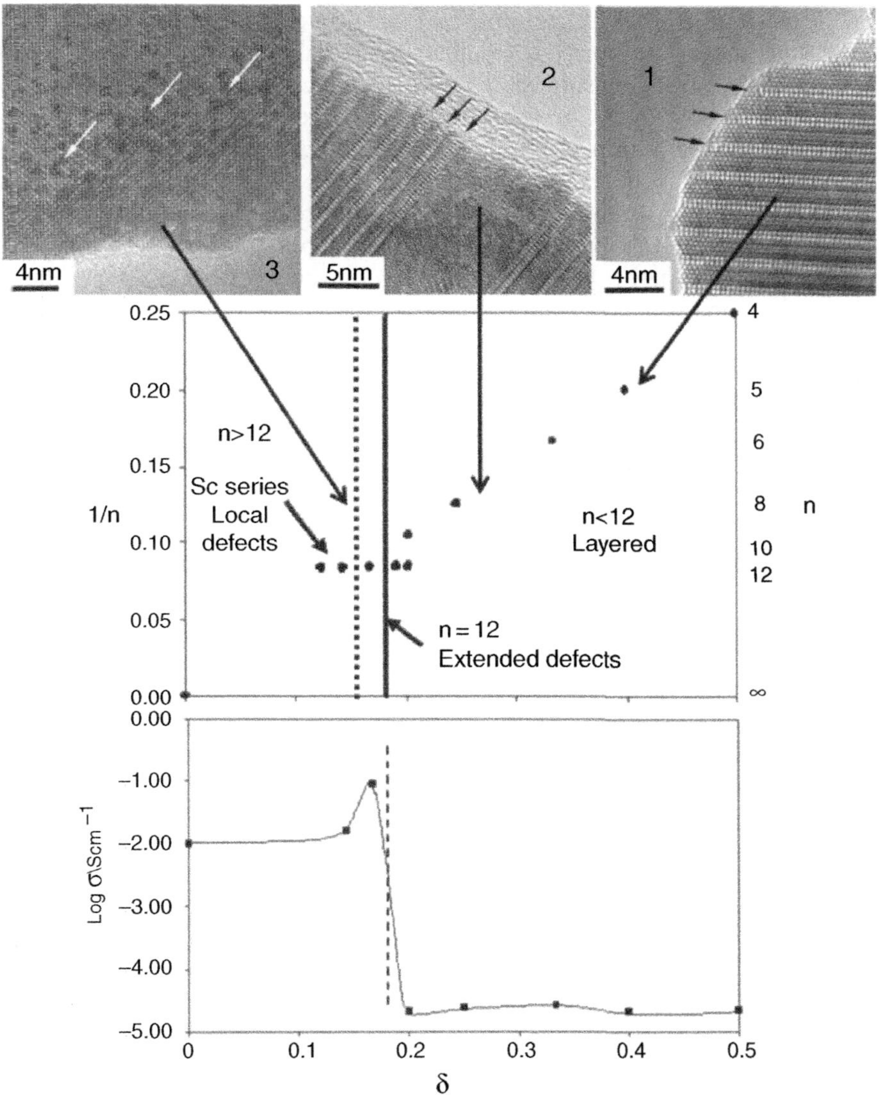

Fig. 8.7 High Resolution Transmission Electron Microscopy (HRTEM) images (*top*) of samples from the "$La_4Sr_{n-4}Ti_nO_{3n+2}$" series varying from ordered extended planar oxygen excess defects (*1*; $n=5$) through random layers of extended defects (*2*; $n=8$) to disordered extended defects (*3*; $n=12$). The *middle picture* shows the location of these phases on the composition map with $1/n$ plotted against oxygen excess, δ, in perovskite unit $ABO_{3+\delta}$. The *bottom diagram* shows defect electronic conductivity (σ/cm^{-1}) of grain component as determined by ac impedance spectroscopy on samples quenched from 1300°C in air, also plotted against δ [31]

SOFC conditions [33]. Furthermore, Mn is known to accept lower coordination numbers in perovskites [34], especially for Mn^{3+}, and thus it may facilitate oxide ion migration. Similarly, Ga is well known to adopt lower coordination than octahedral in perovskite-related oxides. The possibility of mixed ionic/

electronic conduction is very important because it would allow the electro-oxidation process to move away from the three-phase electrode–electrolyte–gas interface onto the anode surface, with considerable catalytic enhancement.

The anode polarization resistance was measured using three-electrode geometry. By optimization of electrode microstructure, polarization resistances in wet H_2 were 0.12 $\Omega\,cm^2$ in wet H_2, 1.5 $\Omega\,cm^2$ in wet 5% H_2, and a remarkably low value, 0.36 $\Omega\,cm^2$, in wet CH_4, at 950°C. These polarization resistances were attained after about 24 h in fuel conditions; initial polarisation resistances were two to three times higher. This long time period to achieve equilibration is fairly typical for donor-doped strontium titanates that are not cation vacancy compensated, and we attribute this to reorganization of a complex defect structure. The open circuit voltages (OCVs) matched the value predicted by the Nernst equation, 0.97 and 1.13 V at 950°C, for wet 5% H_2 and wet H_2, respectively. The OCVs in wet CH_4, for a one-layer 50:50 YSZ:LSTMG anode, were: 1.39 V at 950°C, 1.32 V at 900°C, and 1.36 V at 850°C. These values were reproducible after 2 days of testing in wet 5% H_2, wet H_2 and wet CH_4 [35].

Although $SrVO_3$ shows excellent electronic conductivity of 1000 S/cm at 800°C and an oxygen partial pressure of 10^{-20} atm, it is unstable under a more oxidizing atmosphere [36]. The conductivity in a reducing atmosphere may drop rapidly if strontium at the A site is partially or completely replaced by lanthanum, and neither is the stability in oxidizing condition improved. A material that has both adequate high-temperature conductivity in a reducing atmosphere and redox stability has not been found in the vanadates. In addition, the reduction process of $SrVO_3$ is rather slow once the material is oxidized.

The perovskite oxide $La_{0.6}Sr_{0.4}Co_{0.2}Fe_{0.8}O_3$ was proposed as an SOFC anode at intermediate temperatures (550°–700°C) [37]. The stability of these materials under fuel atmosphere is in doubt, however, even at such low temperatures. $SrFeCo_{0.5}O_x$ exhibits both high electronic and ionic conductivities in air and is applicable for ceramic membrane used for gas separation [38]. It has been reported that mixed conductors $SrFeCo_{0.5}O_x$, $SrCo_{0.8}Fe_{0.2}O_{3-\delta}$ and $La_{0.6}Sr_{0.4}Fe_{0.8}Co_{0.2}O_{3-\delta}$ may be used as SOFC anode materials. Although there remain some questions about the exact structure of $SrFeCo_{0.5}O_x$, these are generally thought to be related perovskite/brownmillerite intergrowths of the Grenier type [39]. The performance was not ideal when using only these mixed conductors as an anode; however, the performance was improved when these mixed conductors were used as an interlayer between an Ni-YSZ anode and YSZ electrolyte. The long-term stability would again be a problem because $SrFeCo_{0.5}O_x$ is unstable under anodic conditions [40].

As stated above, $LaCrO_3$-based materials have been investigated as interconnect materials for SOFCs [8]; however, they are also potential anode materials for SOFCs due to their relatively good stability in both reducing and oxidizing atmospheres at high temperatures [41]. The reported polarization resistance using these materials is too high for efficient SOFC operation, although significant improvements have been achieved using low-level doping of the B site. As no significant weight loss was observed when $LaCrO_3$ was exposed to a reducing atmosphere ($P_{O2} = 10^{-21}$ atm at 1000°C) [42], this indicates that chromium strongly retains its sixfold coordination. Indeed, Cr^{III} is well known to strongly prefer sixfold coordination in its chemistry; thus, it is difficult to introduce the oxygen vacancies that are required for oxygen ion conduction into the $LaCrO_3$ lattice.

When the B sites are doped by other multivalence transition elements that do tolerate reduced oxygen coordination, such as Mn, Fe, Co, Ni, and Cu, oxygen vacancies may be generated at the B-site dopants in a reducing atmosphere at high temperature. Thus, a significant degree of B-site dopant is required to generate a percolation path for oxygen vacancies to achieve high oxygen ion conductivity. Quite a lot of attention has been focused on 3% replacement of Cr by V, and although methane cracking seems to be avoided [43], the polarization resistance is still of the order of 10 Ω cm [2, 21, 38, 44]. The introduction of other transition elements into the B site of $La_{1-x}Sr_xCr_{1-y}M_yO_3$ (M = Mn, Fe, Co, Ni) has been shown to improve the catalytic properties for methane reforming [45]. Of the various dopants, nickel seems to be the most successful, and the lowest polarization resistances have been reported for 10% Ni-substituted lanthanum chromite [46]; however, other workers have found nickel evolution from 10% Ni-doped lanthanum chromites in fuel conditions [47]. Certainly nickel oxides would not be stable in fuel atmospheres, and although the nickel may be stabilized by the lattice in the higher oxidation state, there will always be the suspicion that the activity of nickel-doped perovskites is the result of surface evolution of nickel metal and hence questions about long-term stability. A composite anode of 5% Ni with a 50:50 mixture of $La_{0.8}Sr_{0.2}Cr_{0.8}Mn_{0.2}O_3$ and $Ce_{0.9}Gd_{0.1}O_{1.95}$ was successfully used for SOFCs with different fuels [48].

The combination of both titanium and chromium at the B site of a perovskite has also been investigated [49, 50]. The highest conductivity in 5% H_2/Ar of 5 S/cm at 1000°C was observed with composition $La_{0.4}Ca_{0.6}Cr_{0.2}Ti_{0.8}O_{3-\delta}$; however, the catalytic effect of these materials for oxidization of hydrogen at the anode is possibly not ideal, because a large anode polarization resistance was observed when $La_{0.7}Sr_{0.3}Cr_{0.8}Ti_{0.2}O_3$ was applied as the SOFC anode [49].

8.6 A(B,B′)O_3 Perovskites

As perovskites with one cation occupying the B site have not generally yielded good enough properties for efficient anode operation in SOFCs, an extensive series of studies has considered the possibility of enhancing performance by using two different B-site ions, both with concentration in excess of percolation limit (i.e., >30%). The objective is to obtain complementary functionality from appropriate cation combinations, hopefully without seriously degrading the good properties induced by the individual ions. Not surprisingly, many of the tested combinations did compromise properties, but in some important instances good complementary functionality has been achieved.

Early studies focused on double perovskites with niobium and a first-row transition metal or main group ion occupying the B site. With Nb and Mn occupying the B site, electronic conductivity is fairly low, probably reflecting the rock salt-type ordering of the B cations [51]. Using Cu and Nb somewhat improves conductivity in air, but in reducing conditions copper metal is exolved and conductivity is impaired, as the resultant perovskite is more resistive and the copper does not form a conducting network [52]. Using Ga with Nb again results in an ordered superstucture that impairs electronic conductivity [53].

Improved performance has been obtained with complex perovskites based upon Cr and Mn at the B sites forming compositions $(La,Sr)Cr_{1-x}M_xO_{3-\delta}$ [54]. Previous workers have focused upon doped lanthanum chromite, where doping is used in the solid-state chemical sense of up to 20% dopant on the B site, usually 5% or 10%. $(La_{0.75}Sr_{0.25})Cr_{0.5}Mn_{0.5}O_3$ (LSCM) exhibits comparable electrochemical performance to Ni–YSZ cermets. The electrode polarization resistance approaches 0.2 Ω cm^2 at 900°C in 97% H_2/3% H_2O. Very good performance is achieved for methane oxidation without using excess steam. The anode is stable in both fuel and air conditions and shows stable electrode performance in methane. Thus, both redox stability and operation in low-steam hydrocarbons have been demonstrated, overcoming two of the major limitations of the current generation of nickel zirconia cermet SOFC anodes. Catalytic studies of LSCM demonstrate that it is primarily a direct oxidation catalyst for methane oxidation as opposed to a reforming catalyst [55], with the redox chemistry involving the Mn–O–Mn bonds [56]. Although oxygen ion mobility is low in the oxidized state, the diffusion coefficient for oxide ions in reduced LSCM is comparable to yttria-stabilized zirconia [57].

Another important double perovskite is $Sr_2MgMoO_{6-\delta}$, which has recently been shown to offer good performance, with power densities of 0.84 W/cm^2 in H_2 and 0.44 W/cm^2 in CH_4 at 800°C, and good sulfur tolerance [58]. The molybdenum-containing double perovskite was initially prepared at 1200°C in flowing 5% H_2 and then deposited on top of a lanthanum ceria buffer layer before testing [59].

8.7 Tungsten Bronze Anode Materials

Tungsten bronze-type materials have also been investigated as potential SOFC anodes. The tungsten bronze structure can be obtained from the perovskite by rotation of some of Ti/NbO_6 octahedra: 40% of the A sites (A2 sites) are increased in size from tetracapped square prisms to pentacapped pentagonal prisms, 40% remain essentially unchanged, and the remaining 40% of the sites is decreased in size (Fig. 8.8). The formula may be written as $A_{0.6}BO_3$ when the small-size A sites are left empty. The distortion of the octahedra means that some B–O bonds are extended and some are shorter than the average. The

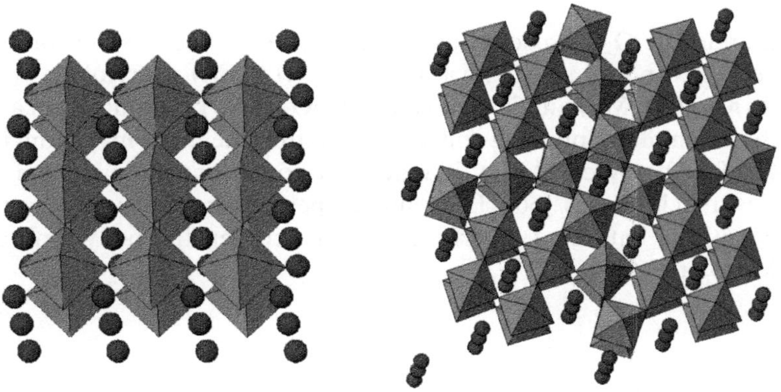

Fig. 8.8 Comparison between perovskite and tetragonal tungsten bronze lattices

connection of the short B–O bond may supply a percolation path for charge transfer, which may lead to high electronic conductivity.

Among the various $(Ba,Sr,Ca,La)_{0.6}M_xNb_{1-x}O_3$ (M = Ni, Mg, Mn, Fe, Cr, In, Sn) compositions, $Sr_{0.2}Ba_{0.4}Ti_{0.2}Nb_{0.8}O_3$ exhibits the highest conductivity (10 S/cm at $P_{O2} = 10^{-20}$ atm at 930°C) [60, 61]. These materials exhibit rather low conductivity in air (~10^{-3} S/cm at 930°C) because there is limited oxygen reduction under this condition. The conductivity increases with decreasing P_{O2} and reaches 1–10 S/cm at a P_{O2} lower than 10^{-17} atm at 930°C when Ti^{4+}/Nb^{5+} was partially reduced, which releases electrons for charge transfer; however, the performance of the $Sr_{0.2}Ba_{0.4}Ti_{0.2}Nb_{0.8}O_3$ tungsten bronze as an SOFC anode is not ideal [61]. Introduction of Mn into the bronze was observed to significantly reduce polarization losses; however, performance was still inferior to better mixed conducting oxides such as titania-doped YSZ [62, 63].

8.8 Anode Materials for All-Perovskite Fuel Cells

The all-perovskite solid oxide fuel cell concept is highly attractive, offering structurally coherent interfaces with good physical and thermal matching. Perovskite electrolytes, normally based upon lanthanum gallate, and perovskite cathodes are already widely utilized in SOFCs; however, perovskite anodes are not so available. A number of possible perovskite anodes have been described in this chapter, and there have been some early successes in implementing these in all-perovskite SOFCs. $(La_{0.75}Sr_{0.25})Cr_{0.5}Mn_{0.5}O_3$ (LSCM) has been utilized successfully in an all-perovskite SOFC with Co-doped LSGM and $La_{0.6}Sr_{0.4}CoO_3$ cathode, achieving a power density of 0.3 Wcm^{-2} at 850°C using a 0.6-mm-thick electrolyte [64]. Using an LSGM electrolyte prepared by tape casting with a thickness of about 120 μm $La_{0.8}Sr_{0.2}MnO_{3-\delta}$ (LSM) and $La_{0.75}Sr_{0.25}Cr_{0.5}Mn_{0.5}O_{3-\delta}$ (LSCM) as cathode and anode material, respectively, good values of power output in a conventional electrolyte-supported cell were achieved, 570 mW/cm^2 using wet hydrogen as fuel and pure O_2 as oxidant at 800°C [65]. An all-perovskite solid oxide fuel cell incorporating (La,Sr)(Ga,Mg)O_3 (LSGM) electrolyte, (La,Sr)(Ga,Mn)O_3 (LSGMn) anode, and (La,Sr)CoO_3 (LSC) cathode fabricated by a combination of dry pressing and screen printing, yielded a maximum power density of 350 mW/cm^{-2} 800°C under test conditions with H_2 and O_2 at the relevant electrodes [66]. Even higher performance was achieved with $Sr_2MgMoO_{6-\delta}$, but using a ceria interlayer between the perovskite anode and electrolyte [58].

8.9 Conclusions

Some materials can match some requirements, but it is very hard to find a material that can match all the stringent requirements for SOFC anodes, particularly redox stability and conductivity. Materials with perovskite structure are promising, such as the redox stable anode $(La_{0.75}Sr_{0.25})Cr_{0.5}Mn_{0.5}O_3$ [54] and $Sr_2MgMoO_{6-\delta}$ [58]. These materials exhibit performance approaching that of the traditional Ni–YSZ anode. Therefore, we can use a nickel-free redox-stable anode for SOFCs. With further optimization of the composition and microstructure, the performance of these materials may be further improved and hopefully replace the traditional Ni–YSZ anode in the future. There have been a number of recent reviews including the topic of oxide anodes where further information may be obtained [67, 68]

Acknowledgments The author thanks his colleagues, especially S.W. Tao and T.D. McColm, for their assistance in developing this manuscript, and thanks ESF, EPSRC, and NEDO (Japan) for financial support for research on perovskite anodes.

References

1. B.C.H. Steele, Solid State Ionics 129, 95 (2000)
2. K. Kuroda, I. Hashimoto, K. Adachi, J. Akikusa, Y. Tamou, N. Komada, T. Ishihara, Y. Takita, Solid State Ionics 132, 199 (2000)
3. J.H. Lee, K.N. Kim, J.W.S.J. Kim, B.K. Kim, H.W. Lee, J. Moon, J. Mater. Sci. 42, 1866 (2007)
4. X.G. Zhang, S. Ohara, H. Okawa, R. Maric, T. Fukui, Solid State Ionics 2001, 139,145.
5. S.W. Tao, J.T.S. Irvine, J.A. Kilner, Adv. Mater. 17, 1734–1737 (2005)
6. J.A. Kilner, S.J. Skinner, T. Ishihara, K. Otsuka, J.T.S. Irvine, T. McColm, Y. Jiang, Electrochem. Soc. Proc.VII, 2001–16, 224–233 (2001)
7. M.F. Hsu, L.J. Wu, J.M. Wu, Y.H. Shiu, K.F. Lin, Electrochem. Solid State Lett. 9, A193 (2006)
8. N.Q. Minh, J. Am. Ceram. Soc. 76, 563 (1993)
9. E.P. Murray, M.J. Sever, S.A. Barnett, Solid State Ionics 148, 27 (2002)
10. Y. Matsuzaki, I. Yasuda, Solid State Ionics 132, 261 (2000)
11. B.C.H. Steele, I. Kelly, H. Middleton, R. Rudkin, Solid State Ionics, 28–30, 1547 (1988)

12. P.R. Slater, D.P. Fagg, J.T.S. Irvine, J. Mater. Chem. 7, 2495 (1997)
13. T.D. McColm, J.T.S. Irvine, Ionics, 7, 116 (2001)
14. N. Peng, J.T.S. Irvine, A.G. Fitzgerald, J. Mater. Chem. 8, 1033 (1998)
15. S. Hui, A. Petric, SOFC VI, Elec. Soc. Proc. 19, 632 (1999)
16. P.R. Slater, J.T.S. Irvine, T. Ishihara, Y. Takita, J. Solid State Chem. 139, 135 (1998)
17. T. Ishihara, H. Matsuda, Y. Takita, Solid State Ionics 79, 147 (1995)
18. T. Ishihara, H. Matsuda, Y. Takita, J. Amer. Chem. Soc. 116, 3801 (1994)
19. H.U. Anderson, Solid State Ionics 52, 33 (1992)
20. H.Y. Tu, Y. Takeda, N. Imanishi, O. Yamamoto, Solid State Ionics 117, 277 (1999)
21. S. Primdahl, J.R. Hansen, L. Grahl-Madsen, P.H. Larsen, J. Electrochem. Soc. 148, A74 (2001)
22. G. Pudmich, B.A. Boukamp, M. Gonzalez-Cuenca, W. Jungen, W. Zipprich, F. Tietz, Solid State Ionics 135, 433 (2000)
23. O.A. Marina, N.L. Canfield, J.W. Stevenson, Solid State Ionics 149, 21 (2002)
24. O.A. Marina, L.R. Pederson, Proc. 5th European Solid Oxide Fuel Cell Forum (ed. Huijsmans J.) pp. 481 (European SOFC Forum, Switzerland, 2002)
25. S.Q. Hui, A. Petric, J. Euro. Ceram. Soc. 22, 1673 (2002)
26. S.Q. Hui, A. Petric, Mater. Res. Bull. 37, 1215 (2002)
27. S.Q. Hui, A. Petric, J. Electrochem. Soc. 149, J1 (2002)
28. S.W. Tao, J.T.S. Irvine, unpublished results.
29. J.T.S. Irvine, P.R. Slater, P.A. Wright, Ionics 2, 213 (1996)
30. J. Canales-Vazquez, M.J. Smith, J.T.S. Irvine, W.Z. Zhou, Adv. Func. Mater. 15, 1000–1008 (2005)
31. J.C. Ruiz-Morales1, J. Canales-Vasquez, C. Savaniu, D. Marrero-Lopez, W. Zhou, J.T.S. Irvine, Nature 439, 568–571 (2006)
32. J. Canales-Vazquez, S.W. Tao, J.T.S. Irvine, Solid State Ionics 159, 159 (2003)
33. P. Holtappels, J. Bradley, J.T.S. Irvine, A. Kaiser, M. Mogensen, Electrochemical characterization of ceramic SOFC anodes. J. Electrochem. Soc. A 148, 923–929 (2001)
34. K.R. Poeppelmeier, M.E. Leonowicz, J.M. Longo, $CaMnO_{2.5}$ and $Ca_2MnO_{3.5}$ – new oxygen-defect perovskite-type oxides. J. Solid State Chem. 44, 89–98 (1982)
35. J.C. Ruiz-Morales, J. Canales-Vazquez, C. Savaniu, D. Marrero-Lopez, P. Nunez, W. Zhou, J.T.S. Irvine, Phys. Chem. Chem. Phys. 9, 1821–1830 (2007)
36. S.Q. Hui, A. Petric, Solid State Ionics 143, 275 (2001)
37. A. Hartley, M. Sahibzada, M. Weston, I.S. Metcafe, D. Mantzavinos, Catalysis Today 55, 197 (2000)
38. U. Balachandran, B. Ma, P.S. Maiya, R.L. Mieville, J.T. Dusek, J.J. Picciolo, J. Guan, S.E. Dorris, M. Liu, Solid State Ionics 108, 363 (1998)
39. J.C. Grenier, G. Schiffmacher, P. Caro, M. Pouchard, P. Hagenmuller, J. Solid State Chem. 20, 365 (1977)
40. B. Ma, U. Balachandran, Mater. Res. Bull. 33, 223 (1998)
41. H. Yokokawa, N. Sakai, T. Kawada, M. Dokiya, Solid State Ionics 52, 43 (1992)
42. T. Nakamura, G. Petzow, L.J. Gauckler, Mater. Res. Bull. 14, 649 (1979)
43. P. Vernoux, M. Guillodo, J. Fouletier, A. Hammou, Solid State Ionics 135, 425 (2000)
44. Y. Matsuzaki, I. Yasuda, Solid State Ionics 132, 261 (2000)
45. J. Sfeir, P.A. Buffat, P. Möckli, N. Xanthopoulos, R. Vasquez, H.J. Mathieu, J.V. Herle, K.R. Thampi, J. Catal. 202, 229 (2001)
46. J. Sfeir, J.V. Herle, R. Vasquez, Proc. 5th European Solid Oxide Fuel Cell Forum (ed. Huijsmans J.) pp. 570 (European SOFC Forum, Switzerland, 2002)
47. A.-L. Sauvet, J.T.S. Irvine, Proc. 5th European Solid Oxide Fuel Cell Forum (ed. Huijsmans J.) pp. 490 (European SOFC Forum, Switzerland, 2002)
48. J. Liu, B.D. Madsen, Z.Q. Ji, S.A. Barnett, Electrochem. Solid State Lett. 5, A122 (2002)
49. G. Pudmich, B.A. Boukamp, M. Gonzalez-Cuenca, W. Jungen, W. Zipprich, F. Tietz, Solid State Ionics 135, 433 (2000)

50. V. Vashook, L. Vasylechko, H. Ullmann, U. Guth, Solid State Ionics 158, 317 (2003)
51. S.W. Tao, J.T.S. Irvine, J. Mater. Chem. 12, 2356 (2002)
52. S.W. Tao, J.T.S. Irvine, Solid State Ionics 154–155, 659 (2002)
53. T.C. McColm, J.T.S. Irvine, Solid State Ionics 152–153, 615 (2002)
54. S.W. Tao, J.T.S. Irvine, Nature Mat. 2, 320 (2003)
55. S. Tao, J.T.S. Irvine, S.M. Plint, J. Phys. Chem. 110, 21771–21776 (2005)
56. S.M. Plint, P.A. Connor, S. Tao, J.T.S. Irvine, Solid State Ionics, 177, 2005–2008 (2006)
57. E.S. Raj, J.A. Kilner, J.T.S. Irvine, Solid State Ionics, 177, 1747–1752 (2006)
58. Y.H. Huang, R.I. Dass, Z.L. Xing, J.B. Goodenough, Science, 2006, **312**, 254–257
59. Y.H. Huang, R.I. Dass, **J.C. Denyszyn,** J.B. Goodenough, J. Electrochem. Soc. **153,** A1266–A1272 (2006)
60. P.R. Slater, J.T.S. Irvine, Solid State Ionics 120, 125 (1999)
61. A. Kaiser, J.L. Bradley, P.R. Slater, J.T.S. Irvine, Solid State Ionics 135, 519 (2000)
62. J.T.S. Irvine, P.R. Slater, A. Kaiser, J.L. Bradley, P. Holtappels, M. Mogensen, Proceedings 4th European SOFC Forum, 2000, Volume 2, p. 471.
63. P. Holtappels, J. Bradley, J.T.S. Irvine, A. Kaiser, M. Mogensen, J. Electrochem. Soc. 148, A923 (2001)
64. S. Tao, J.T.S. Irvine, J.A. Kilner, Adv. Mater. 17, 1734–1737 (2005)
65. J. Pena-Martinez, D. Marrero-Lopez, J.C. Ruiz-Morales, C. Savaniu, P. Nunez, J.T.S. Irvine, Chem. Mater. B18B, 1001–1006. (2006)
66. M.F. Hsu, L.J. Wu, J.M. Wu, Y.H. Shiu, K.F. Lin, Electrochem Solid State Lett. 97, 789 (2006)
67. S.W. Tao, J.T.S. Irvine, The Chemical Record, 4, 83–95 (2004)
68. A. Atkinson, S. Barnett, R.J. Gorte, J.T.S. Irvine, A.J. McEvoy, M. Mogensen, S.C. Singhal, J. Vohs, Nature Mater. 3, 17–27 (2004)

Chapter 9
Intermediate-Temperature Solid Oxide Fuel Cells Using $LaGaO_3$

Taner Akbay

9.1 Introduction

Among other types of fuel cells, deployment of solid oxide fuel cells (SOFCs) into the distributed energy sector of the consumer market is fast becoming a reality, predominantly because technological advancements are making them more reliable and relatively cost competitive. In addition, there has been a concerted effort in designing and manufacturing novel ceramic materials for high-performance cell components [1–4]. Over the decades, a large collection of materials have been systematically investigated for possible application of SOFC electrolyte and electrodes. In addition, reduced temperature operation of SOFCs has been attracting attention due to numerous advantages offered by this mode of operation. Some of these advantages can be listed as follows: (1) less expensive metallic materials can be used as interconnects, (2) less degradable components imply higher stability and durability, and (3) responses to start-up and shut-down procedures are faster. Even internal reforming of hydrocarbonaceous fuels remains possible at intermediate temperatures. Overall, the common denominator impetus is to achieve higher area (or volume)-specific power densities at lower temperatures.

Mitsubishi Materials Corporation, The Kansai Electric Power Co., Inc., and Kyushu University have been working together on multiple projects for developing intermediate-temperature SOFCs. Substantial technical advancement has been made in manufacturing SOFC units ranging from a 1-kW class module to a scalable 10-kW class combined heat and power (CHP) system. The core technology is based on high-performance cells operable in a temperature window between 600° and 800°C.

Lanthanum gallate-based perovskite oxide is selected as the electrolyte material for high-performance cells. Strontium is used as a dopant to the A site, and magnesium together with cobalt are doped to the B site of the

T. Akbay (✉)
Mitsubishi Materials Corporation, Tokyo, Japan
e-mail: akbay@mmc.co.jp

T. Ishihara (ed.), *Perovskite Oxide for Solid Oxide Fuel Cells*,
Fuel Cells and Hydrogen Energy, DOI 10.1007/978-0-387-77708-5_9,

lanthanum gallate. Double doping the B site with cobalt in addition to strontium actually causes a decrease in oxide ion transport number at high temperatures. However, this property is regarded as not detrimental for the intermediate-temperature operation of the cells. In addition, a unique cell stacking technology is developed by using metallic separator plates that eliminate the use of seals.

In this chapter, the key competences of our technology toward the development of commercial SOFC power generation units utilizing cation-doped lanthanum gallate-based electrolyte are summarized.

9.2 Cell Development

9.2.1 Electrolyte

The primary requirement of a fully dense (i.e., impermeable) electrolyte is to possess sufficient levels of ionic conduction under SOFC operating conditions. In addition to the chemical compatibility with the electrode materials, electrolyte materials are also expected to be stable in dual atmospheres of considerably different oxygen partial pressures. On the other hand, mechanical strength is a dominant criterion in deciding the cell geometries (i.e., electrode or self-support).

9.2.1.1 Doped Lanthanum Gallate

High oxide ion conductivity of doped lanthanum gallate compounds at intermediate temperatures makes them favorable materials as electrolytes for self-supported SOFC designs [5–10]. In our manufacturing facilities, the self-supported cells are exclusively fabricated in disk-type planar geometries. Typical dimensions of the electrolyte support are 120 mm in diameter and 200 μm in thickness.

A conventional solid-state reaction technique is employed for the synthesis of the electrolyte materials. Commercially available powders of La_2O_3, $SrCO_3$, Ga_2O_3, MgO, and CoO are mixed, ball-milled. and calcined in air to obtain the final composition of $La_{0.8}Sr_{0.2}Ga_{0.8}Mg_{0.15}Co_{0.05}O_{3-\delta}$ (LSGMC). The calcined mixture is then reground and mixed with an organic binder to produce the LSGMC slurry. As the most prevalent method for fabricating relatively thick layers of cell components, tape casting is employed for the preparation of green sheets. Hence, the LSGMC slurry is tape cast into green sheets of appropriate thickness by using a doctor blade. Green densities are measured as almost 50%. Disks are then cut out of the green sheet and fired in air at 1400°–1500°C for 6 h. As for the relative densities of the sintered disks, 98%–99% of the theoretical values are confirmed.

Because of the inclusion of cobalt as an additional B-site dopant, LSGMC possesses a certain amount of electronic (electron or hole) conductivity, which, in turn, causes a decrease in the open circuit voltage (OCV) of the cell. Reducing the electrolyte thickness is generally thought to be the most favorable way to increase the cell performance via decreasing the ohmic losses. However, one needs to be careful in reducing the thickness of mixed conducting electrolytes, such as LSGMC, because the electronic leak current would cause considerable decrease in the cell potential. Our estimations suggests that, at intermediate temperatures, the optimum LSGMC electrolyte thickness is around 100 μm [11].

For self-supported cells, the other factor that indirectly dictates the electrolyte thickness is the mechanical strength of the electrolyte. As the electrolyte layer gets stronger, the possibility of manufacturing thinner self-supported cells becomes higher. As powder characteristics and processing conditions determine the final microstructure, densely packed small-grained electrolytes can be manufactured by using fine-sized spherical powders together with lower sintering temperatures. Figure 9.1 shows the scanning electron microscopy (SEM) images of the dense LSGMC microstructures obtained by sintering at different temperatures. The electrochemically active cell can then be manufactured by attaching air- and fuel-side electrodes to the electrolyte, as shown in Fig. 9.2.

9.2.2 Anode

The level of catalytic activity of anode determines how efficiently the fuel can be electrochemically oxidized in a fuel cell. In anode conditions, i.e., reducing atmospheres, most noble and transition metals may provide the desired activity. However, at intermediate temperatures only a selected set of metals can meet the whole set of criteria, such as high morphological and dimensional stability and low thermal expansion mismatch. Currently, nickel is the metal most commonly used as an anode material in SOFCs. The total electrical conductivity of a desired anode material should possess electronic as well as ionic components. Ionic conduction is widely accepted as the necessary conduction mechanism to enlarge the reaction zone.

9.2.2.1 Nickel/Rare Earth Metal-Doped Ceria Cermet

It is well known that the ionic conductivity of aliovalent-doped ceria solutions show a maximum at a certain dopant concentration and cation radius. Compared to divalent cation doping, however, trivalent dopants are observed to contribute higher conductivity values in ceria. Among trivalent rare earth metals, samarium and gadolinium are accepted as the most effective dopants. Therefore, we have selected a cermet made up of nickel and 20 at% samarium-doped ceria (SDC) as the anode material in our standard cells [12–14].

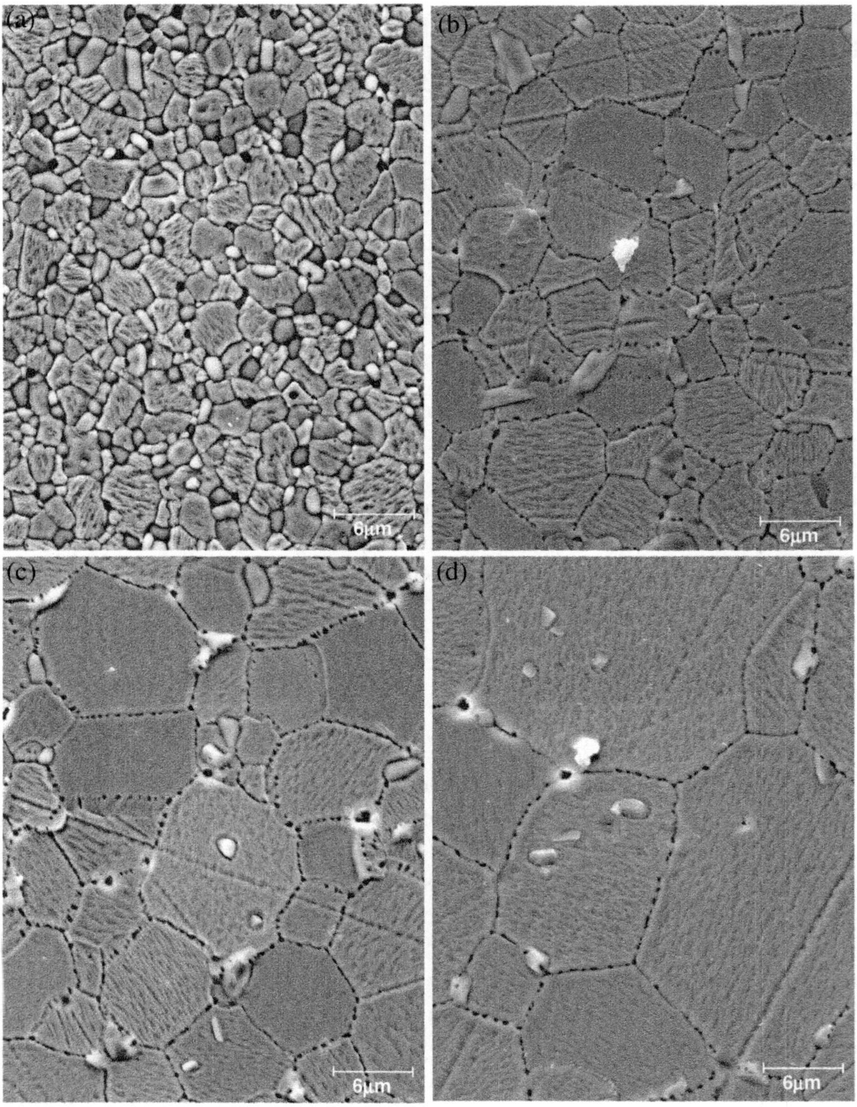

Fig. 9.1 Scanning electron microscopy (SEM) images of $La_{0.8}Sr_{0.2}Ga_{0.8}Mg_{0.15}Co_{0.05}O_{3-\delta}$ (LSGMC) sintered at (**a**) 1350°C, (**b**) 1400°C, (**c**) 1450°C, and (**d**) 1500°C

For preparation of the anode, a slurry composed of NiO and $Ce_{0.8}Sm_{0.2}O_{2-\delta}$ (SDC) mixture is screen printed onto the electrolyte. The firing process involves treatment of the screen-printed anode layer on the electrolyte at 1100°–1300°C for 3 h in air. The relative amounts and particle size ratios of the mixture of NiO and SDC are optimized in such a way that the resulting microstructure is resistant to coarsening of reduced nickel particles under the operating

Fig. 9.2 Cross section of the standard cell

conditions of the fuel cell. Figure 9.3(a) shows the typical air-fired microstructure of the Ni-SDC anode. The average thickness of the porous anode is about 30–50 μm.

Apart from nickel coarsening, additional causes of the inadmissible increase in anodic overpotentials can be thought of as blockage of the nickel network by the presence of SDC particles, poor adhesion between the anode and the electrolyte, slow in situ steam reforming reaction of methane on nickel, insufficient active length of triple-phase boundaries, and so on. To overcome these

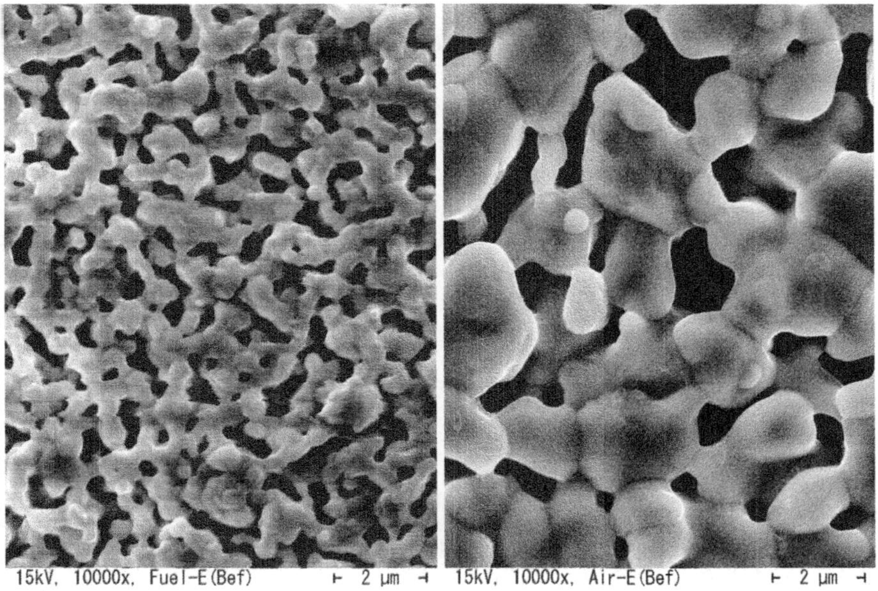

(a) (b)

Fig. 9.3 SEM images of (**a**) anode and (**b**) cathode

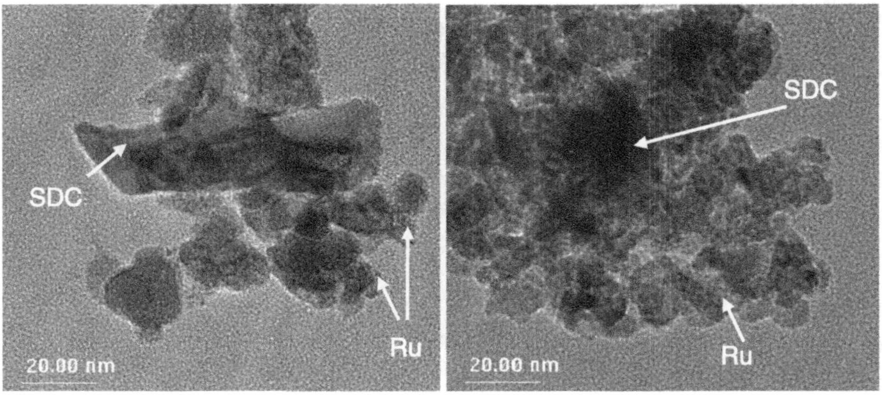

(a) (b)

Fig. 9.4 Ru-dispersed *SDC* (NiO and $Ce_{0.8}Sm_{0.2}O_{2-\delta}$) particles with the concentration of (**a**) 1 wt% (**b**) and 10 wt%

problems, nano-sized particles of certain transition metals may be incorporated in the anode to modify its morphology. For this, we have selected ruthenium to be dispersed in the cermet of nickel and SDC.

The Ni-Ru-SDC anode is prepared by using a modified batch of SDC powder. Relatively coarse particles of SDC are mixed with ruthenium chloride to form a suspension. Following the reduction of ruthenium cation, the resulting powder is separated and dried. Figure 9.4 shows the transmission electron microscopy (TEM) images of 1 wt% (a) and 10 wt% (b) Ru-dispersed SDC particles obtained by the aforementioned preparation method. The optimum value, on the other hand, is determined in such a way that when mixed with NiO to form the anode, the concentration of ruthenium should be around 1 wt%, as estimated from the compositions of widely used precious metal catalysts [15].

Among trivalent rare earth metals, gadolinium is also known to make a solid solution with ceria to increase its ionic conductivity. Catalytic activity of the anode, on the other hand, can be further improved by embedding highly dispersed nickel particles in the ceramic matrix [15]; this should improve the long-term stability of the anode through delayed sintering of nickel particles.

9.2.3 Cathode

Similar to the basic requirements for anodes, a suitable cathode material should also possess high catalytic activity to electrochemically reduce oxygen at operating conditions of SOFCs. At intermediate temperatures, however, mixed conducting oxides and certain noble metals may be utilized as cathode materials. Although most of the candidates have intrinsic thermal expansion mismatch and incompatibility with other cell components, a subset of doped oxides are found to be suitable and widely used as cathode materials in a range of intermediate- to high-temperature fuel cells.

9.2.3.1 Strontium-Doped Samarium Cobaltite

Strontium-doped samarium cobaltite (SSC) is selected as the air electrode material of our standard cells. The slurry of $Sm_{0.5}Sr_{0.5}CoO_{3-\delta}$ powder is screen printed onto the electrolyte disk with an already sintered anode. The final sintering of the cathode is performed in air at a temperature range of 1000°–1200°C for 3 h. The average thickness of the porous cathode is about 30–50 μm. Figure 9.3(b) shows the electron microscope image of the air-fired microstructure of the air electrode of our standard cell. It has been demonstrated that the cathodic overpotential of the cell is very small [16].

9.2.3.2 Lanthanum-Doped Barium Cobaltite

The A site of barium cobaltite ($BaCoO_3$) is partially substituted by lanthanum ions to improve the catalytic and electrical properties of this particular oxide. We have performed a series of performance measurements by varying the lanthanum content between 30 at% and 50 at% and determined the optimal value as 50 at%, that is, an equal amount to the barium content on the A site of the barium cobaltite. Figure 9.5 shows a comparison between I-V-P characteristics of cells made up of two different cathode materials, namely, $Ba_{0.5}La_{0.5}CoO_{3-\delta}$ (BLC) and SSC measured at 750°C using pure hydrogen as fuel. It is found that the cells made up of the BLC cathode have a very similar performance to those utilizing SSC as a cathode material [17].

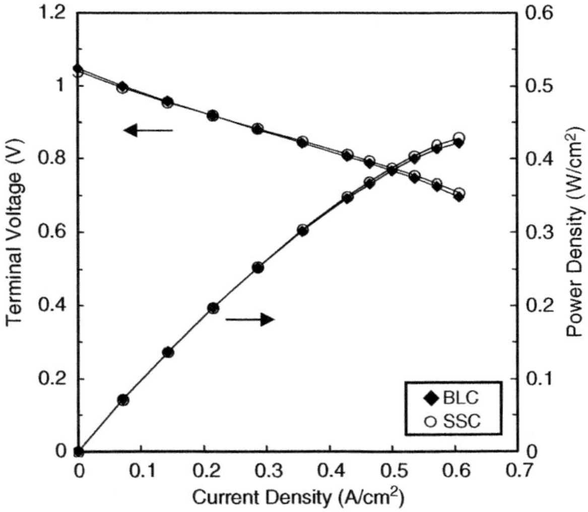

Fig. 9.5 I-V-P characteristics of single cells using strontium-doped samarium cobaltite (SSC) and $Ba_{1-x}La_xCoO_{3-\delta}$ (BLC) as cathode

9.3 Stack Development

In fuel cells, as in batteries, electrochemically active cells should be electrically connected in series to produce usable power. Stacking is the term most commonly used for connecting cells in electrical series for fuel cells. Basically, the stack design addresses the delivery of gaseous species to the cells together with loss-free collection of the electrical current and management of heat. These, in principle, have direct influence on the electrochemical performances of the individual cells. Successful heat management, on the other hand, is critically important for the structural integrity of the components used in a stack assembly.

Compared to alternative stacking configurations proposed for planar cells, interconnecting cells without the need for using seals offer a great deal of simplicity [18]. Our disk-type cell stacking concept exploits the sealless design concept and consists of a number of repeat units, each of which is made up of only cells, porous current collectors, and metallic separators. Figure 9.6 illustrates the conceptual drawing of the single cell stack unit. Air and fuel gas are supplied to the unit through the channels inside the separators. The openings at the center of separators serve as gas inlets to the porous current collectors. The uniform flow regimen of gas along the radial direction over the electrodes is maintained by the help of the porous current collectors. The thicknesses and the porosity values of the current collectors are carefully optimized to obtain the highest rates of axial diffusion of gases. Ferritic stainless steel is used as a material for the separator plates. As there are no circumferential seals attached to the cells, the unutilized fuel and depleted air are freely mixed and burned around the stack unit. The combustion heat is further utilized for the heat management of the entire stack and the balance of plant components around it (steam generator, pre-reformer, and heat exchangers, etc.).

Figure 9.7 shows the actual cell stack unit assembly for the fourth-generation 1-kW class module. The individual separator plates are designed

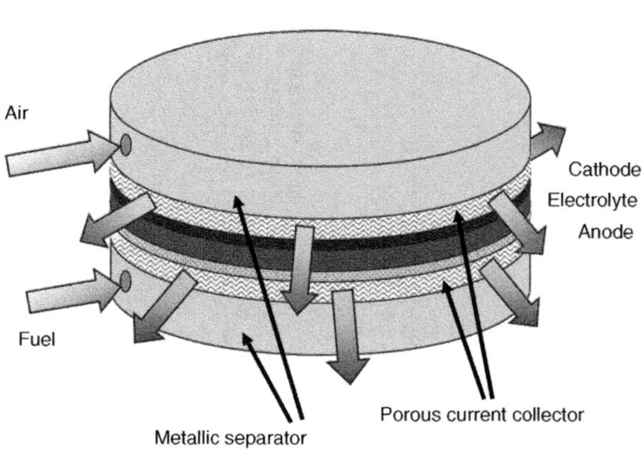

Fig. 9.6 Conceptual illustration for the sealless stack unit

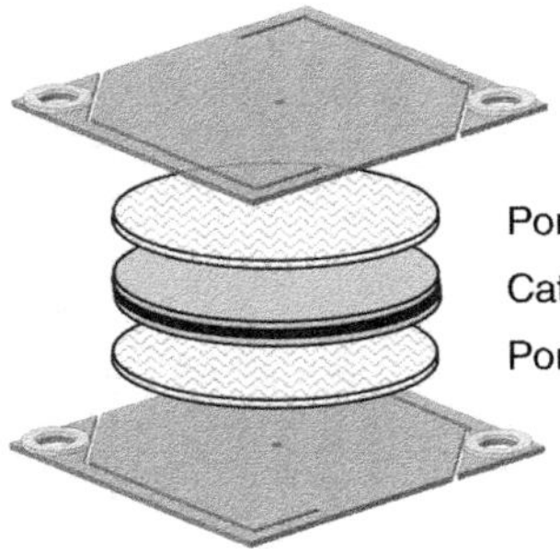

Fig. 9.7 Cell stack unit assembly for the fourth-generation 1-kW class module

by using a rather unique concept of internal gas manifolding. Holes on opposite corners serve as gas manifolds when separator plates are stacked by placing ceramic rings over them. Inner gas channels in the separator plate connect the gas manifolds to two openings located at the center of opposite sides of the separator. The fuel gas supplied from the central opening flows through the anode-side current collector and undergoes internal steam reforming and anode electrode reaction. Meanwhile, the air supplied from the opening on the opposite side of the separator flows through the cathode-side current collector and takes part in the cathode electrode reaction.

Apart from supplying gases to the cells and serving as electrical connection between individual cells, the separator plate in our stack design has an additional functionality for isolating the compressive forces on the disk-type cells placed at their center and the ceramic rings used for the gas manifolds. This unique functionality is accomplished by attaching flexible arms to the separator plates. The manifold ends of the separator arms and ceramic rings must be tightened by bolts and nuts to make hermetic seals. On the other hand, the interconnection parts of the separators where cells and current collectors are placed require certain levels of pressure to minimize electric contact resistance between them. The milder load requirement on the interconnection parts is mainly exerted by a weight at the top of the cell stack.

The fourth-generation 1-kW class stack shown in Fig. 9.8 consists of 46 cells connected in electrical series. Electrically insulated clear hole flanges are attached to the extreme ends of the stack and tightened by using stud bolts to fixate the entire assembly. The air inlet for the stack is attached to the air manifold at the mid-height of the stack while the fuel inlets are attached to top and bottom flanges that have built-in openings to the fuel manifolds. To enhance the heat exchange between the stack and balance of plant components, an additional radiator plate is inserted at the mid-height of the stack.

Fig. 9.8 CAD representation of the fourth-generation 1-kW class SOFC stack

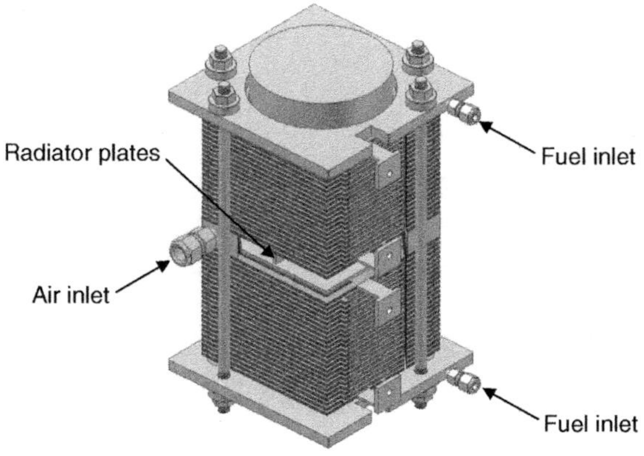

9.4 Module Development

The previously mentioned sealless stacking concept is highly pertinent to electrical power scaling by means of altering the number of cells. In other words, the number of cells can easily be increased for higher power output per stack. However, there is an intrinsic limit for the maximum number of standard-sized (120 mm in diameter) cells that can actually be stacked due to the very nature of the sealless stacking design. As can be imagined, the larger the number of cells, the taller the stack, which exerts its own weight as increased amount of pressure on the porous current collectors and cells toward the bottom of the stack. To eliminate a possible mechanical failure in cells, we have decided to build a generic-sized stack to be connected electrically in series for manufacturing larger-sized SOFC power production units.

As illustrated in Fig. 9.9, we often use the term module for a DC power generation unit composed of generic SOFC stack(s) together with all the rest of the hot balance of plant components efficiently packed inside a thermally insulated lining. The module utilizes desulfurized town gas and deionized water for internal steam reforming. Air is preheated before entering to the stack. The relative positions of the components inside the module are carefully designed for optimal heat management.

9.4.1 A 1-kW Class Single-Stack Module

Single-stack modules capable of generating around 1 kW electrical power output have been developed as test platforms and continuously evolved up to the current design, named the fourth generation. In all generations, the internal steam reforming concept is adopted for thermally self-sustained operation, except for the first-generation module, which is designed for hydrogen fuel. In each design iteration, the SOFC stack together with the balance of plant

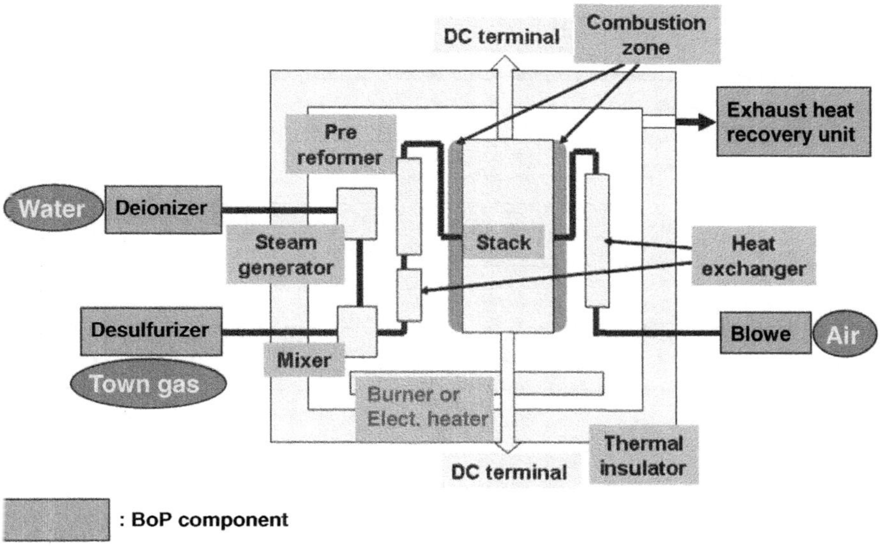

Fig. 9.9 Concept of the module

components is optimized for obtaining higher electrical conversion efficiencies and long-term stability.

A constant power output durability test of the fourth-generation 1-kW class module is performed over 4200 h. The operating conditions are listed in Table 9.1. The data recorded during the entire test period are plotted as a graph (shown in Fig. 9.10). The operation was interrupted briefly at 1000 h to correct the erratic behavior of the stack that started around 600 h. After the restart, the remaining period of operation was trouble free. The electrical efficiency remained about 50% Higher Heating Value (HHV) throughout the test. The voltage degradation after the restart was calculated as 0.5% per 1000 h.

A cyclic power output durability test is also conducted for more than 1000 cycles. The DC power output of the stack is alternately cycled between 100% and 10% for the frequency of 6 cycles per hour (Fig. 9.11). The stack behavior was rather stable, while the degradation rate for certain cells was slightly higher than the constant power output test case.

Table 9.1 Operating conditions for the durability test of the fourth-generation 1-kW module

Fuel	Town gas (13 A)
Total current	33.9 A
Current density	0.3 A/cm^2
Fuel utilization	71%
S/C	3.0
Maximum separator temperature	790°C

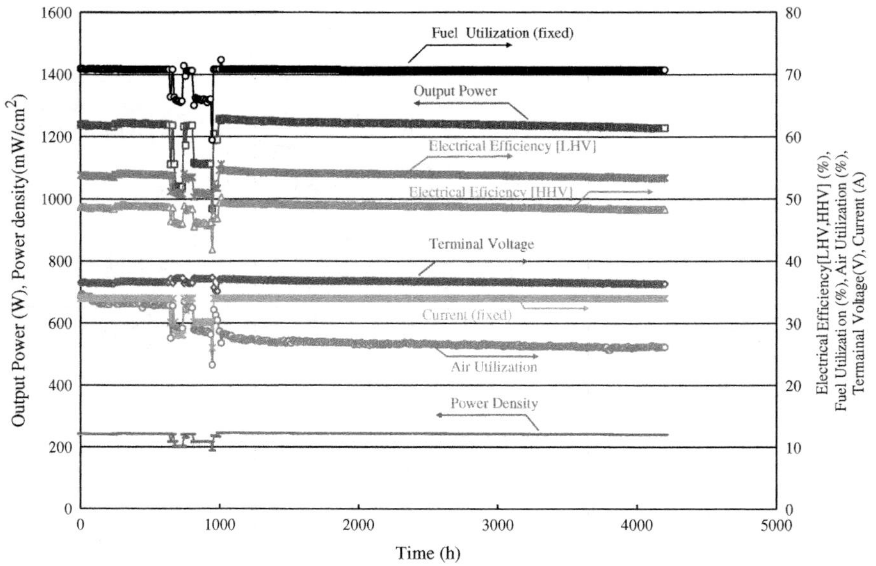

Fig. 9.10 1-kW module durability test over 4200 h

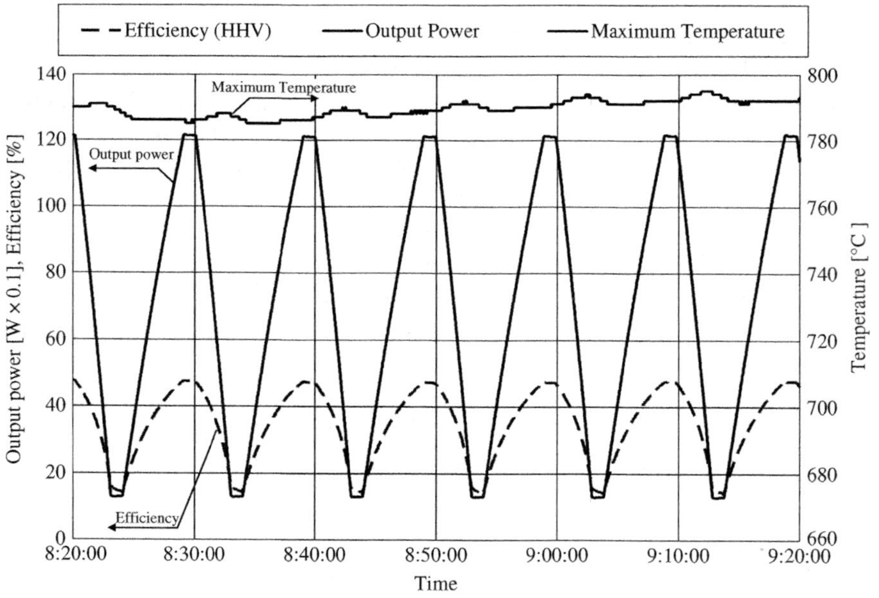

Fig. 9.11 Cyclic durability test results for the 1-kW module

9.4.2 A 10-kW Class Multi-Stack Module

Figure 9.12 depicts the conceptual drawing of the 10-kW class intermediate-temperature SOFC module. The outer dimensions of the module are about 1 m (W) × 1 m (D) × 2 m (H). A three-dimensional array of 16 generic stacks (2 × 2 × 4) that are connected electrically in series is used to produce the output power of 12.6 kW DC with a gross electrical efficiency of 50% (HHV) at an operation temperature below 800°C. A flat-plate box-type steam reformer is designed and positioned vertically between the stacks to keep its temperature as high as possible. Fuel gas and air streams introduced to the module are passed through dedicated plate-type heat exchangers before being distributed to the SOFC stacks in the array. Start-up burners that utilize town gas are positioned inside the insulator linings near the stacks to heat up the module. Table 9.2 summarizes the design specifications of the 10-kW class module.

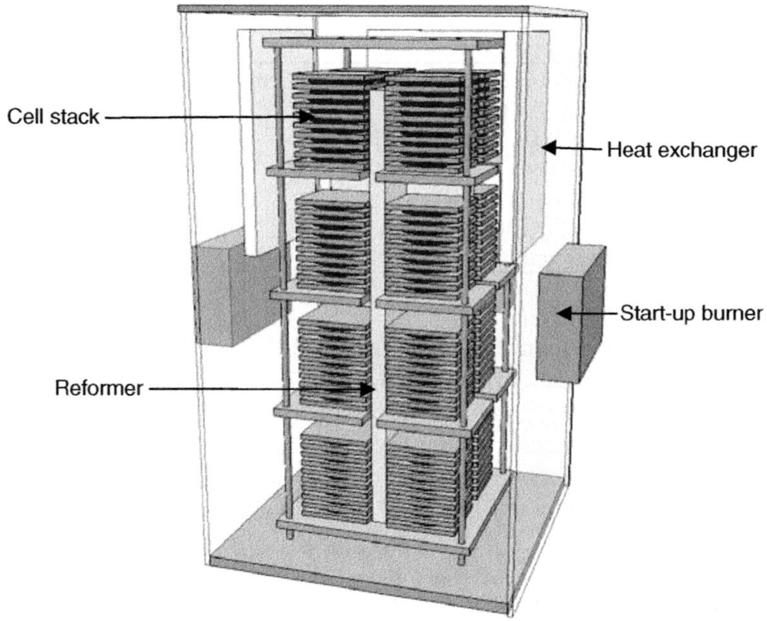

Fig. 9.12 Conceptual view of a 10-kW class module

Table 9.2 Specifications of the 10-kW class module

Fuel	Town gas (13 A)
Output power	12.6 kW DC
Electrical efficiency	50% HHV
Maximum separator temperature	<800°C

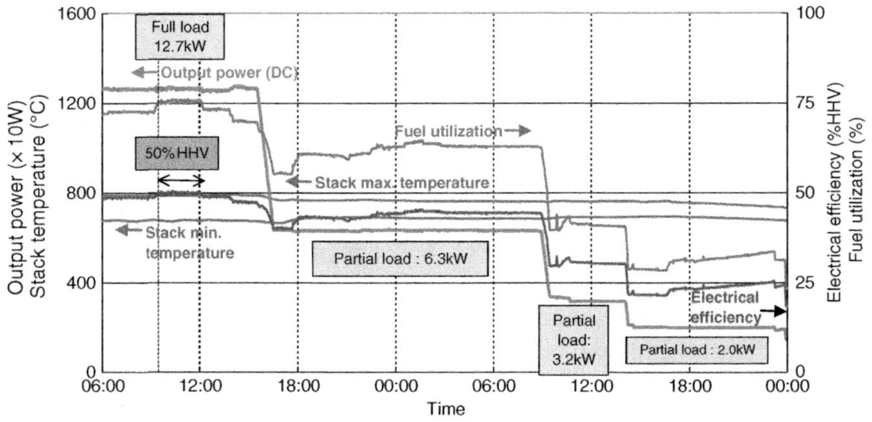

Fig. 9.13 Operation data of the first 10-kW class module

Table 9.3 Typical performance of the 10-kW class module

DC output power (kW)	12.7
Average area-specific power density (W/cm^2)	0.208
DC terminal voltage (V)	415
Efficiency at DC terminal (%HHV)	50
Fuel utilization (%)	76
Air utilization (%)	53
Average cell voltage (V)	0.77
Stack temperature (°C)	675–787

A short-term performance test demonstrated that the 10-kW class module is capable of meeting the design specifications for full load and partial load operations. Figure 9.13 shows the various data recorded during the test. The DC power output of 12.7 kW is obtained under full load operation at 76% fuel utilization, analogous to the electrical efficiency of 50% HHV. The operational characteristics during the thermally self-sustained operation are listed in Table 9.3.

9.5 System Development

While demonstrating the 10-kW class module capability in meeting the DC target specifications, a combined heat and power (CHP) system is developed for utilizing the module within a complete system configuration [19, 20]. The photograph shown in Fig. 9.14 is for the actual 10-kW class CHP system installed at Rokko Testing facilities of The Kansai Electric Power Co., Inc. The system consists of a 10-kW class SOFC module, a unit for gas and water supply equipment, a unit for power electronics, a control unit, and an exhaust heat recovery unit.

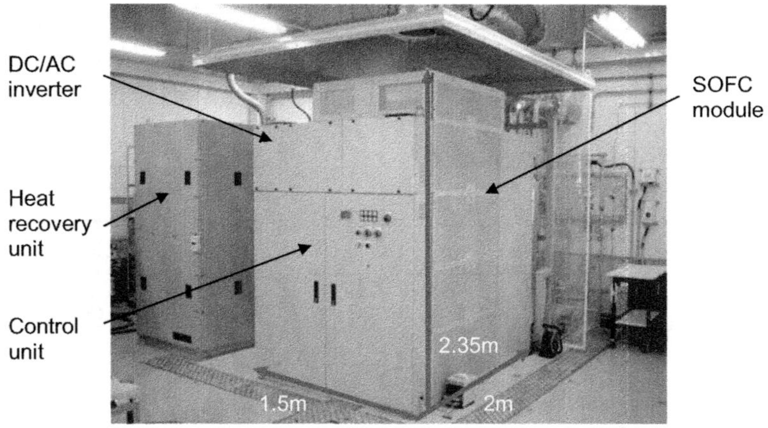

Fig. 9.14 The 10-kW class CHP system

The control unit of the system is specifically designed for automated start-up, power generation, hot-standby, and scheduled or emergency shut-down processes. For the power electronics, a DC/AC inverter is attached to the module for grid connection. The heat recovery unit is attached to the module's exhaust to generate hot water at a temperature between 60° and 90°C. The volumetric capacity of the hot water tank is selected as 370 liters with a suitably sized water pump. The condensed water recycling unit is attached to the system for possible utilization of pure water for internal steam reforming.

An initial performance test is performed on the system to validate the design specifications. The test results are summarized in Table 9.4. The AC power output of the system is obtained as 10.1 kW with a corresponding AC electrical conversion efficiency of 41% HHV. The overall system efficiency, on the other hand, is recorded as 82% HHV when the module exhaust heat is recovered as hot water at 60°C. These results clearly demonstrate that an intermediate-temperature SOFC CHP system of this size is perfectly capable of generating electric power as well as quality heat with high efficiency figures.

Figure 9.15 shows the data obtained during the long-term stability test performed over an accumulated period of 3400 h. The system is stably operated during

Table 9.4 Test results of the 10-kW CHP system

Fuel	Town gas (13 A)
AC output power	10.1 kW
AC electrical efficiency	41% HHV
Overall efficiency[a]	82% HHV
Maximum separator temperature	776°C

CHP, combined heat and power.
[a]Exhaust heat is recovered as hot water at 60°C.

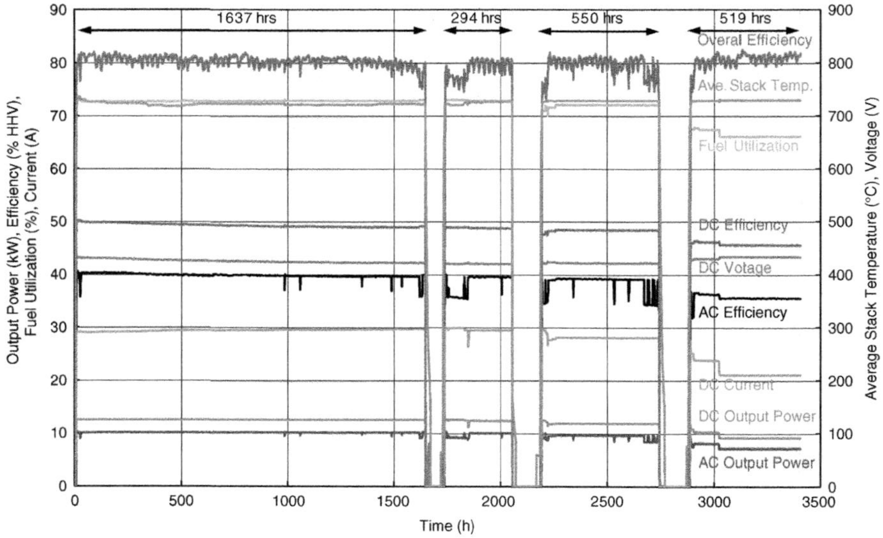

Fig. 9.15 Long-term stability test of the 10-kW CHP system

the comprehensive portion of the test period except for some unexpected electrical trips caused by malfunctioning gas flow controllers and temperature sensing system. The fluctuation in the overall system efficiency is attributed to a lag between the heat recovery control and the module's inlet water flow rate control.

9.6 Stack Modeling

As numerous parameters affect the operational characteristics of fuel cells, the search for an optimum combination can be very difficult. In this respect, a mathematical modeling approach is widely used in fuel cell research and development to perform various analyses ranging from estimation to design iteration. Predictive models, however, are only useful when they are verified by experimental evidence.

Among various levels of modeling, particular attention is paid to the cell and stack models for supporting our stack development. In this section, the discussion is be restricted to a standard computational fluid dynamics (CFD) analysis coupled with rigorous electrochemistry of the generic 34-cell stack. The standard approach includes solving the governing equations of mass and momentum conservation (Navier–Stokes) together with energy, species, and charge transport phenomena. Chemical and electrochemical reaction kinetics are also included to evaluate necessary source or sink terms for the transport equations. A commercially available CFD package is used for solving the coupled set of transport equations [21]. Detailed explanation of the multiphysics model, which is outside the scope of this section, may be found elsewhere [22, 23].

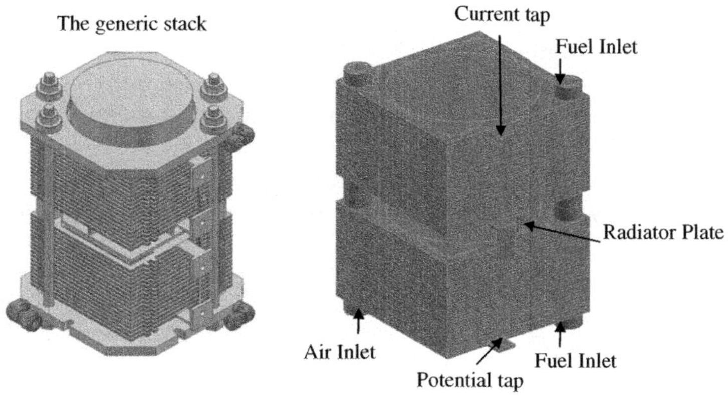

Fig. 9.16 Computational domain prepared for the generic stack

The computational domain (Fig. 9.16) is prepared by generating a grid for the entire stack. Due to diverse length scales along the axis of the stack, grid optimization is done to make the numerical results independent of the mesh size.

At fuel manifold inlets, gaseous species concentrations are specified as equilibrium compositions of the town gas reformate at 650°C. Steam-to-carbon ratio is kept as 3.06 for this particular steady-state analysis. Both fuel and air gas manifold inlet conditions are summarized in Table 9.5. Mixed convective and radiative heat transfer boundary conditions are applied to the side surfaces of the stack to accurately model the heat exchange with the balance of plant components. Top and bottom surfaces, on the other hand, are assigned with

Table 9.5 Independent variables of the generic stack simulation

			Volume fraction						
	Inlet temp.(°C)	Flow rate (liter/min)	O_2	N_2	CH_4	CO_2	CO	H_2	H_2O
Air	570	32.0	0.21	0.79					
Fuel	650	6.93			0.0148	0.0840	0.0759	0.5347	0.2906

Table 9.6 Physical properties used in the generic stack simulation

	Thermal conductivity (W/mK)	Electrical resistivity (Ω m)	Viscous resistance (1/m²)	Porosity	Tortuosity
Separator	26.3	5.88×10^{-8}	–	–	–
Fuel-side current collector	2.47	4.26×10^{-7}	$1 \times 10^{+9}$	0.9	1.0
Anode	10.7	7.09×10^{-7}	$1 \times 10^{+12}$	0.4	1.5
Electrolyte	1.32	1.02×10^{-1}	–	–	–
Cathode	3.2	1.26×10^{-5}	$1 \times 10^{+12}$	0.5	1.5
Air-side current collector	5.67	4.76×10^{-8}	$1 \times 10^{+9}$	0.9	1.0

known heat flux values. All external surfaces are kept at zero current, except those labeled as "potential tap" and "current tap," for which a fixed electrical current value (29.95 A, corresponding to 72.5% fuel utilization) is assigned. The potential tap is kept at the ground potential. Physical properties used in the model are listed in Table 9.6.

Figure 9.17 shows the species concentration profiles along the radial axis of cells numbered 1, 9, and 17. Although the maximum temperatures estimated at

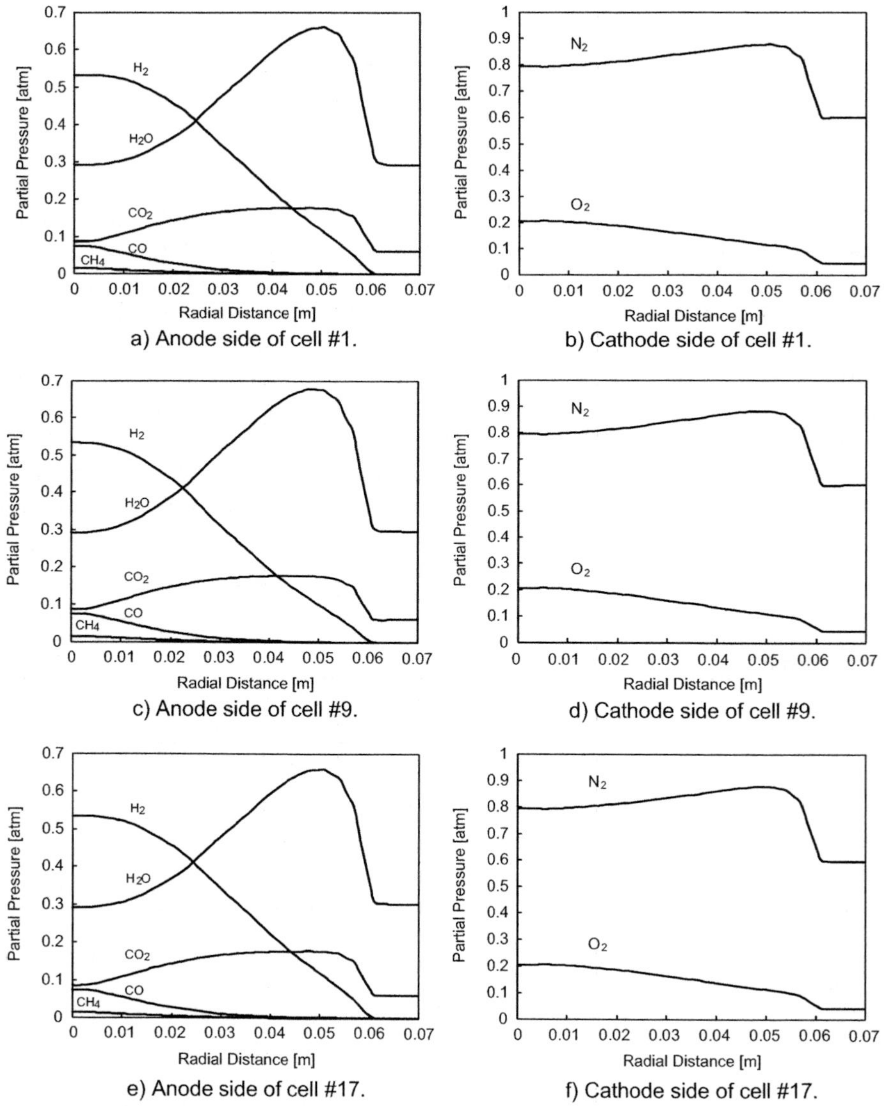

Fig. 9.17 Gaseous species concentrations on selected cells

the center of these three cells are quite different, the gas concentrations remain uniform, implying that the manifolds are capable of contributing to stable gas dynamics in the stack. In addition, the model successfully captures the phenomenon of back-diffusion of air against the depleted fuel stream over the anode side of the cells. It can be noticed that methane undergoes the in situ steam reforming process over the cell and becomes 100% converted at a distance corresponding to two-thirds of the cell radius.

The surface temperature distribution of the stack is shown in Fig. 9.18. The cooling effect of the air is slightly more noticeable than that of the fuel. Because the stack loses extra heat from the top and bottom surfaces as well as the surfaces adjacent to the radiator plate, there are unavoidable hot spots on the front and back surfaces. On the other hand, the maximum temperature reaches about 800°C at the center of the separator plates adjacent to cells numbered 9 and 26 (Fig. 9.19).

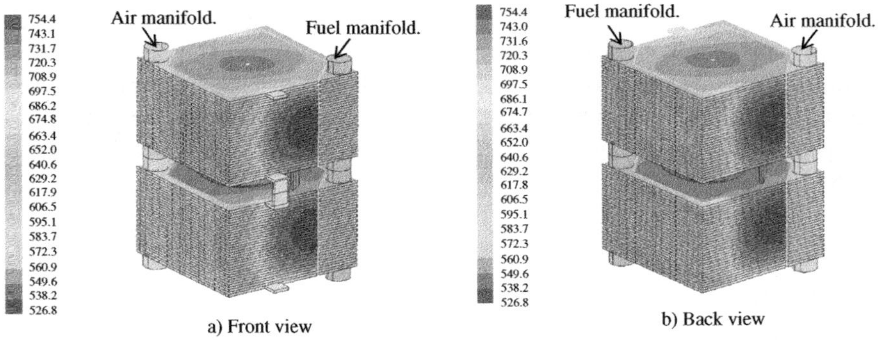

Fig. 9.18 Temperature distribution (°C) on the generic stack surface

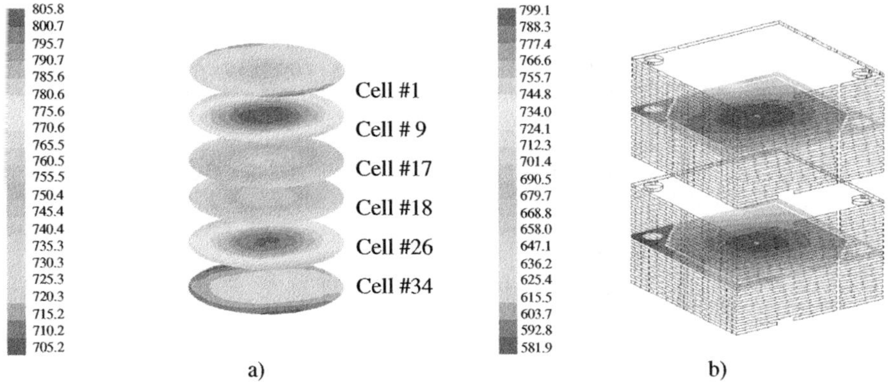

Fig. 9.19 Temperature distribution (°C) on **(a)** selected cells and **(b)** selected separators

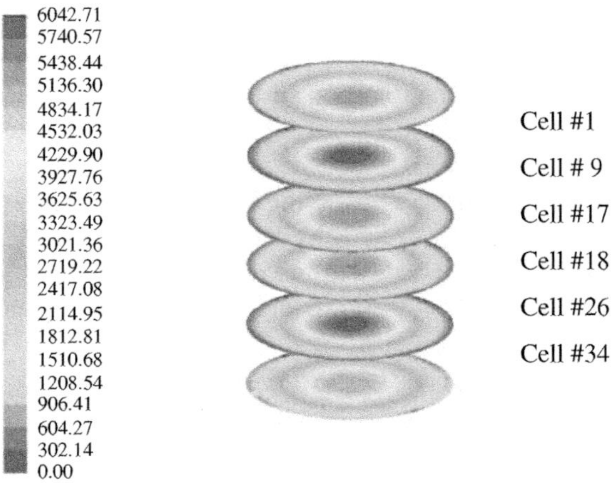

Fig. 9.20 Current density distribution (A/m^2) on selected cells in the generic stack

Figure 9.20 shows the current density distribution on the cells numbered 1, 9, 17, 18, 26, and 34, starting from the top of the stack. The current density assumes a bell-shaped profile on the surface of each cell. The maximum current density reaches 0.6 A/cm^2 at the center of the cells that observe the highest temperature. The average cell potential, on the other hand, is calculated as 0.785 V.

References

1. B.C.H. Steele, J. Power Sources 49, 1 (1994)
2. N.Q. Minh, T. Takahashi, Science and Technology of Ceramic Fuel Cells, Elsevier, Amsterdam (1995)
3. N.Q. Minh, Solid State Ionics 174, 271 (2004)
4. S.C. Singhal, Solid State Ionics 152–153, 405 (2002)
5. T. Ishihara, H. Matsuda, Y. Takita, J. Am. Chem. Soc. 116, 3801 (1994)
6. T. Ishihara, H. Matsuda, Y. Takita, Solid State Ionics 79, 147 (1995)
7. M. Feng, J.B. Goodenough, Eur. J. Solid State Inorg. Chem. 31, 663 (1994)
8. K. Huang, R.S. Tichy, J.B. Goodenough, J. Am. Ceram. Soc. 81, 2565 (1998)
9. P. Huang, A. Petric, J. Electrochem. Soc. 142, 1644 (1996)
10. T. Ishihara, T. Akbay, H. Furutani, Y. Takita, Solid State Ionics 113–115, 585 (1998)
11. T. Ishihara, S. Ishikawa, C. Yu, T. Akbay, K. Hosoi, H. Nishiguchi, Y. Takita, Phys. Chem. Chem. Phys. 5, 2257 (2003)
12. T. Inagaki, H. Yoshida, K. Miura, S. Ohara, R. Maric, X. Zhang, K. Mukai, T. Fukui, Proc. Electrochem. Soc. Vol. 2001–16, 963 (2001)
13. X. Zhang, S. Ohara, R. Maric, KI. Mukai, T. Fukui, H. Yoshida, M. Nishimura, T. Inagaki, K. Miura, J. Power Sources 83, 170 (1999)
14. S. Ohara, R. Maric, X. Zhang, K. Mukai, T. Fukui, H. Yoshida, T. Inagaki, K. Miura, J. Power Sources 86, 455 (2000)
15. S. Awatsu, H. Iwasaki, K. Isono, N. Chitose, G. Uozumi, T. Inagaki, T, Ishihara, ECS Trans. 7(1), 1583 (2007)

16. T. Ishihara, M. Honda, T. Shibayama, H. Minami, H. Nishiguchi, Y. Takita, J. Electrochem. Soc. 145, 3177 (1998)
17. M. Kawasaki, N. Chitose, J. Akikusa, T. Akbay, H. Eto, T. Inagaki, T. Ishihara, ECS Trans. 7(1), 1229 (2007)
18. J. Akikusa, K. Adachi, K. Hoshino, T. Ishihara, Y. Takita, J. Electrochem. Soc. 148, 1275, (2001)
19. F. Nishiwaki, T. Inagaki, J. Kano, J. Akikusa, N. Murakami, K. Hosoi, J. Power Sources 155, 809, (2006)
20. M. Shibata, N. Murakami, T. Akbay, H. Eto, K. Hosoi, H. Nakajima, J. Kano, F. Nishiwaki, T. Inagaki, S. Yamasaki, ECS Trans. 7(1), 77 (2007)
21. Fluent, 2006, FLUENT 6.3 User's Guide, Fluent Incorporated, Canterra Resource Park, Cavendish Court, Lebanon, NH, USA
22. T. Akbay, N. Chitose, T. Miyazawa, N. Murakami, K. Hosoi, F. Nishiwaki, T. Inagaki, Proc. The Fourth Int. Conf. Fuel Cell Sci. Eng. Technol. June 19–21 (2006)
23. T. Akbay, N. Chitose, T. Miyazawa, M. Shibata, F. Nishiwaki, and T. Inagaki, Proc. of The Fifth Int. Conf. on Fuel Cell Sci. Eng. and Tech., June 18–20 (2007)

Chapter 10
Quick-Start-Up Type SOFC Using $LaGaO_3$-Based New Electrolyte

Akira Kawakami

10.1 Introduction

TOTO is the top sanitary ware manufacturer in Japan. The company also has much experience in making both traditional and advanced ceramic products. Solid oxide fuel cells (SOFCs) are mainly composed of ceramics, and our fabrication technology has been utilized to produce high-performance SOFCs at a low cost. TOTO commenced research and development of the tubular-type cell (whose diameter is 16 mm) using a zirconia-based electrolyte for stationary power applications in 1989 (Table 10.1). We have demonstrated more than 10,000 h of continuous operation in single-cell tests, and our modules have accumulated more than 3,000 h of operation as of 2004 with electrical efficiency (Lowe heating value, LHV) of 55% [1]. Since 2004, TOTO has been collaborating with Hitachi, Ltd. on a new NEDO (New Energy and Industrial Technology Development Organization) project. We are developing a 20-kW class co-generation system by integration with TOTO's stacks (Fig. 10.1). Meanwhile, TOTO also started developing micro-tubular SOFCs using an $LaGaO_3$-based new electrolyte for portable power applications in 2002.

A micro-tubular SOFC has potential for portable and transportation applications because of its advantages, including high volumetric power density and high thermal shock resistance, which enable rapid start-up [2]. We started developing the micro-SOFCs using micro-tubular cells (whose diameter is less than 5 mm) in 2002 under the sponsorship of the NEDO project [3]. The objectives of this project are to lower the operating temperature to 500°–700°C and to develop a compact stack that will enable both downsizing of the system and rapid start-up (in minutes or less) for the SOFC system. To achieve high power density at lower temperatures, the anode-supported design with a thin film of lanthanum gallate with strontium and magnesium doping was selected. The single cells and cell stacks were evaluated using various

A. Kawakami (✉)
TOTO Ltd., Chigasaki, Kanagawa 253-8577 Japan
e-mail: akira.kawakami@toto.co.jp

T. Ishihara (ed.), *Perovskite Oxide for Solid Oxide Fuel Cells*,
Fuel Cells and Hydrogen Energy, DOI 10.1007/978-0-387-77708-5_10,

Table 10.1 Materials used in TOTO tubular cell with zirconia-based electrolyte

Components	Materials
Cathode tube	$(La,Sr)MnO_3$
Electrolyte	ScSZ
Anode	Ni/YSZ
Interconnector	$(La,Ca)CrO_3$

Fig. 10.1 Appearance of TOTO stacks for stationary power applications

practical fuels, i.e., hydrogen, simulated reformate gas of liquid petroleum gas (LPG), and direct fueling of dimethyl ether (DME).

In this chapter, we summarize the current status of micro-tubular SOFC for portable power applications at TOTO.

10.2 Micro-Tubular Cell Development

Table 10.2 shows materials and fabrication processes of the TOTO micro-tubular cell. The anode substrate tube made of $NiO/(ZrO_2)_{0.9}(Y_2O_3)_{0.1}$ (NiO/YSZ) was formed by extrusion molding. The anode interlayer and the electrolyte were subsequently coated onto the anode substrate by slurry coating and co-fired. These techniques allow us to significantly reduce the cost of SOFCs. An $NiO/(Ce_{0.9}Gd_{0.1})O_{1.95}$ (NiO/GDC10) anode interlayer was inserted between the substrate and the electrolyte to enhance the anode performance at lower temperatures. $La_{0.8}Sr_{0.2}Ga_{0.8}Mg_{0.2}O_{2.8}$ (LSGM) was selected for the electrolyte material, which is being considered as a promising material for the intermediate-temperature operating SOFC [4–5]. For the cathode material, $(La_{0.6}Sr_{0.4})(Co_{0.8}Fe_{0.2})O_3$(LSCF) was coated on the LSGM layer by slurry coating and then fired. An important point in cell fabrication is the insertion of a dense $(Ce_{0.6}La_{0.4})O_{1.8}$ (LDC40) layer between the LSGM layer and NiO/GDC10 layer to prevent undesirable nickel diffusion from the anode to the LSGM during the co-firing procedure [6]. However, LDC40 is a material of extremely low sintering property, so that it is difficult to densify the

LDC40 layer. Therefore, we investigated the various additives for the sintering promotion of LDC40 powder. As a result, it was found that the addition of a small amount of Ga_2O_3 to LDC40 was effective in obtaining a fully dense LDC40 layer below a sintering temperature of 1300°C (Fig. 10.2), and it also improved electrical conductivity of the LDC40 itself.

Table 10.2 Materials and fabrication process of micro-tubular cell

Components	Material	Fabrication	Firing
Anode tube	NiO/YSZ	Extrude molding	
Anode interlayer	NiO/GDC10		Co-firing
Electrolyte	LDC40(Ga_2O_3)-LSGM (double layered)	Slurry coating	
Cathode	LSCF		Firing

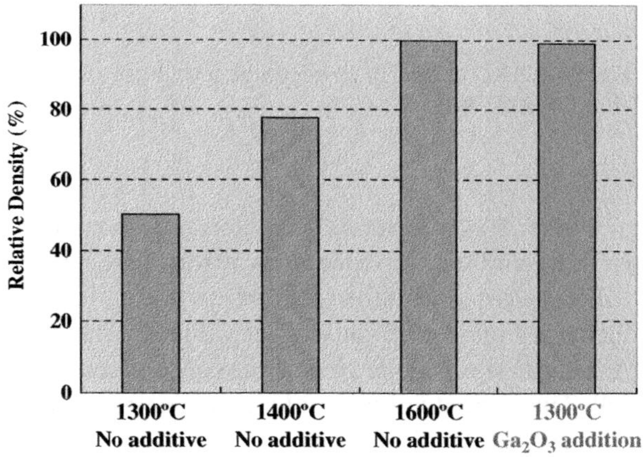

Fig. 10.2 Effect of small addition of Ga_2O_3 to LDC40

According to electron probe micro-analyzer (EPMA) observations of the cross sections of micro-tubular cells, it was seen that the reaction between the LSGM layer and the anode was avoided by the introduction of the LDC40 layer, and there was no obvious element diffusion between the LSGM layer and the LDC40 layer. Furthermore, it was found that LDC40 is a preferable material as the insertion layer between the LSGM layer and the anode because it is not very reactive with LSGM. Shown in Fig. 10.3 are the crystal phases identified by the X-ray diffraction (XRD) method after heat-treating the mixture at 1400°C, which is obtained by uniformly mixing equivalent weights of the LSGM powder and each of the cerium-containing oxide powders. In the case of LDC40, no other phase was observed in a matrix, whereas in the case of GDC10, a large amount of reaction generating phase $SrLaGa_3O_7$ was observed. The reaction-generating phases of $SrLaGa_3O_7$ are electrically highly resistant, which suggests that when these reaction-generating phases are developed between the LSGM layer and the cerium-containing layer, the electric power-generating capacity is lowered.

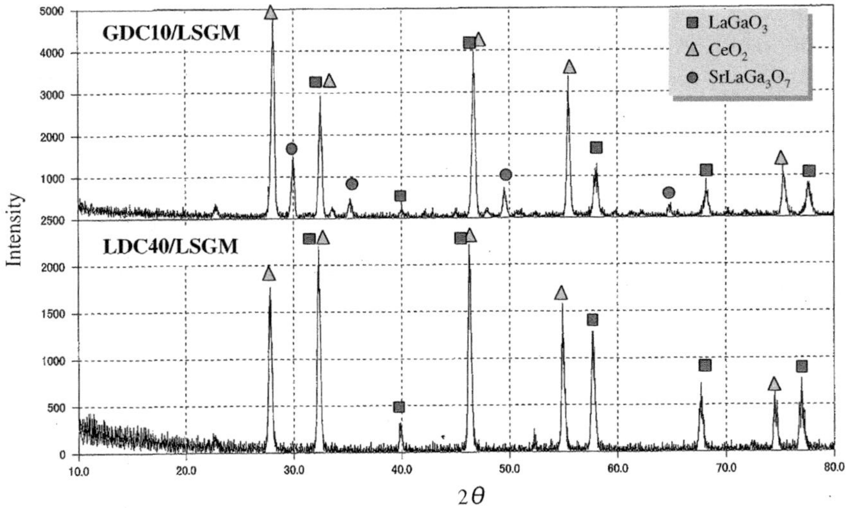

Fig. 10.3 XRD pattern of powder mixture consisting of LSGM and cerium-containing oxide after firing at 1400°C

Figure 10.4 is a picture of a micro-tubular single cell. The diameter of the cell is 5 mm, and the active length is 50 mm. The single cell was joined to the current collector cap with silver-brazed metal, and its performance was tested in a furnace. Figure 10.5 shows the evaluation method for single-cell performance. A fuel gas was supplied inside the cell, and air was supplied to the outside of the cell. The current voltage and impedance of the single cells were measured using a potentiostat and a frequency response analyzer in the temperature range from 500° to 700°C.

Fig. 10.4 TOTO micro-tubular cell

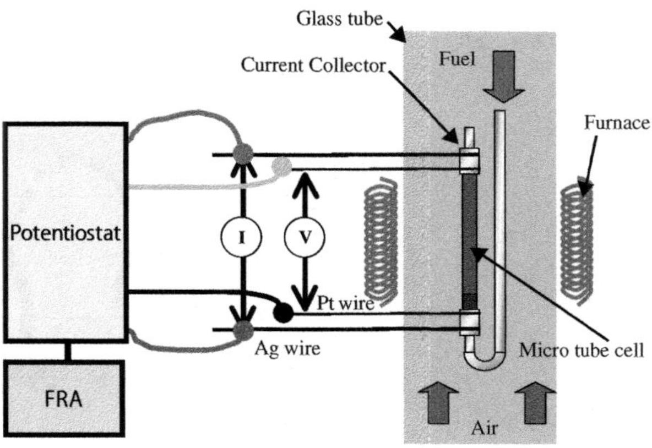

Fig. 10.5 Evaluation method for single-cell performance

Figure 10.6 shows the typical I-V curves of a micro-tubular single cell using dry H_2 in N_2 as fuel. H_2 flow was fixed at 0.12 L/min. The open circuit voltage (OCV) was close to the theoretical value. The results indicated that the electrolyte has a good gas tightness, and the chemical reaction between LSGM and Ni is effectively avoided by the layer of LDC40 with Ga_2O_3. (In the case of LDC40 without Ga_2O_3, the OCV was 0.85 V.) We obtained the maximum power densities of 0.85, 0.70, and 0.24 W/cm^2 at 700°, 600°, and 500°C, respectively. Figure 10.7 shows the impedance spectra of a micro-tubular cell measured under 0.125 A/cm^2 at various temperatures. It has

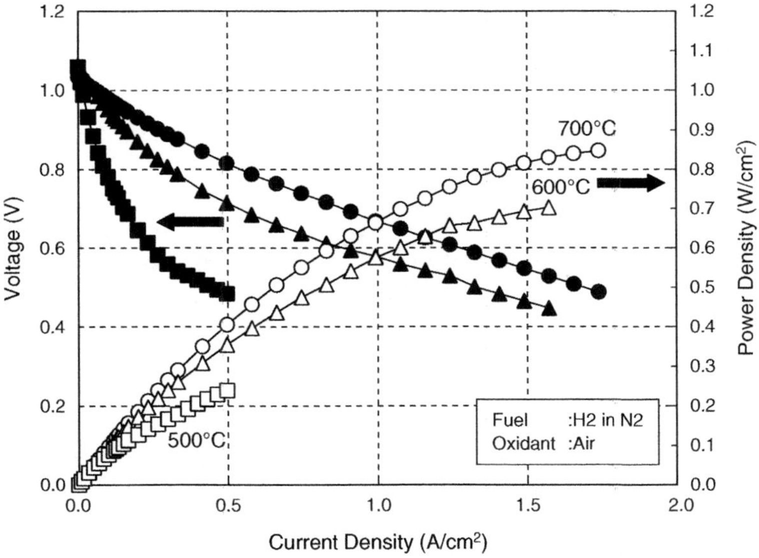

Fig. 10.6 I–V characteristic of micro-tubular single cell

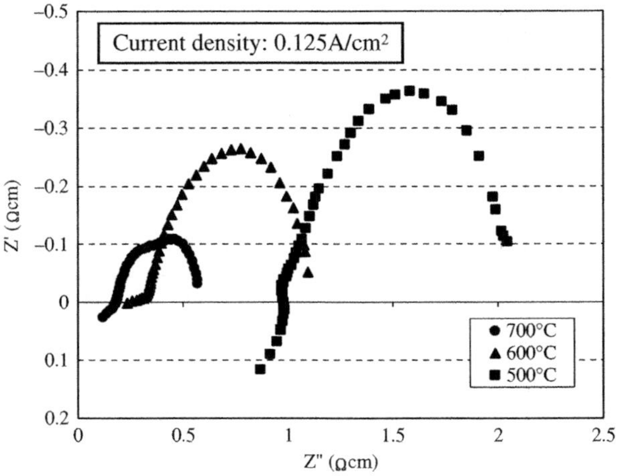

Fig. 10.7 Impedance of micro-tubular cell

been generally assumed that the intercept with the real axis at the highest frequency represents the ohmic resistance, and the width of the low frequency arc represents the electrode resistance. The electrode resistance increased significantly with decreasing operation temperature, and ohmic resistance at 500°C was very high. The most likely cause of the high resistance is the low ionic conductivity of LDC40. Therefore, it is expected that the cell performance can be improved by optimizing the anode electrode and the thickness of the LDC40 layer. Figure 10.8 shows the fuel utilization effects on micro-tubular cell performance measured under 0.125 A/cm^2 at 700° and 600°C. The observed cell voltage was close to the theoretical value calculated by the Nernst equation, indicating that the micro-tubular cell can be operated at a high efficiency.

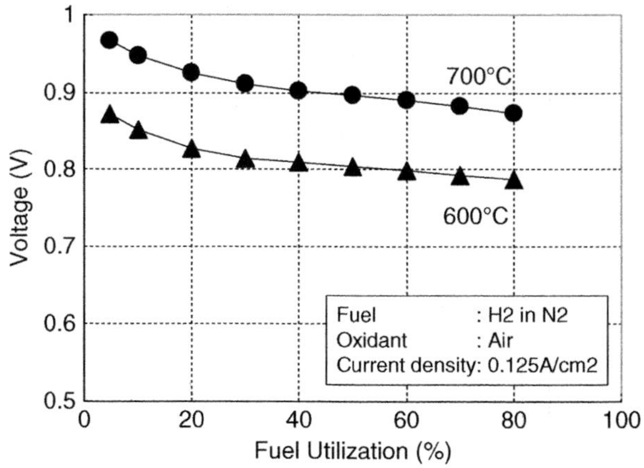

Fig. 10.8 Fuel utilization effect on micro-tubular cell performance

10.3 Rapid Thermal Cycling

The thermal cycling performance of the micro-tubular single cell was evaluated by directly inserting and removing the cell from a high-temperature furnace (530°C) for 1,000 times. The cell was heated from room temperature to 500°C within 5 min, and the OCV was generated over 1.0 V within 2 min (Fig. 10.9). The 1,000 thermal shocks were completed successfully with no sign of delamination or crack formation, which indicates that there is a strong possibility of being able to develop a micro-tubular cell capable of performing quick start-up and shut-down. In this test, the voltage degradation rate at 0.125 A/cm^2 was 1.89% for each 100 thermal cycles (Fig. 10.10).

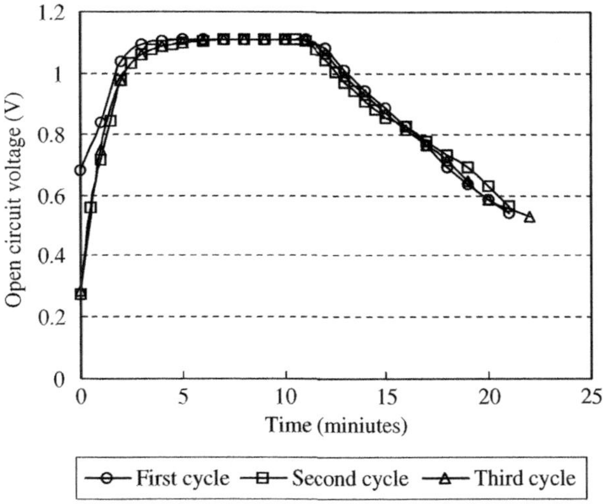

Fig. 10.9 Start-up performance of micro-tubular cell

10.4 Fuel Flexibility

One of the significant advantages of SOFCs is that they are compatible with many different kinds of fuels, in contrast to other fuel cells. The single cells were evaluated using practical fuels such as LPG and DME that can be easily handled and stored in portable cartridges.

Cell performance using H_2 in N_2 (1:1) gas mixture or simulated reformate gas is compared in Fig. 10.11. The composition of simulated reformate was 32% H_2, 13% CO, 5% CO_2, and 50% N_2, based on the preliminary experiment of the catalytic partial oxidation (CPOX) reforming of LPG. As shown in the figure, the differences in cell performance were smaller at lower current densities. However, performance using reformate gas was lower at higher current densities at temperatures of 600° and 700°C. The

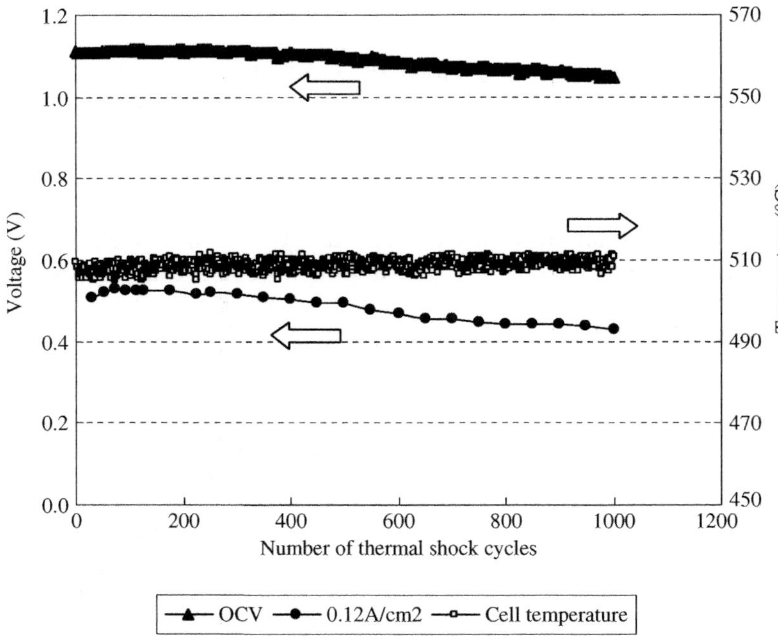

Fig. 10.10 Thermal shock cycling test: 1000 cycles

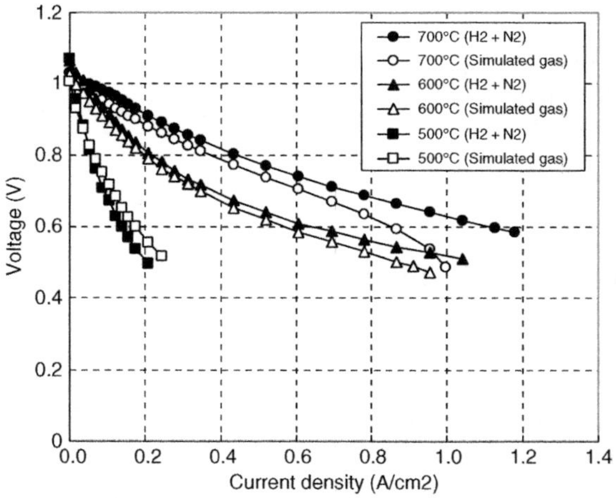

Fig. 10.11 I–V curve tested on H_2 in N_2 and simulated reformate gas

differences became significant with increasing operating temperatures. To identify the differences, the impedance spectra were measured under a current density of 0.8 A/cm^2 at 700°C (Fig. 10.12). The electrode resistance on simulated reformate was higher than that on H_2 in N_2, and it was

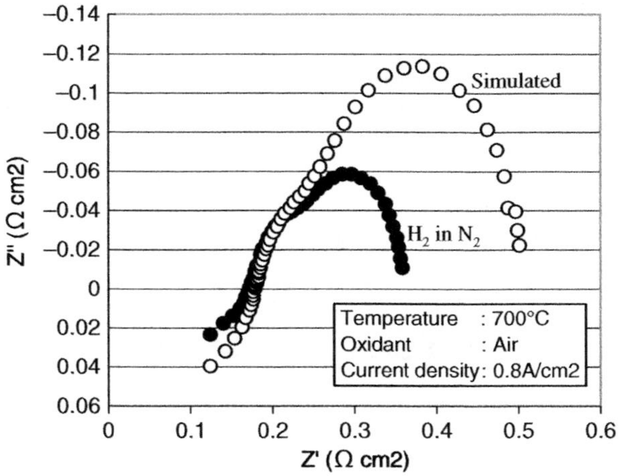

Fig. 10.12 Impedance tested on H_2 and N_2 and simulated reformate gas

thought that this difference was caused by the CO transport resistance from the anode in the high current density area. Therefore, we are now trying to improve the anode performance.

Figure 10.13 shows cell performance using DME + air mixture as fuel at 550°C. The DME flow rate was fixed at 85 mL/min, and the excess air ratios (air–fuel ratio/theoretical air–fuel ratio) were 0.1, 0.2, 0.3, and 0.4, respectively. Direct use of fuel without a reformer in SOFCs will greatly simplify the system, and this is important for SOFCs, especially in portable and transportation applications. DME is an attractive fuel because it is highly active and easily liquefied and stored. As shown in the figure, the use of DME + air as fuel resulted in higher performance than that of H_2 in N_2, and also no carbon deposition was observed during operation. Figure 10.14 shows the DME

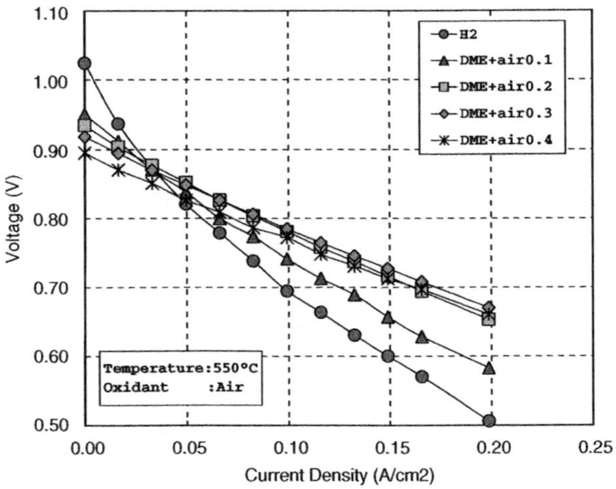

Fig. 10.13 I-V curve tested on DME + air mixture as fuel at 550°C

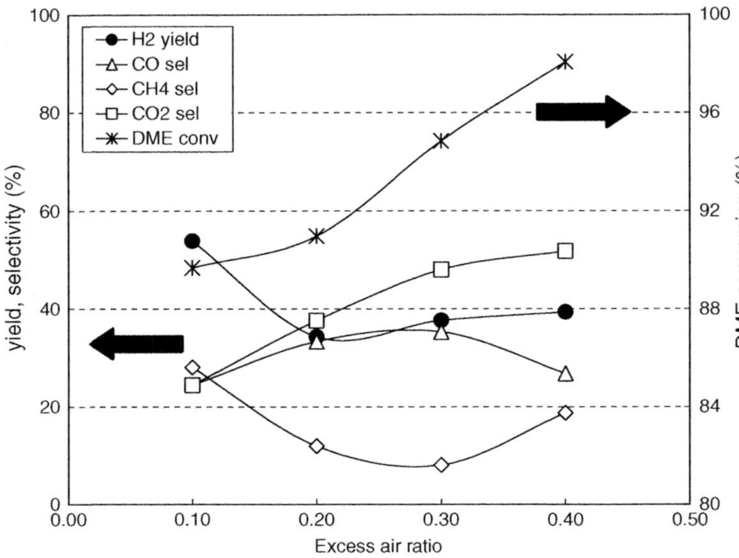

Fig. 10.14 DME conversion rate and exhaust gas composition tested on DME + air mixture. as fuel

conversion rate and the exhaust gas composition analyzed by gas chromatography. (The water content was not measured.) The DME conversion rate and CO_2 content increased with an excess air ratio. Therefore, the increased performance achieved by using the DME + air mixture is probably the result of the increased cell-surface temperatures caused by the decomposition and combustion of DME. (The furnace temperature was kept at 550°C.) These results demonstrated the high possibility of micro-tubular cells being used for direct fueled operations.

10.5 Stack Development

To evaluate the performance of cells in a bundle, we built a stack consisting of 14 micro-tubular cells (Fig. 10.15). These 14 micro-tubular cells were arranged in 7 parallel and 2 serial rows. This stack was evaluated in a furnace using hydrogen as a fuel, and it successfully demonstrated 43 W, 37 W and 28 W power generation at temperatures of 700°, 600°, and 500°C, respectively (Fig. 10.16). Table 10.3 summarizes the results we have obtained from the stack evaluation. The maximum stack power densities were 478 W/L and 239 W/kg at 700°C. These results demonstrated that micro-tubular SOFCs have a high potential for use as compact portable fuel cell generators.

Fig. 10.15 Appearance of micro-tubular SOFC stack

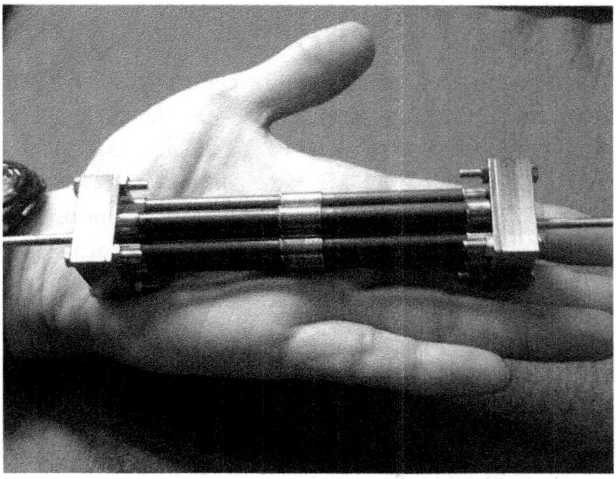

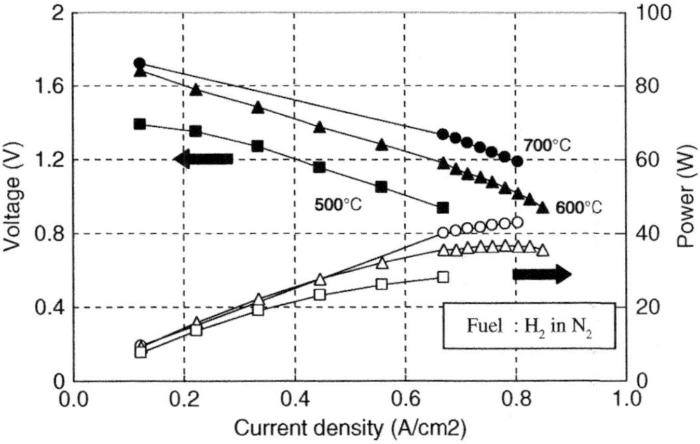

Fig. 10.16 Performance of micro-tubular SOFC stack

Table 10.3 Performance of micro-tubular SOFC stack at 700°C

Maximum power (W)	43
Stack volume (liters, L)	0.09
Stack weight (kg)	0.18
Stack power density (W/L)	478
Stack power density (W/kg)	239

10.6 Summary

Anode-supported micro-tubular cells with a thin film of lanthanum gallate with strontium and magnesium doping were developed. The single cells and the cell stack were tested using various fuels, such as hydrogen, simulated reformate gas of LPG, and direct fueling of DME. Excellent performance was shown at lower temperatures ranging from 500° to 700°C. These results demonstrated that micro-tubular SOFCs have a strong potential for portable and transportation applications. Further research into durability, quick start-up, and compactness of the stack is underway.

TOTO has been promoting the use of tubular cells and stacks for stationary applications since 2004, and that of micro-tubular cells for portable applications since 2006, to several system integrators. We intend to become a leading supplier of SOFCs for stationary and portable power generators providing high-performance and low-cost SOFC cells and stacks to power system integrators. One of the system integrators to which TOTO started promoting micro-tubular cells is Protonex Technology Corporation. Their generators are based on sub-kilowatt SOFC and are designed to operate on practical fuels, including propane, LPG, and kerosene. These generators are more compact, quiet, and have higher efficiency than conventional internal combustion engine generators. Also, they also run longer than batteries, making them ideal for remote power needs for commercial and industrial applications [7].

Acknowledgments The development was supported by NEDO in Japan.

References

1. T. Saito, T. Abe, K. Fujinaga, M. Miyao, M. Kuroishi, K. Hiwatashi, A. Ueno, Development of Tubular SOFC at TOTO. SOFC-IX, Electrochemical Society Proceedings Vol. 2005-07, 133–140 (2005)
2. J. Van Herle, J. Sfeir, R. Ihringer, N. M. Sammes, G. Tompsett, K. Kendall, K. Yamada, C. Wen, M. Ihara, T. Kawada, J. Mizusaki, Improved Tubular SOFC for Quick Thermal Cycling. Proceedings of the Fourth European Solid Oxide Fuel Cell Forum, 251–260 (2000)
3. A. Kawakami, S. Matsuoka, N. Watanabe, A. Ueno, T. Ishihara, N. Sakai, K. Yamaji, H. Yokokawa, Development of low-temperature micro tubular type SOFC. Proceedings of the 13th Symposium on Solid Oxide Fuel Cells in Japan, 54–57 (2004)
4. T. Ishihara, H. Matsuda, Y. Takita, Doped $LaGaO_3$ perovskite type oxide as a new oxide ionic conductor. J. Am. Chem. Soc 116, 3801–3803 (1994)
5. M. Feng, J.B. Goodenough, A superior oxide-ion electrolyte. Eur. J. Solid State Inorg. Chem. 31, 663–672 (1994)
6. K. Huang, J.H Wan, J.B. Goodenough, Increasing power density of LSGM-based solid oxide fuel cells using new anode materials. J. Electrochem. Soc 148(7), A788–A794 (2001)
7. C. Martin, J. Martin, Portable 250 W solid oxide fuel cell battery charger operating on liquid fuels. 2006 Fuel Cell Seminar 96 (2006)

Chapter 11
Proton Conductivity in Perovskite Oxides

Truls Norby

11.1 Introduction

Perovskite oxides offer in almost all respects a wide variety of properties because of the structure's ability to host varying cations, substitutions, non-stoichiometry, and defects of many kinds. Proton conduction, resulting from oxide ability to dissolve protons from water vapor or hydrogen, is no exception: Some perovskites contain virtually no protons at all and are thus barriers to protons, hydrogen, and water vapor. Other perovskites are dominated by proton defects to high temperatures and are predominantly proton conductors.

The first significant report of proton conduction in perovskites refers to $LaYO_3$ and $SrZrO_3$.[1] Throughout the 1980s, Iwahara and coworkers revealed higher proton conduction in a number of II-IV perovskites, such as $SrCeO_3$ [2] and $BaCeO_3$.[3] These findings were soon accompanied by results on proton conduction for I-V perovskites such as $KTaO_3$ [4] and, later on, complex perovskites such as $Sr_2(GaNb)O_6$ [5] and $Ba_3CaNb_2O_9$.[6]. In all these cases, the oxide was acceptor doped to facilitate charge compensation by protons dissolved from water vapor. Throughout the 1990s and 2000s, proton solubility and conductivity have been identified in a large number of perovskite-related oxides, as we shall see.

The proton conductivity can—if relatively pure—be used in galvanic sensors for hydrogen or water vapor activity, usually at elevated temperatures. These sensors utilize hydrogen activity gradients between the tested environment and a reference electrode. An example is In-doped $CaZrO_3$ used for probing hydrogen in molten aluminium [7].

In many perovskites, the proton conductivity is high enough to be utilized in high-drain applications, such as fuel cells, steam electrolyzers, hydrogen pumps, and various hydrogenation/dehydrogenation electrochemical reactors [8].

T. Norby (✉)
Department of Chemistry, Centre for Materials Science and Nanotechnology, University of Oslo, FERMiO, Gaustadalléen 21, NO-0349 Oslo, Norway
e-mail: t.e.norby@kjemi.uio.no

T. Ishihara (ed.), *Perovskite Oxide for Solid Oxide Fuel Cells*, Fuel Cells and Hydrogen Energy, DOI 10.1007/978-0-387-77708-5_11,

A prominent example comprises acceptor-doped $BaCeO_3$, which has a bulk proton conductivity of the order of 0.01 S/cm in wet atmospheres in the temperature range 400°–1000°C. A variety of laboratory-scale fuel cells have been operated with this as electrolyte, with high performance in many cases. For instance, a thin electrolyte supported on dense Pd gave high power densities all the way down to 400°C with hydrogen as fuel [9].

Proton-conducting electrolytes provide interesting advantages over their oxide ion-conducting counterparts. With increasing load, the oxide ion-conducting fuel cell produces water vapor on the anode side, which lowers the cell voltage and threatens the stability of Ni in the anode cermet; this further reduces the fuel utilization and requires considerable fuel circulation. A proton-conducting fuel cell, on the other hand, produces water at the cathode. The high air flow usually used anyway takes care of this, while the anode gas remains undiluted by water vapor, keeping the Nernst voltage high. The fuel utilization can be high, and the anode is not in danger. While particularly beneficial for hydrogen as fuel, it is also advantageous for carbon-based fuels such as methane [10].

Perovskites and other oxidic proton conductors furthermore do not require water molecules as proton vehicles, and a major water management is thus not necessary, in contrast to low-temperature polymer electrolytes, for example.

If the material possesses mixed ambipolar conduction, it becomes permeable to a gas species and can be used as a gas separation membrane. This condition is well known and has been investigated for mixed oxide ion and electron conductors, which are oxygen permeable. However, mixed proton and electron conductors also exist: they are hydrogen permeable and can be used in hydrogen separation processes [11]. These conductors are potentially useful for separating hydrogen from syngas in fossil-fueled power plants with precombustion CO_2 capture, for hydrogen purification, etc. In comparison with hydrogen-permeable metals such as Pd, or microporous membranes, the dense mixed conducting ceramics have the potential advantages of higher structural integrity and longevity, high chemical stability (e.g., against evaporation), lower price, and, most importantly, thermal integration because of the higher operating temperature. Examples of perovskites demonstrated to perform hydrogen separation by mixed proton electron conduction include $SrCeO_3$ and $BaCeO_3$.

The major problem with ceramic proton conductors is that those that exhibit high proton conductivity, such as the ones we have exemplified so far, are rather basic in nature because Sr and Ba are main components. They are thus vulnerable to destructive reaction with acidic gases such as CO_2 or SO_2/SO_3, especially at moderate temperatures, and may also react with water at moderate and low temperatures to form hydroxides.

All in all, proton conductivity in oxides is a matter of compromise in composition and temperature between high concentration of protons (favorable hydration kinetics), high proton mobility, and chemical robustness. In this contribution, we concentrate on a description of protonic defects and their thermodynamics in various perovskite-related oxides, give an overview of the resulting proton conductivities, make a short trip to non-perovskites for comparison, and round off with a brief status on fuel cells with proton-conducting oxides.

11.2 Proton Conductivity in Acceptor-Doped Perovskites

11.2.1 Protons in Oxides

Hydrogen is a donor in oxides and will, under most conditions, be ionized to a proton. This proton will be located in the electron cloud of an oxide ion, such that the species is a hydroxide ion on the site of an oxide ion, being effectively positive and in Kröger–Vink notation written as $OH_O^{\bullet}$. The proton may diffuse by jumping to a neighboring oxide ion, and this process is considered to dominate over diffusion of the OH^- ion as such (which usually needs an oxygen vacancy). Because it is the free proton that moves, it is also relevant and acceptable to denote the protonic defect as an interstitial proton, $H_i^{\bullet}$, although the interstitial site is not a regular one.

11.2.2 Hydration of Acceptor-Doped Perovskites

Practical cases of proton conduction in perovskites almost exclusively involve acceptor-doped systems with charge-compensating oxygen vacancies in the dry state. The doping reaction for an acceptor Mf on the B site of a II–IV perovskite ABO_3 in dry atmospheres can in this case be written as follows:

$$\mathrm{Mf_2O_3(s) + 2AO(s) = 2A_A^x + 2Mf_B' + 5O_O^x + v_O^{\bullet\bullet}} \tag{11.1}$$

In the presence of water vapor, protons can be incorporated directly as charge-compensating defects during the synthesis according to this equation:

$$\mathrm{Mf_2O_3(s) + 2AO(s) + H_2O(g) = 2A_A^x + 2Mf_B' + 4O_O^x + 2OH_O^{\bullet}} \tag{11.2}$$

One may, from this, note that the solubility of acceptors increases with increasing p_{H_2O} when protons are dominating charge-compensating defects. However, this is difficult to utilize in practice, because synthesis, annealing, and sintering usually are done at higher temperatures, where protons are in the minority.

Oxygen vacancies and protons are in equilibrium through the important hydration reaction:

$$\mathrm{H_2O(g) + O_O^x + v_O^{\bullet\bullet} = 2OH_O^{\bullet}} \tag{11.3a}$$

$$K_3 = \exp\left(\frac{\Delta S_3^0}{R}\right)\exp\left(\frac{-\Delta H_3^0}{RT}\right) = \frac{[\mathrm{OH_O^{\bullet}}]^2}{[\mathrm{O_O^x}][\mathrm{v_O^{\bullet\bullet}}]p_{H_2O}} \tag{11.3b}$$

The expression requires that the concentrations are all given in the same units, so that the standard concentrations cancel, and that the partial pressure of water is given in bars. The electroneutrality condition in our case reads:

$$2[v_O^{\bullet\bullet}] + [OH_O^{\bullet}] = [Acc'] = \text{constant} \tag{11.4}$$

where Acc′ hereafter denotes acceptors in general. Note that an electroneutrality such as this comprises volume or molar concentrations, not site fractions. The foregoing electroneutrality has two limiting cases: if protons dominate (at low temperatures), $[OH_O^{\bullet}] \approx [Acc'] = \text{constant}$. The constancy arises from the acceptors being frozen in or being all dissolved so that varying solubility is not effective.

On the other hand, if protons are in the minority so that $2[v_O^{\bullet\bullet}] \approx [Acc'] = \text{constant}$, the proton concentration becomes:

$$[OH_O^{\bullet}] = ([O_O^x][v_O^{\bullet\bullet}]p_{H_2O}K_3)^{1/2} = [O_O^x]^{1/2}[Acc']^{1/2}p_{H_2O}^{1/2} \exp\frac{\Delta S_3^0}{2R} \exp\frac{-\Delta H_3^0}{2RT} \tag{11.5}$$

If one operates with small defect concentrations, we may assume that $[O_O^x]$ equals the concentration of oxide ion sites, and for molar concentrations this equals 3 in perovskites.

We can also obtain an analytical solution to the full electroneutrality. If we combine the equilibrium expression with the electroneutrality condition and assume $[O_O^x] = [O] = \text{constant} >> [v_O^{\bullet\bullet}] + [OH_O^{\bullet}]$, we obtain, for the concentration of protons:

$$[OH_O^{\bullet}] = \frac{[O]K_3 p_{H_2O}\left(-1 + \sqrt{1 + \frac{8[Acc']}{K_3 p_{H_2O}[O]}}\right)}{4} \tag{11.6a}$$

If we instead assume the more general $[O_O^x] + [v_O^{\bullet\bullet}] + [OH_O^{\bullet}] = [O] = \text{constant}$, we obtain:

$$[OH_O^{\bullet}] = \frac{[O]K_3 p_{H_2O}\left(-1 + \sqrt{1 - \frac{2[Acc']}{[O]} + \frac{[Acc']^2}{[O]^2} + \frac{8[Acc']}{K_3 p_{H_2O}[O]} - \frac{4[Acc']^2}{K_3 p_{H_2O}[O]^2}}\right)}{4 - K_3 p_{H_2O}} \tag{11.6b}$$

which is a reorganized version of the same equation given by Kreuer et al. [12] At moderate dopant and defect concentrations, the two solutions provide the same concentration for practical purposes. Note that the proton concentration comes out in molar fraction or volume concentration corresponding to the unit that is used for [O] and [Acc′]. In molar concentration, [O] = 3 for perovskites.

The concentration of oxygen vacancies is now obtained simply from the concentration of protons and the electroneutrality (Eq. (11.4)). The concentration of other (minority) defects can be found by linking to the vacancy concentration by appropriate defect-chemical reactions.

K_3 and its thermodynamic parameters are important in that they determine whether the material is primarily dominated by oxygen vacancies or by protons. This point can be investigated by studying the proton concentration versus temperature using, for example, IR spectroscopy, thermogravimetry, or conductivity, and this has been done for many perovskites. It turns out that the entropy change ΔS_3^0 ends up around -120 J/mol K, as expected empirically for the loss of 1 mole of gas, while the enthalpy change ΔH_3^0 varies widely. Some perovskites, such as $BaCeO_3$, have large negative values (exothermic) of more than -150 kJ/mol, and they are thus dominated by protons in wet atmospheres, and it takes a high temperature to shift the equilibrium to the left. Others, such as $SrTiO_3$, have moderate negative enthalpies and are dominated by protons only at relatively low temperatures. Finally, there are perovskites such as $LaGaO_3$ in which protons are never observed under any conditions and where modeling verifies that the enthalpy of hydration is actually positive [13].

In an attempt to find correlations between hydration thermodynamics and other materials properties, Norby et al. [14] noted that the best so far encompasses the difference in electronegativity between the B-site and A-site constituents of the perovskite. Figure 11.1 shows an update of this correlation plot. Although other correlations to electronegativity differences are in use [15], ours is yet not rationalized to any extent, and probably represents only a first or rough approximation, judged from the scatter. A linear regression of the data yields the following:

$$\Delta H_3(\mathrm{kJ/mol}) = -173(9) + 370(42)\Delta X_{B-A} \qquad (11.7)$$

We have recently tried to challenge the correlation and reduce the scatter that arises from the uncertain extraction of enthalpies from "equilibrium" measurements by measuring the enthalpy directly in calorimeters, notably combined DSC/TG instruments where the water exchange and associated enthalpy can be recorded simultaneously. It has so far turned out that combination instruments lack the isothermal stability to obtain significant results.

We are also examining hydration of some Pb-based perovskites such as $PbZrO_3$ where the electronegativity difference, the entry on the x-axis in Fig. 11.1, is negative. The correlation then predicts these perovskites to have very large negative hydration enthalpies and to be very strongly hydrated. The results so far, both by experiments and by density functional theory (DFT) simulations [16], suggest that the hydration is significant but moderate and that the correlation must be modified. It may be, for instance, that it is the absolute value of the electronegativity difference which must be applied.

Some perovskites with smaller band gaps have acceptor dopants compensated by electron holes rather than oxygen vacancies, especially at lower temperatures and, of course, high oxygen activities. In principle, this will depress the tendency of proton dominance; $BaPrO_3$ is predicted to have a large negative enthalpy of hydration of oxygen vacancies (Eq. 11.3), but the dominance of

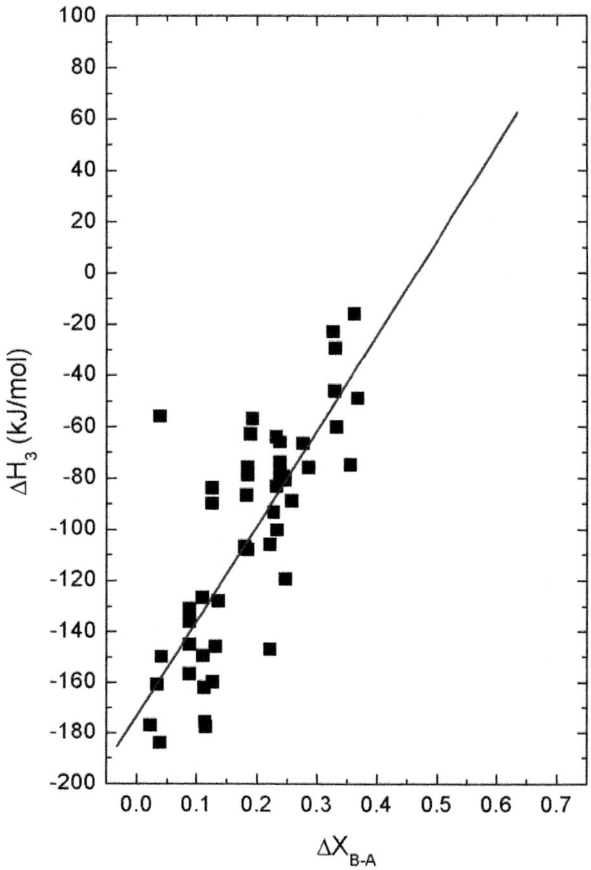

Fig. 11.1 Hydration enthalpy vs. difference in Rochow–Allred electronegativities between B- and A-site constituents in perovskites. The linear regression gives ΔH_3(kJ/mol) = $-173(9) + 370(42)\,\Delta X_{B-A}$

holes suppresses the protons [17]. Put differently, the hydration of the electron holes, which can be written:

$$H_2O(g) + 2O_O^x + 2\,h^{\bullet} = 2OH_O^{\bullet} + \tfrac{1}{2}O_2(g) \tag{11.8}$$

is less favorable than the hydration of oxygen vacancies.

11.2.3 Proton Diffusion

The protons, always residing on oxide host ions, exhibit thermal rotational and stretching vibrations. They may be located in a variety of local energy minima, depending on which neighboring oxide ion(s) they are directed toward. Depending on the O–O distance, the protons may set up hydrogen bonds, OH–O, between the two oxide ions, decrease their distance somewhat, and affect the structure slightly.

The protons rotate around their oxide ion hosts. The activation barrier for rotational diffusion is generally low so that these rotations are easy [18], but they lead to no long-range proton migration. The stretching vibrations, on the other hand, may lead to a jump to the next oxide ion. The diffusivity of protons may be expressed:

$$D_{H^+} = (1 - x_{H^+})aZs_{OH-O}^2\nu_{H^+} \exp\frac{\Delta S_{m,H^+}}{k} \exp\frac{-\Delta H_{m,H^+}}{kT}$$
$$= (1 - x_{H^+})D_{0,H^+} \exp\frac{-\Delta H_{m,H^+}}{kT} \quad (11.9)$$

where $x_{H^+}$is the fraction of protons over oxide ion sites so that $1 - x_{H^+}$ is the probability that the target of the jump is free to accept a proton (assuming that all sites are occupied by an oxide ion with or without one proton), a is a factor arising from geometry, Z is the number of neighbors, s_{OH-O} is the jump distance (effectively the O–O distance as long as the rotation is tacitly assumed part of the jump), ν_{H^+} is the effective attempt vibrational frequency, and $\Delta S_{m,H^+}$ and $\Delta H_{m,H^+}$ are the entropy and enthalpy of activation of migration, respectively. We may normally take x_{H^+} to be small and thus $(1 - x_{H^+}) \approx 1$.

In normal three-dimensional lattice diffusion (for non-protonic species), $a = 1/6$, and the vibrational frequency is defined such that $aZ\nu_{H^+}$ is the effective frequency of attempts in any direction. For protons in oxides, it is more reasonable to let $a = 1$, $Z = 1$, and $\nu_{H^+} = 10^{14}\ \mathrm{s}^{-1}$. If $s_{OH-O} = 2.81$ Å, and we additionally let $\Delta S_{m,H^+} = 0$, then we get, for the pre-exponential of proton diffusivity, $D_{0,H^+} = s_{OH-O}^2\nu_{H^+} = 7.8 \times 10^{-6}\ \mathrm{m^2/s} = 7.8 \times 10^{-2}\ \mathrm{cm^2/s}$, but, as we shall see below, this is not obtained in practice.

Proton transport exhibits a large isotope effect between protons and deuterons. This is often of the order of $\sqrt{2} \sim 1.4$ and is thus attributed to the classical ratio of attempt vibrational frequencies based on the reduced mass of OH versus OD. However, it is now fairly well accepted that it is more reasonable to attribute it to the semiclassical difference in zero-point energy of the proton versus the deuteron, that gives the proton an activation energy of the final jump (when the oxide ions are close during their vibrations) which is 0.04–0.06 eV lower than for deuterons. Thus, while most of the activation energy is assigned to the O–O vibrations, a small part is still attributed to the proton jump, and this latter part has a 0.04–0.06 eV lower energy for protons than for deuterons. This difference happens to give a ratio of 1.4 or slightly higher at temperatures where proton conduction is typically investigated (a few hundred degrees centigrade). The pre-exponential of proton migration may still hold the classical isotope effect, but probably has also other isotope effects (such as the so-called sticking probability) that may actually cancel the classical one or even reverse it [19]. The pre-exponential of proton migration should, in addition to the effect of the low sticking probability, be lowered by the frequency of oxide ion vibrations. For simplicity, one might think of the oxide ion vibration frequency of around 10^{13} Hz as the first estimate entering in the mobility of protons rather than the proton's own 10^{14} Hz. When mobility is extracted from proton conductivities, it usually comes out a factor 10–100 lower than estimated from classical proton jumps, in accordance with the above. Thus, it is realistic to expect D_{0,H^+} to be in the range 10^{-7}–10^{-6} m^2/s or 10^{-3}–10^{-2} cm^2/s.

The activation energy for a proton to jump to the next oxide ion is dependent on the O–O distance, because the electron density decreases and potential increases in between. The activation energy is thus high in unpolarizable, stiff lattices, whereas it decreases considerably and temporarily during O–O vibrations in softer lattices. These vibrations comprise the diminishment of the O–O distance but also the linearization of the bent OH–O hydrogen bond required to facilitate the proton transfer [18]. For this reason, the activation energy of proton migration is well above 1 eV in close-packed lattices such as Al_2O_3, is intermediate at 0.7–1 eV in lattices with larger cations, such as rare earth sesquioxides, and usually as low as 0.5 ± 0.1 eV in perovskites.

It is interesting to note that the proton activation energy, seemingly regardless of class of oxide, is usually around two-thirds of the activation energy for oxygen vacancy migration. This point may be understood by the fact that the proton is totally reliant on the oxide lattice vibrations. The same activation barrier has to be overcome, but while the oxide ion must make its way all the way through the barrier saddle point to arrive in the vacancy, the proton can jump or tunnel when the oxide ion on which it sits has made it most of the way (two-thirds of the barrier height).

The low activation enthalpy of proton mobility in perovskites may be attributed to the large A-site cations, which allow considerable dynamics in and between the BO_6 octahedra, and to the fact that the octahedra are corner sharing so that jumps can take place within the octahedra as soon as the proton has rotated into the appropriate direction from the previous jump. The low activation energy of migration, combined with the low defect energies (acceptance of large concentrations of dopants and charge-compensating defects), is what makes the perovskites such good proton conductors.

It appears that cubic perovskites exhibit higher diffusivities of protons than less symmetrical lattices. This difference is most easily rationalized by the oxide ion sites being equivalent, so that there are no sites that act as traps by requiring a higher activation energy for the liberating jump.

11.2.4 Charge Mobility and Conductivity of Protons

The charge mobility of protons is given via the Nernst–Einstein equation, as follows:

$$u_{H^+} = \frac{e}{kT} D_{H^+} = \frac{e}{kT} D_{0,H^+} \exp\frac{-\Delta H_{m,H^+}}{kT} = u_{0,H^+}\frac{1}{T}\exp\frac{-\Delta H_{m,H^+}}{kT} \quad (11.10)$$

where we have again assumed $(1 - x_{H^+}) \approx 1$. Ideally, the pre-exponential $u_{0,H^+} = \frac{e}{k} D_{0,H^+}$ is approximately 0.1 m^2 K/Vs = 1000 cm^2 K/Vs for proton migration, but the real attempt frequency and sticking probability lower this to 0.001–0.01 m^2 K/Vs = 10–100 cm^2 K/Vs.

The conductivity σ_i of any species i is given by its charge $z_i e$, volume concentration of charge carrier particles c_i, and mobility u_i:

$$\sigma_i = z_i e c_i u_i \tag{11.11a}$$

If the concentration is, instead, expressed as volume concentration of moles of charge carriers $c_{m,i}$, the conductivity becomes:

$$\sigma_i = z_i F c_{m,i} u_i \tag{11.11b}$$

For protons, $z = 1$, and it is convenient to obtain the conductivity in terms of the fraction of protons over oxide ion sitesx_{H^+} and the molar concentration of oxide ion sites $c_{m,O^{2-}}$:

$$\sigma_{H^+} = F x_{H^+} c_{m,O^{2-}} u_{H^+} \tag{11.12a}$$

or, in terms of the molar fraction of protons x_{m,H^+} and the molar concentration of the oxide c_m:

$$\sigma_{H^+} = F x_{m,H^+} c_m u_{H^+} \tag{11.12b}$$

In regions where the proton concentration is given by the acceptor dopants, $x_{m,H^+} = x_{m,Acc'} =$ constant and the activation energy of proton conduction is given by that of the proton mobility. In regions where the protons are minor defects, and oxygen vacancies instead compensate the acceptors, the concentration term has a temperature dependency given by $\Delta H_3^0/2$ so that the overall activation enthalpy of proton conduction is $\Delta H_{\sigma_{H^+}} = \Delta H_{m,H^+} + \Delta H_3^0/2$. As ΔH_3^0 is negative for proton-conducting oxides, and most often $\Delta H_3^0/2$ is larger in magnitude than $\Delta H_{m,H^+}$, the conductivity decreases with increasing temperature at high temperatures where the protons have become a minority. All in all, the proton conductivity typically goes through a maximum as a function of temperature.

11.2.5 Proton Conductivity in Acceptor-Doped Simple Perovskites, ABO_3

Proton conductivity is found in a range of perovskites, ABO_3, covering all combinations of valence of the A and B cations. The protonic defects have to be compensated by some effectively negative defect, most often acceptor substituents. The next prerequisite seems to be, as discussed earlier, that the electronegativity difference between B and A is not too large to favor hydration of the oxygen vacancies.

Figure 11.2 shows the partial proton conductivity for a number of acceptor-doped perovskites, calculated from data for proton mobility and thermodynamics of hydration.

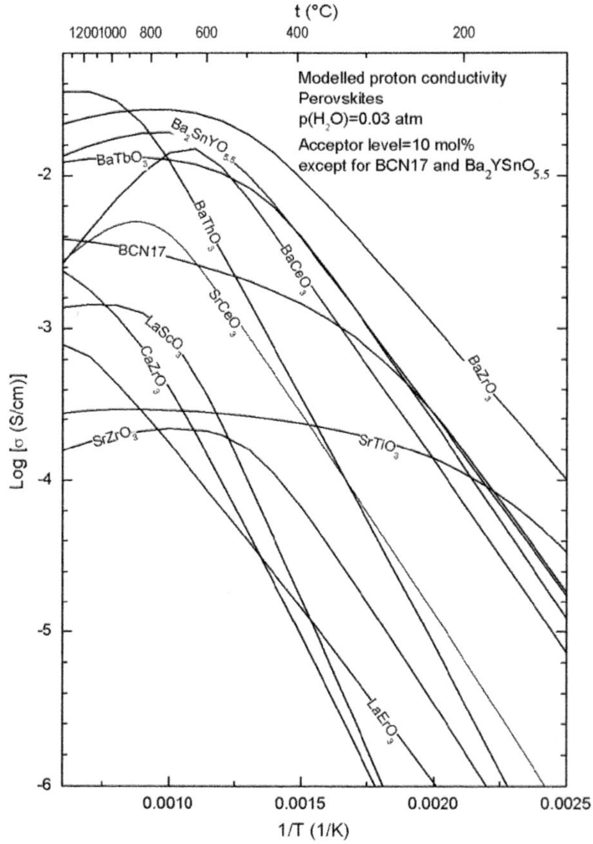

Fig. 11.2 Partial proton conductivity vs. 1/T for bulk (grain interior) of selected perovskites at pH_2O = 0.03 atm, calculated on basis of proton mobility and hydration thermodynamics parameters. Acceptor level is chosen as 10 mol% (3.33% of oxide ions are protonated at full hydration) for $BaZrO_3$ [20], $BaCeO_3$ [12], $BaTbO_3$ [21], $BaThO_3$ [21], $SrCeO_3$ [22], $SrTiO_3$ [20, 23], $SrZrO_3$ [20], $CaZrO_3$ [24, 25], $LaScO_3$ [26], and $LaErO_3$ [27]. Data for BCN17 ($Ba_3Ca_{1.17}Nb_{1.83}O_{8.745}$) [28] and $Ba_2YSnO_{5.5}$ [29] are calculated with acceptor levels as in formula. Curves should represent reasonable estimates of partial proton conductivity in temperature range of measurement, but may otherwise contain large errors caused by correlations between the parameters used

Among III–III perovskites, La^{3+} is the only common A-site cation. $LaScO_3$ hydrates well but exhibits medium proton mobility and a proton conductivity peaking just above 10^{-3} S/cm. $LaYO_3$ and $LaErO_3$ are also perovskite related, but with lower tolerance factors and lower conductivity.

$LaCrO_3$, $LaMnO_3$, $LaFeO_3$, and $LaCoO_3$ have increasingly acidic B-site cations and decreasing band gaps, and reliable findings of protons or proton conductivity are not reported for these oxides. For $LaMnO_3$ and $LaFeO_3$ [30], we have ourselves actively looked for evidence of effects of protons but found none.

Sr- and Mg-substituted $LaGaO_3$ (LSGM) is another member where no protons are found [31]. Lattice modeling of this material has indeed predicted the hydration enthalpy to be positive (endothermic) [32].

Among II–IV perovskites we find the best proton conductors. $BaZrO_3$, $BaCeO_3$, $BaTbO_3$, and $BaThO_3$ all exhibit large negative hydration enthalpies and small activation energies for proton migration. The proton conductivity peaks above 10^{-2} S/cm. They have large grain boundary resistances, especially for $BaZrO_3$, which has the largest band gap. It seems that the grain boundary resistance decreases with decreasing band gaps and increasing partial electronic conduction, but it is unclear whether this reflects the proton transport itself or variations in electronic transport properties across the grain boundary.

$BaTiO_3$ does not hydrate much and exhibits only a small proton conductivity. Combining Ba with more acidic tetravalent cations is not known to lead to proton uptake or proton conductivity.

Moving to strontium-based perovskites, we find the same trend as for the barium members. However, the conductivity is smaller, peaking between 10^{-3} and 10^{-2} S/cm for the best ones ($SrZrO_3$ and $SrCeO_3$). $SrCeO_3$ is one of the best studied proton conductors, but the tolerance factor is low and the material is on the verge of decomposition into the binary oxides. It is therefore very vulnerable to reaction with CO_2, for example.

Also, a few calcium-based perovskites are proton conducting, notably $CaZrO_3$, which is used in hydrogen sensors for molten aluminium. It may be noted that $CaCeO_3$ does not exist. Both $CaTiO_3$ and $SrTiO_3$ hydrate, but only significantly at relatively low temperatures, and thus exhibit modest proton conductivities.

Along the same scheme as for Ba perovskites, combinations with acidic and/or ambivalent B-site cations, e.g., in $SrMnO_3$ or $SrFeO_3$, do not lead to hydration.

I–V perovskites such as $KTaO_3$ were early on found to exhibit proton conduction [4], but the conductivity was low, and fundamental properties such as hydration thermodynamics have not been investigated in detail.

One may consider that WO_3 and ReO_3 represent 0–VI perovskites. Hydration as such is not well described for WO_3, but it dissolves hydrogen that ionizes to protons and electrons and forms a bronze. It is interesting to note that the mineral bernalite, $Fe(OH)_3 \bullet nH_2O$, possesses a ReO_3-like perovskite structure in which Fe occupies the B sites, OH the "oxygen" sites, and water molecules a fraction (typically 25%) of the A sites.

Complex perovskites employ usually two B-site cations of different valence, typically in simple ratios to form certain valence sums. Many of these are used as electroceramics in which the B-site cation ordering is essential, but the valence sum is made up such that the oxygen sublattice is kept full. An example of this is $BaCa_{1/3}Nb_{2/3}O_3$ (BCN), often written as $Ba_3CaNb_2O_9$. By increasing the Ca content at the cost of Nb, charge-compensating oxygen vacancies are introduced, for instance, in the classical "BCN18" or $Ba_3Ca_{1.18}Nb_{1.82}O_{8.73}$. (Fig. 11.2 includes data for BCN17.) An example of another class is

$Sr_2Sc_{1+x}Nb_{1-x}O_{6-x}$ [33] and analogues [34]. Both the hydration thermodynamics and proton mobility and conductivity of such complex perovskites can be said to follow the systematics of the simple perovskites. It has been argued that the complex perovskites have the advantage that the doping can be done simply by shifting the ratio of the two B-site cations and not introducing another dopant ion. However, it is the present author's view that the extra cation is already there by having two B-site cations, and that little or nothing is gained.

The most interesting aspect of complex perovskites probably lies in the possibility to order the B-site cations that are present in a simple ratio, e.g., 1:2 or 1:1, so as to avoid carrier traps, and at the same time have a moderate oxygen deficiency (lower than the 1/6 of the Brownmillerite-type stoichiometries). One of these classes is represented by $Ba_2YSnO_{5.5}$ (or $Ba_4(Y_2Sn_2)O_{11}$). Here, 1 in 12 of the oxygen positions is vacant. The material hydrates, probably helped by the basicity of the Y^{3+} ion, and is a good proton conductor (see Fig. 11.2). Another class is exemplified by $Ba_4(Ca_2Nb_2)O_{11}$, the "BCN50" end member of Ca-enriched BCN. At sufficiently low temperature and high pH_2O they may go through a phase transformation into an ordered or disordered oxyhydroxide, as we shall come back to later.

Ruddlesden–Popper-phases, with general composition $A_{n+1}B_nO_{3n+1}$, are more basic than the normal perovskite end members ABO_3 and in principle more attractive to protons, but only modest proton conductivities have been found, notably in Sr_2TiO_4 [35].

11.2.6 Effects of Defect–Acceptor Interactions

Oxygen vacancies are often found to be effectively associated to acceptor dopants according to the following equilibrium:

$$v_O^{\bullet\bullet} + Acc' = (v_O Acc)^{\bullet} \tag{11.13}$$

The hydration of this associate, according to the following:

$$H_2O(g) + (v_O Acc)^{\bullet} + O_O^x = 2OH_O^{\bullet} + Acc' \tag{11.14}$$

can perhaps be expected to be less favorable (less exothermic) than that of the free vacancies, provided that the protons dissolved are less associated to the acceptor. Although this is probably an important factor in many systems, it is mainly neglected in interpretation of data for hydration. The defect chemistry and "master curves" of proton concentration (similar to the foregoing Eqs. (11.6a) and (11.6b)) can be derived for hydration of associated vacancies, but this is not further covered here.

Eventually, at sufficiently low temperatures, protons will themselves tend to be trapped by acceptors according to this equation:

$$OH_O^{\bullet} + Acc' = (OH_OAcc)^x \tag{11.15}$$

This reaction is not studied as well as the corresponding reaction for oxygen vacancies. However, Kreuer et al. [20] found that $BaZrO_3$ doped with Y had a somewhat lower activation energy and higher conductivity than with the same amount of Sc, and they attributed this to oxide ions coordinated to Sc becoming more electron rich (basic), thus bonding protons more strongly. The stronger trapping by Sc comes despite Sc^{3+} having a better size fit to the Zr^{4+} sites, emphasizing, according to Kreuer et al., that the acid–base character of the cations (including electronegativity) plays a large role for mobility as well as for hydration. The effect on the activation energy is nevertheless modest (0.43 eV for Y vs. 0.50 for Sc). Although it gives two orders of magnitude higher proton conductivity at 100°C in the work by Kreuer et al., the effect will be much reduced at operating temperatures.

All in all, it seems that there are effects of dopants, but trapping in associates such as those described by Eq. (11.11) are hardly evidenced experimentally in conductivity measurements so far.

11.2.7 Grain Boundaries

State-of-the-art proton-conducting perovskite oxides are troubled by remarkably high grain boundary impedances. This problem is most prominent for $BaZrO_3$, where for a long time the true bulk impedance was too small to be detected in impedance spectroscopy [36]. It is likely that the grain boundary impedance for protons has the same origin as that for oxygen vacancies in many oxygen ion-conducting oxides, such as Y-stabilized ZrO_2 (YSZ) and acceptor-doped CeO_2. The grain boundary core attracts an excess of oxygen vacancies because this minimizes the energy of the misfits between the two lattices. Possibly, protons act as terminators of "dangling bonds" in the grain boundary core and are also enriched there. The core thus attains a net positive charge, which is balanced by a net negative charge of the space charge layer (SCL) around the core [37]. This negative charge comes about by a deficiency in positive species (such as oxygen vacancies, protons, and holes) and a surplus of negative species (such as electrons and acceptor dopants).

Much work has been invested in understanding and reducing the grain boundary proton resistance, especially with regard to the effects of A-site nonstoichiometry and impurities. We shall not treat this here but just mention that, somewhat surprisingly, the grain boundary resistance disappeared in Y-substituted $BaZrO_3$ by long-term annealing at very high temperatures [38].

In some systems, n-type electronic conductivity under reducing conditions is enhanced relative to protonic over grain boundaries, which is expected from the effect of the space charge layer [39]. The result is that the effective protonic transport number for a DC application may be less in the grain boundaries and thus less overall than in the bulk; this may lead to electrochemical reactions at the edge of the space charge regions, where protons are transformed into neutral hydrogen by consuming an electron over the grain boundary.

11.3 Proton Conduction in Inherently Oxygen-Deficient Perovskites

11.3.1 Hydration of Ordered Oxygen Deficiency

In many perovskite-related oxides, the number of oxygen vacancies can be great enough to order; this is, for instance, the case in $Ba_2In_2O_5$ and $Ba_4(Ca_2Nb_2)O_{11}$. These oxides can also hydrate, but in this case the water reacts with vacant interstitial sites rather than with charged vacancies. Moreover, the resulting protons may prefer the interstitial sites or regular sites, so that a number of hydration reactions are possible:

$$H_2O(g) + v_i^x = (H_2O)_i^x \tag{11.16}$$

$$H_2O(g) + O_O^x + v_i^x = OH_i' + OH_O^{\bullet} \tag{11.17}$$

$$H_2O(g) + 2O_O^x + v_i^x = O_i'' + 2OH_O^{\bullet} \tag{11.18}$$

These hydration reactions are probably less favorable than the hydration of regular oxygen vacancies because the ordered vacancies are lower in energy than the free vacancy. At sufficient hydration, the protons appear to order, and the material becomes an oxyhydroxide, in the above cases $Ba_2In_2O_4(OH)_2$ and $Ba_4(Ca_2Nb_2)O_{10}(OH)_2$. The new ordered phase may form with or without step changes in thermogravimetry versus T or p_{H_2O}, for example, indicating a change in the structure or not, respectively. A valuable treatment of phase relationships and nonstoichiometry (water deficiency) in the $Sr_4(Sr_2Ta_2)O_{10}(OH)_2$–$Sr_4(Sr_2Ta_2)O_{10}(OH)_2$ system has been provided by Schober [40].

Also in this class there are large differences between different oxides. Although $Ba_2In_2O_5$ hydrates to the oxyhydroxide around 300°C in wet atmospheres, the analogous Brownmillerite $Sr_2Fe_2O_5$ does not have this tendency. It thus seems that we are again back at those with proton affinity being basic and thus chemically vulnerable.

11.3.2 Nomenclature and Hydration of Disordered Intrinsic Oxygen Deficiency

At sufficiently high temperatures, $Ba_2In_2O_5$, $Sr_2Fe_2O_5$, and $Ba_4(Ca_2Nb_2)O_{11}$ will disorder, so that free oxygen vacancies will be the ones to hydrate. But how is the defect chemistry done in this case? Let us take $Ba_2In_2O_5$ as an example: there are clearly oxygen vacancies around, but what are the charge-compensating defects? One might say that the material is $BaZrO_3$ doped 100% with In, or $LaInO_3$ doped 100% with Ba. This idea works but feels unsatisfying because no doping is involved. Consider instead that the disordered system is the perfect structure, so that each oxygen site is supposed to be filled with 5/6 oxide ions. We denote this a $\frac{5}{6}O$ site. It has, ideally, the charge of -2 5/6 = -5/3. This notation means that both the vacancy and the oxide ion constitute defects, with their real charges differing from the ideal charge of the partially occupied site. The electroneutrality condition and total site restriction then read:

$$\tfrac{1}{3}[O_{\frac{5}{3}O}^{\frac{1}{3}\prime}] = \tfrac{5}{3}[v_{\frac{5}{3}O}^{\frac{5}{3}\bullet}] \tag{11.19}$$

$$[O_{\frac{5}{3}O}^{\frac{1}{3}\prime}] = [v_{\frac{5}{3}O}^{\frac{5}{3}\bullet}] = 6 \tag{11.20}$$

As both species—the ion and the vacancy—are defective, they both carry effective charge and thus contribute to conductivity; this allows further interaction with other defects and elucidation of the behavior of minority defects. Hydration can be written:

$$H_2O(g) + O_{\frac{5}{3}O}^{\frac{1}{3}\prime} + v_{\frac{5}{3}O}^{\frac{5}{3}\bullet} = 2OH_{\frac{5}{3}O}^{\frac{2}{3}\bullet} \tag{11.21}$$

The entry in the electroneutrality, equilibrium constant, and solution with respect to, e.g., T or p_{H_2O} is done in the usual manner. One obtains results familiar from ordinary defect chemistry but without having to resort to "100%" doping or other even more incorrect approaches.

The same type of defect chemistry can be applied to other kinds of intrinsic disorder. Examples comprise the disordered polymorphs of δ-Bi_2O_3 ($\frac{3}{4}O$ sites), $Ba_4(Ca_2Nb_2)O_{11}$ ($\frac{11}{12}O$ sites), $Ca_{12}Al_{14}O_{33}$ ($\frac{1}{6}O$ sites), $LaNb_3O_9$ ($\frac{1}{3}La$ sites), $Ba_3La(PO_4)_3$ ($\frac{3}{4}Ba + \frac{1}{4}La$ sites), and $CsHSO_4$ ($\frac{1}{4}H$ sites).

11.3.3 Order–Disorder Reactions Involving Hydrated Inherently Oxygen-Deficient Perovskites (Oxyhydroxides)

There are, according to the foregoing discussion, several possibilities of ordered phases of hydrated inherently oxygen-deficient perovskites. The order refers to the vacant oxygen sites where water molecules, hydroxide ions, or oxide ion interstitials reside. However, also protons distributed randomly may order, and this seems indeed to happen when these materials are hydrated fully at relatively low temperatures; the protons are ordered and the proton conduction is accordingly small. The ordering of protons may be as water molecules in interstitial sites or as hydroxide ions in particular lattice sites.

Such ordered oxyhydroxides may form thermal defect pairs of proton vacancies and proton interstitials:

$$OH_{OH}^{x} + O_{O}^{x} = O_{OH}^{/} + OH_{O}^{\bullet} \quad (11.22)$$

Moreover, they will probably have a water deficiency by proton vacancies and hydroxide or oxide ion vacancies, e. g.:

$$2OH_{OH}^{x} = O_{OH}^{/} + v_{OH}^{\bullet} + H_2O(g) \quad (11.23)$$

At a sufficiently high temperature, provided the oxyhydroxide is still stable, the ordered protons may disorder fully. This is then a phase transformation into a phase of partial proton occupancy on oxide ions. In $Ba_4(Ca_2Nb_2)O_{10}(OH)_2$ we may have, for instance, $O + \frac{1}{6}H$ sites with real charge of $-11/6$ as "perfect." A proton is then $OH_{O+\frac{1}{6}H}^{\frac{5}{6}\bullet}$ and an oxide ion is $O_{O+\frac{1}{6}H}^{\frac{1}{6}/}$.

The proton conductivities of the fully hydrated $Sr_4Sr_2Nb_2O_{10}(OH)_2$ and $Sr_4Sr_2Ta_2O_{10}(OH)_2$ appear to have activation energies of about 100 kJ/mol and 70 kJ/mol, respectively. A number of other alkaline earth niobates and tantalates have similar proton conductivities and activation energies. The question remains whether some of these oxyhydroxides can be disordered. If they can, they would benefit from larger proton content and no acceptor–proton trapping as compared to ordinary acceptor-doped simple perovskites. The peaks in conductivity of BCN50 [41] and similar perovskites [42] upon heating the hydrated material (oxyhydroxide) might be taken to indicate that this may be possible. However, a recent conductivity study of $Sr_4(Sr_2Nb_2)O_{10}(OH)_2$ in 1 atm H_2O in our laboratory indicates that decomposition of such basic materials results in hydroxides such as $Sr(OH)_2$ and $Ba(OH)_2$ that melt at 400°–500°C. Their high ionic conductivities in the molten state should then not be mistaken for high solid-state proton conduction in the disordered perovskite-related oxyhydroxide.

11.4 Hydration of Undoped Perovskites

Let us return to the normal ABO_3 perovskites and, for academic and educational reasons, look briefly at hydration of undoped materials. Materials may contain nonstoichiometry in the A-to-B ratio as a result of inaccuracies in the synthesis or by evaporation loss during annealing or sintering. Also, some perovskites preferentially form cation vacancies on one site, normally the A site, dissolving the overshooting cations on the other site or precipitating or evaporating them. All these processes end up with a constant frozen level of cation vacancies, acting as acceptors and adding, often unnoticeably, to the level of deliberate dopants or impurities, compensated by oxygen vacancies, holes, or protons.

But what about the perfectly pure, undoped, and stoichiometric perovskite in equilibrium with water vapor? Higher charged defects such as oxygen vacancies benefit from higher doping levels, but those singly charged such as protons benefit at lower doping levels. It is thus possible to show that any oxide dominated by protons as positive defect-compensating acceptors will also be dominated by protons in the undoped case under the same conditions. Thus, all the acceptor-doped perovskites we know that are dominated by protons will also dissolve protons in the undoped case. The negative defect will then have to be oxygen interstitials, metal vacancies, or electrons. It is reasonable to believe that, for perovskites, metal vacancies will be the negative defect in nominally undoped samples. The hydration reaction for a II–IV-perovskite is then written:

$$3H_2O(g) + 3O_O^x = 6OH_O^{\bullet} + v_A^{//} + v_B^{////} \quad (11.24)$$

or, for vacancies preferentially on the A site, e.g.:

$$H_2O(g) + ABO_3 = 2OH_O^{\bullet} + v_A^{//} + B_B^x + O_O^x + AO(s) \quad (11.25)$$

Although such equilibria for undoped oxides are, as already said, necessarily a basis for dissolution of protons in the doped cases, their thermodynamics—a combination of hydration of oxygen vacancies and Schottky equilibrium—is little studied.

11.5 Proton Conductivity in Selected Classes Of Non-Perovskite Oxides and Phosphates

Proton solubility and conduction are known also for many non-perovskite classes of oxides, comprising mainly fluorite-related structures and structures with oxide ion tetrahedra.

Among the fluorite-related oxides, the MO_2 oxides (M = Zr, Hf, Ce) exhibit very little bulk solubility of protons, and the acceptor-doped oxides remain pure oxide ion conductors with negligible proton conductivity at all temperatures. Rutile TiO_2, on the other hand, exhibits significant proton solubility and conductivity. Some acceptor-doped pyrochlores, such as $La_2Zr_2O_7$, exhibit quite high and pure proton conductivity in wet atmospheres [43], of the order of magnitude of the perovskite $LaScO_3$ (Fig. 11.3). Rare earth oxides also exhibit proton conductivity [44] (see Fig. 11.3).

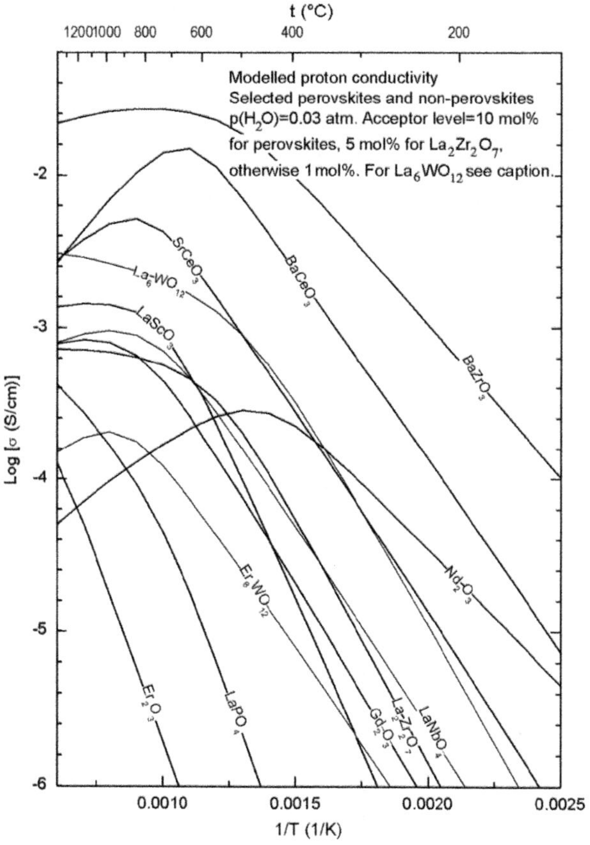

Fig. 11.3 Partial proton conductivity vs. 1/T for bulk (grain interior) of selected perovskites and non-perovskites, calculated on basis of proton mobility and hydration thermodynamics parameters. Acceptor level is chosen as 10 mol% (3.33% of oxide ions are protonated at full hydration) for perovskites, 5 mol% for $La_2Zr_2O_7$, and 1 mol% for other non-perovskites. The curve for La_6WO_{12} is for undoped material but is modeled with an effective acceptor level of 1.38 mol% arising from thermogravimetry of water uptake. $pH_2O = 0.03$ atm. Parameters and references are for $BaZrO_3$ [20], $BaCeO_3$ [53], $SrCeO_3$ [22], $LaScO_3$ [26], $La_2Zr_2O_7$ [54], Nd_2O_3 [55], Gd_2O_3 [44], Er_2O_3 [44], $LaPO_4$ [56], $LaNbO_4$ [47], Er_6WO_{12} [57], and La_6WO_{12} [58]. Curves should reflect partial proton conductivity in temperature range of measurement, but may otherwise contain large errors caused by correlations between the various parameters used

In the 1990s it was discovered that acceptor-doped $LaPO_4$ was hydrated and conducted protons much the same way as oxides [45]. However, the solubility of dopants was low, limiting the achievable proton content and conductivity (see Fig. 11.3). The limited solubility can be assigned to the high energy of oxygen deficiency (oxygen vacancies or diphosphate groups) and hydrogen phosphate groups. It should be mentioned that phosphates are probably not suited for use in high-temperature fuel cells because of volatility of reduced P-containing compounds such as PH_3.

A few years ago, we started to investigate niobates and tantalates with tetrahedral oxide ion arrangements. Acceptor-doped rare earth orthoniobates and orthotantalates, $LnNbO_4$ and $LnTaO_4$, proved to be proton conductors, with $LaNbO_4$ being the one with the highest proton conductivity (see Fig. 11.3), peaking at around 0.001 S/cm [46–48]. Also, La_3NbO_7 shows a similar proton conduction [49]. $LaNb_3O_9$, on the other hand, which is a bronze-type perovskite-related compound, neither dissolves significant amounts of protons nor exhibits any measurable proton conductivity.

Although acceptor-doped $LaGaO_3$ perovskite dissolves no protons, $LaBaGaO_4$ with an oxygen tetrahedral dissolves protons and exhibits considerable proton conductivity [50]. Modeling of proton transport in this material shows that transfer of protons between tetrahedra is easy, whereas the between-oxide ions within one tetrahedron—necessary for long-range transport—is more difficult and is responsible for the experimental activation energy [51]. Probably, this comparatively high intrapolyhedron proton diffusion barrier is a general reason for the low proton mobilities of non-perovskites, unless rotational reorientation of the tetrahedra makes intratetrahedral jumps unnecessary, as in $CsHSO_4$ [52].

All in all, the non-perovskite oxides and phosphates with oxygen tetrahedral suffer from limiting solubility of acceptors and consequently low proton concentrations. In all cases, the hydration enthalpy becomes more negative as the oxides get more stable and close packed, but this is accompanied by higher activation energies of proton migration. Thus, for the non-perovskites the proton conductivity seems to be a compromise between concentration and mobility, in contrast to the perovskites, where the two aspects mainly go together. However, although the best perovskite proton conductors are chemically and mechanically weak, the non-perovskite proton-conducting oxides comprise many very stable and robust materials.

There has recently been much interest around proton conduction in condensed phosphates. Lanthanum metaphosphate ($La(PO_3)_3$) exhibits a modest proton conductivity [59] whereas diphosphates of tetravalent metals, e.g., SnP_2O_7 and TiP_2O_7, appear to exhibit a high proton conduction peak at intermediate temperatures (around 200°C). The effect is reportedly enhanced by substituting In^{3+} for the tetravalent cation, and the conductivity can exceed 0.1 S/cm [60]. It is uncertain what is the defect or doping mechanism behind these behaviors. The same materials exhibit a lower, temperature-dependent conductivity above 400°C, tentatively attributed to protons from hydrolysis of the diphosphate groups [61].

Also, the disilicate $La_2Si_2O_7$ has been shown to be a proton conductor, although modest [62]. This may play a role in oxides with silicon impurities, where this phase may form in the grain boundaries [63].

Acceptor-doped $LaBO_3$ also displays modest proton conductivity [64]. Recently, it was found that lanthanum oxyborate, $La_{26}O_{27}(BO_3)_8$, contains 1 vacant oxygen position among the 28 and that this can be hydrated [65]. The material thus becomes a proton conductor at intermediate and low temperatures.

11.6 Developments of Proton-Conducting SOFCs

Demonstrations of proton-conducting fuel cells (PCFCs), aside from traditional polymer membrane fuel cells (PEMFCs), are made using proton-conducting electrolytes that are perovskites, solid acids, or high-temperature polymers. The solid acid fuel cells (SAFCs) are based on $CsHSO_4$ or CsH_2PO_4 and show good performances because of high proton conductivity and attractive operating temperatures around 200°C [66], where supporting components and some electrode technologies from PEMFCs can be applied. Challenges are related to the low melting or decomposition temperatures, solubility in liquid water, stability toward reduction and evaporation in the fuel-side atmosphere, and general softness of the materials. High-temperature polymer membrane fuel cells based on sulfonated or phosphonated polybenzimidazole (PBI) operate typically at 200°C, and as the membranes are now being made available to research laboratories and industries, we will probably see an interesting development in this area.

Within proton-conducting solid oxide fuel cells, all significant demonstrations until now have been made on acceptor-doped perovskites, notably those Sr- or Ba based. Most often these efforts have stranded on the limiting mechanical and chemical stability of the electrolyte because of reaction with acidic gases under operation, cycling, or storage. Additional problems are related to the insufficient knowledge and research on electrodes for these new fuel cells. For instance, Ni-cermet anodes are often used with the proton-conducting electrolyte as part of the cermet, but the commonly used $BaCeO_3$ reacts with the NiO–Ni pair under production and operation, and this destroys the electronic conduction of the anode cermet and reduces the ionic transport number of the electrolyte [67]. Remarkably few tests conclude with grain boundary resistance as a major problem, a problem that would be anticipated from conductivity studies.

In 1995, Bonanos et al. [68] reviewed the literature and their own results and tabulated performances reported for a number of H_2-O_2 or H_2-air fuel cells with 0.4- to 0.5-mm-thick, 10%–20% Gd- or Nd-doped $BaCeO_3$ electrolytes, Pt anodes, and Pt or Ag cathodes, operated at 800°C. At 700 mV cell voltage, they delivered from 70 to 285 mA/cm^2 of current density, corresponding to 50–200 mW/cm^2 of power density and effective electrolyte conductivities of around 10–40 mS/cm. These conductivities are, to a first approximation, in agreement with bulk (grain interior) conductivity data for doped $BaCeO_3$s, indicating that electrode and grain boundary impedances were small.

Some years later, the company Protonetics Inc. pursued commercialization of fuel cells based on $BaCeO_3$ and reported their own data on running the cells on hydrogen or methane [69]. Power densities in the lower range of those mentioned above were obtained.

Kreuer [70] reported a test of a $Ba(Ce,Zr)O_3$ fuel cell and obtained a modest power density. It was shown that this was much lower than expected from bulk (grain interior) conductivity and in agreement with expectations from electrical properties, including the grain boundary resistance.

In the aforementioned studies, Coors and Kreuer bring up the issue of mixed proton–oxide ion conduction of $BaCeO_3$-based materials: while proton conduction is beneficial for efficiency in hydrogen-burning fuel cells, some oxide ion transport is beneficial for providing water vapor on the anode side for reforming and shifting carbon-based fuels such as methane.

Ito et al. [9] made a laboratory fuel cell with a 0.7-μm-thick, $BaCeO_3$-based electrolyte deposited on a Pd anode substrate and with a noble metal cathode. The cell showed high power densities, more than 1 W/cm^2 at 600°C and about 0.7 W/cm^2 at as low as 400°C, but substantial lifetimes were not reported.

NASA (National Aeronautics and Space Administration) has also pursued ceramic proton conductors as part of their SOFC program by fabricating and characterizing thin $BaCeO_3$-based [71] and other proton-conducting ceramic films.

Meng et al. [72] report power densities of up to around 300 mW/cm^2 for cells using 50-μm films of Gd-doped $BaCeO_3$ (BGCO) as electrolyte, $La_{0.5}Sr_{0.5}CoO_3$-BGCO cercers as cathode, Ni-BGCO cermets as anode, and H_2 or industrial ammonia as fuel. This work is of course very interesting and promising and shows how the otherwise unstable $BaCeO_3$ may be used with a basic fuel such as ammonia. However, the apparent coexistence of Ni and BCGO and some other conclusions in the referred paper, such as protonic conduction in LSGM, are questionable.

Meulenberg et al. have reported preparation of thin films of Y-doped $BaCeO_3$ [73] and other proton-conducting films on porous cermet substrates by first making a dense film of doped ZrO_2 or CeO_2, for example, and then applying a film of a Ba-containing compound (e.g., $BaCO_3$) and letting the two layers react by a solid-state reaction.

There are few reports of fuel cells with $BaZrO_3$-based (e.g., BZY) electrolytes. Acceptor-doped mixed barium zirconate cerates $Ba(Zr,Ce)O_3$ have been investigated for their better stability of the zirconate combined with the better grain boundary conduction of the cerate [74], but fuel cell tests have not given convincing results [70, 75].

11.7 Conclusions

Perovskite oxides dissolve protons from water vapor at high temperatures and may in some cases become proton conductors. The use of acceptor dopants increases the proton concentration, and in addition to the mobility of those protons, the thermodynamics of incorporating the effectively positive proton defects at the cost of the dry-state positive defects are the crucial parameters that determine the proton conductivity. Perovskites exhibit the highest mobilities of protons in oxides, and the hydration thermodynamics is increasingly favorable (exothermic) as the difference in electronegativity between the B- and A-site cations decreases. Because of the concerted action of mobility and dehydration, the proton conductivity most often goes through a maximum with increasing temperature. The maximum is typically found around 800°C and comprises conductivities of 10^{-2} S/cm for the best Ba-based perovskites, somewhat lower for the Sr-based, and 10^{-3} or below for the non-Ba-containing ones. Unfortunately, the best proton conductors are chemically unstable and vulnerable to CO_2, for example. They are also troubled by high grain boundary resistances.

The advances of proton conductivity in acceptor-doped perovskites relies on developing materials with better chemical stability, higher doping levels, higher proton mobility by structural optimizations and smaller doping–proton association energies, and finally by understanding and minimizing grain boundary resistances.

In this chapter, I have treated the defect chemistry and thermodynamics of hydration. I have, moreover, put special emphasis on inherently—not acceptor-doped—oxygen-deficient systems and their hydration in the ordered or disordered states. Finally, I have discussed the new oxyhydroxide compounds that arise from such hydration and advocated that radically new and better proton conductors may rely on understanding how to stabilize, disorder, dope, and in general control the defect chemistry of such oxyhydroxides.

Fuel cells based on proton conductors have clear potential advantages over oxide ion-conducting SOFCs, and perovskites constitute our best class of proton conductors. However, the difficulty of identifying stable and highly proton-conducting materials has so far limited the performance of demonstration cells of this kind, and commercialization of ceramic proton-conducting fuel cells thus still lies ahead of us.

Acknowledgments The author wishes to acknowledge the results, input, and help from colleagues and students at the University of Oslo over the years that have enabled the writing of this chapter. Also, I am grateful for support from several projects funded by the Research Council of Norway (RCN) and The University of Oslo/FUNMAT@UiO that have contributed to the same.

References

1. T. Takahashi, H. Iwahara, Solid-state ionics: protonic conduction in perovskite type oxide solid solutions. Rev. Chim. Miner. 17(4), 243–253 (1980)
2. H. Iwahara, T. Esaka, H. Uchida, N. Maeda, Proton conduction in sintered oxides and its application to steam electrolysis for hydrogen production. Solid State Ionics 3–4, 359–363 (1981)

3. H. Uchida, K. Ogaki, H. Iwahara, High temperature hydrogen sensor and steam sensor using barium cerium oxide ($BaCeO_3$)-based proton conducting ceramics. Proc. Electrochem. Soc. 87–9 (Proc. Symp. Chem. Sens.), 172–179 (1987)
4. W.-K. Lee, A. S. Nowick, L. A. Boatner, Protonic conduction in acceptor-doped $KTaO_3$ crystals. Solid State Ionics, 18–19, 989–993 (1986)
5. K.C. Liang, A.S. Nowick, High-temperature protonic conduction in mixed perovskite ceramics. Solid State Ionics 61, 77–81 (1993)
6. K.C. Liang, Yang Du, A.S. Nowick, Fast high-temperature proton transport in nonstoichiometric mixed perovskites. Solid State Ionics 69, 117–120 (1994)
7. T. Yajima, K. Koide, H. Takai, N. Fukatsu, H. Iwahara, Application of hydrogen sensor using proton conductive ceramics as a solid electrolyte to aluminum casting industries. Solid State Ionics 79, 333–337 (1995)
8. H. Iwahara, Y. Asakura, K. Katahira, M. Tanaka, Prospect of hydrogen technology using proton-conducting ceramics. Solid State Ionics 168, 299–310 (2004)
9. N. Ito, M. Iijima, K. Kimura, S. Iguchi, New intermediate temperature fuel cell with ultra-thin proton conductor electrolyte. J. Power Sources 152, 200–203 (2005)
10. J. Hartvigsen, S. Elangovan, A. Khandkar, AICHE Meeting, St. Louis, MO, USA, Aug. 1993
11. T. Norby, R. Haugsrud, Dense ceramic membranes for hydrogen separation. In: Non-porous Inorganic Membranes (A. F. Sammells, M.V. Mundschau, eds.). Wiley-VCH, Weinheim, 2006, pp. 1–48.
12. K.D. Kreuer, W. Münch, M. Ise, T. He, A. Fuchs, U. Traub, J. Maier, Defect interaction in proton conducting perovskite-type oxides. Ber. Bunsenges. Phys. Chem. 101(9), 1344 (1997)
13. M.S. Khan, M.S. Islam, D.R. Bates, Dopant substitution and ion migration in the $LaGaO_3$-based oxygen ion conductor. J. Phys. Chem. B 102(17), 3099–3104 (1998)
14. T. Norby, M. Widerøe, R. Gløckner, Y. Larring, Hydrogen in oxides. Dalton Trans. (19), 3012–3018 (2004)
15. D. F. Shriver, P.W. Atkins, C.H. Langford, Inorganic Chemistry, 2nd edn. Oxford University Press, Oxford (1994), pp. 165–166
16. T.S. Bjørheim, A. Kuwabara, I. Ahmed, R. Haugsrud, S. Stølen, T. Norby, A combined conductivity and DFT study of protons in $PbZrO_3$ and the alkaline earth zirconates. Solid State Ionics, under publication.
17. A. Magrasó, R. Haugsrud, M. Segarra, T. Norby, Defects and transport in Gd-doped $BaPrO_3$. J. Electroceramics, in print. DOI 10.1007/s10832-008-9541-z.
18. W. Münch, K.D. Kreuer, G. Seifert, J. Maier, A quantum molecular dynamics study of proton diffusion in $SrTiO_3$ and $CaTiO_3$. Solid State Ionics 125, 39–45 (1999)
19. A.S. Nowick, A.V. Vaysleyb, Isotope effect and proton hopping in high-temperature protonic conductors. Solid State Ionics 97, 17–26 (1997)
20. K.D. Kreuer, S.T. Adams, W. Münch, A. Fuchs, U. Klock, J. Maier, Proton conducting alkaline earth zirconates and titanates for high drain electrochemical applications. Solid State Ionics 145, 295–306 (2001)
21. K.A. Furøy, R. Haugsrud, M. Hänsel, A. Magrasó, T. Norby, Role of protons in the electrical conductivity of acceptor-doped $BaPrO_3$, $BaTbO_3$, and $BaThO_3$. Solid State Ionics 178 (7–10), 461–467 (2007)
22. P.I. Dahl, R. Haugsrud, H.L. Lein, T. Grande, T. Norby, M.-A. Einarsrud, Synthesis, densification and electrical properties of strontium cerate ceramics. J. Eur. Ceramic Soc. 27, 4461–4471 (2007)
23. J.H. Yu, J.-S. Lee, J. Maier, Formation of protonic defects in perovskite-type oxides with redox-active acceptors: case study on Fe-doped $SrTiO_3$. Phys Chem Chem Phys, 7, 3560–3564 (2005)
24. J.B. Smith, M.Sc. Thesis (in Norwegian). Dept. of Chemistry, University of Oslo, 1999
25. S. Yamaguchi, K. Kobayashi, T. Higuchi, S. Shin, Y. Iguchi, Electronic transport properties and electronic structure of $InO_{1.5}$-doped $CaZrO_3$. Solid State Ionics, 136–137, 305–311 (2000)

26. D. Lybye, N. Bonanos, Proton and oxide ion conductivity of doped $LaScO_3$. Solid State Ionics, 125, 339–344 (1999)
27. Y. Larring, T. Norby, Protons in $LaErO_3$. Solid State Ionics 70–71, 305–10 (1994)
28. K.D. Kreuer, Aspects of the formation and mobility of protonic charge carriers and the stability of perovskite-type oxides. Solid State Ionics 125, 285–302 (1999)
29. P. Murugaraj, K.D. Kreuer, T. He, T. Schober, J. Maier, High proton conductivity in barium yttrium stannate $Ba_2YSnO_{5.5}$. Solid State Ionics 98, 1–6 (1997)
30. R. Glöckner, Thesis, University of Oslo (2000)
31. C. Haavik, E.M. Ottesen, K. Nomura, J.A. Kilner, T. Norby, Temperature dependence of oxygen ion transport in Sr + Mg substituted $LaGaO_3$ (LSGM) with varying grain sizes. Solid State Ionics 174 (1–2), 233–243 (2004)
32. M. Cherry, M.S. Saiful, J.D. Gale, C.R.A. Catlow, J. Phys. Chem. 99, 14614–14618 (1995)
33. A.S. Nowick, K.C. Liang, Effect of non-stoichiometry on the protonic and oxygen-ionic conductivity of $Sr_2(ScNb)O_6$: a complex perovskite. Solid State Ionics 129, 201–207 (2000)
34. A.S. Nowick, Y. Du, K.C. Liang, Some factors that determine proton conductivity in nonstoichiometric complex perovskites. Solid State Ionics 125, 303–311 (1999)
35. T. Shimura, K. Suzuki, H. Iwahara, Conduction properties of Mg-, Fe- or Co-substituted Sr_2TiO_4 at elevated temperatures. Solid State Ionics 125, 313–318 (1999)
36. K.D. Kreuer, Aspects of the formation and mobility of protonic charge carriers and the stability of perovskite-type oxides. Solid State Ionics 125, 285–302 (1999)
37. F. Iguchi, N. Sata, T. Tsurui, H. Yugami, Microstructures and grain boundary conductivity of $BaZr_{1-x}Y_xO_3$ (x = 0.05, 0.10, 0.15) ceramics. Solid State Ionics 178, 691–695 (2007)
38. S.B.C. Duval, P. Holtappels, U.F. Vogt, E. Pomjakushina, K. Conder, U. Stimming, T. Graule, Electrical conductivity of the proton conductor $BaZr_{0.9}Y_{0.1}O_{3-\delta}$ obtained by high temperature annealing. Solid State Ionics 178, 1437–1441 (2007)
39. C. Kjølseth, H. Fjeld, P.I. Dahl, C. Estournés, R. Haugsrud, T. Norby, Space Charge theory applied to the grain boundary impedance of proton conducting $BaZr_{0.9}Y_{0.1}O_{3-\delta}$. under publication.
40. T. Schober, Phase diagrams in the proton conductor systems $Sr_6Ta_2O_{11}\times nH_2O$ and $Sr_{5.92}Ta_{2.08}O_{11.12}\times nH_2O$. Solid State Ionics 177(5–6), 471–474 (2006)
41. I. Animitsa, A. Neiman, N. Kochetova, B. Melekh, A. Sharafutdinov, Proton and oxygen-ion conductivity of $Ba_4Ca_2Nb_2O_{11}$. Solid State Ionics 162–163, 63–71 (2003)
42. I. Animitsa, T. Norby, S. Marion, R. Glöckner, A. Neiman, Incorporation of water in strontium tantalates with perovskite-related structure. Solid State Ionics 145, 357–364 (2001)
43. T. Shimura, M. Komori, H. Iwahara, Ionic conduction in pyrochlore-type oxides containing rare earth elements at high temperature. Solid State Ionics 86–88, 685–689 (1996)
44. Y. Larring, T. Norby, Protons in rare earth oxides. Solid State Ionics 77, 147–151 (1995)
45. T. Norby, N. Christiansen, Proton conduction in Ca- and Sr-substituted $LaPO_4$. Solid State Ionics 77, 240–243 (1995)
46. R. Haugsrud, T. Norby, Proton conduction in rare earth ortho-niobates and ortho-tantalates. Nat. Mater. 5, 193–196 (2006)
47. R. Haugsrud, T. Norby High-temperature proton conductivity in acceptor-doped $LaNbO_4$. Solid State Ionics 177(13–14), 1129–1135 (2006)
48. R. Haugsrud, T. Norby, High-temperature proton conductivity in acceptor-substituted rare-earth ortho-tantalates, $LnTaO_4$. J. Am. Ceram. Soc. 90(4), 1116–1121 (2007)
49. T. Shimura, Y. Tokiwa, H. Iwahara, Protonic conduction in lanthanum strontium aluminate and lanthanum niobate-based oxides at elevated temperatures. Solid State Ionics 154–155, 653–658 (2002)
50. F. Schönberger, E. Kendrick, M.S. Islam, P.R. Slater, Investigation of proton conduction in $La_{1-x}Ba_{1+x}GaO_{4-x/2}$ and $La_{1-x}Sr_{2+x}GaO_{5-x/2}$. Solid State Ionics 176, 2951–2953 (2005)
51. E. Kendrick, J. Kendrick, K.S. Knight, M.S. Islam, P. Slater, Cooperative mechanisms of fast-ion conduction in gallium-based oxides with tetrahedral moieties. Nat. Mater. 6, 871–875 (2007)

52. W. Münch, K.D. Kreuer, U. Traub, J. Maier, A molecular dynamics study of the high proton conducting phase of $CsHSO_4$. Solid State Ionics 77, 10–14 (1995)
53. K.D. Kreuer, W. Münch, M. Ise, T. He, A. Fuchs, U. Traub, J. Maier, Defect interaction in proton conducting perovskite-type oxides. Ber. Bunsenges. Phys. Chem. 101 (9), 1344 (1997)
54. T. Omata, K. Ikeda, R. Tokashiki, S. Otsuka-Yao-Matsuo, Proton solubility for $La_2Zr_2O_7$ with a pyrochlore structure doped with a series of alkaline-earth ions. Solid State Ionics 167, 389–397 (2004)
55. Y. Larring, T. Norby, The equilibrium between oxygen vacancies, water vapour and protons in rare earth oxides. Solid State Ionics 97, 523–528 (1997)
56. K. Amezawa, Y. Tomii, N. Yamamoto, High temperature protonic conduction in $LaPO_4$ doped with alkaline earth metals. Solid State Ionics 176, 135–141 (2005)
57. R. Haugsrud, H. Fjeld, K. R. Haug, T. Norby, Mixed ionic and electronic conductivity of undoped and acceptor doped Er_6WO_{12}. J. Electrochem. Soc. 154 (1), B77–B81 (2007)
58. R. Haugsrud, Defects and transport properties in Ln_6WO_{12} (Ln = La, Nd, Gd, Er). Solid State Ionics 178 (7–10), 555–560
59. K. Amezawa, Y. Kitajima, Y. Tomii, N. Yamamoto, M. Widerøe, T. Norby, Protonic conduction in acceptor-doped LaP_3O_9. Solid State Ionics 176 (39–40), 2867–2870 (2005)
60. M. Nagao, T. Kamiya, P. Heo, A. Tomita, T. Hibino, M. Sano, Proton conduction in In^{3+}-doped SnP_2O_7 at intermediate temperatures. J. Electrochem. Soc. 153, A1604–A1609 (2006)
61. V. Nalini, T. Norby, A.M. Anuradha, Protonic conduction in TiP_2O_7. Proc. 10th Asian Conf. Solid State Ionics, B.V.R. Chowdari, M.A. Careem, M.A.K.I. Dissanayake, R.M. G. Rajapakse, V.A. Seneviratne (eds.), Kandy, Sri Lanka, June 12–16, 2006. World Scientific, Singapore, pp 321–328
62. K. Amezawa, J. Yamada, N. Kitamura, Y. Tomii, T. Handa, N. Yamamoto, High temperature protonic conduction in Sr-doped $La_2Si_2O_7$. Solid State Ionics 176, 341–347 (2005)
63. H. Fjeld, R. Haugsrud, A.E. Gunnæs, T. Norby, Proton and oxide ion conductivity in the grain interior and grain boundaries of Ca-doped $Er_2Ti_2O_7$ with Si impurities. Solid State Ionics, 179(33–34), 1849–1853 (2008)
64. K. Amezawa, N. Takahashi, N. Kitamura, Y. Tomii, N. Yamamoto, High temperature protonic conduction in $LaBO_3$ with the aragonite-type structure. Solid State Ionics 175, 575–579 (2004)
65. S. Noirault, S. Célérier, O. Joubert, M.T. Caldes, Y. Piffard, Incorporation of water and fast proton conduction in the inherently oxygen-deficient compound $La_{26}O_{27}(BO_3)_8$. Adv. Mater. 19, 867–870 (2007)
66. D.A. Boysen, T. Uda, C.R.I. Chisholm, S.M. Haile, High-performance solid acid fuel cells through humidity stabilization. Science 303 (5654), 68–70 (2004)
67. V. Agarwal, M.L. Liu. Electrochemical properties of $BaCe_{0.8}Gd_{0.2}O_3$ electrolyte films deposited on Ni-$BaCe_{0.8}Gd_{0.2}O_3$ substrates. J. Electrochem. Soc. 144(3), 103 (1997)
68. N. Bonanos, K.S. Knight, B. Ellis, Perovskite solid electrolytes: Structure, transport properties and fuel cell applications. Solid State Ionics 79, 161–170 (1995)
69. W.G. Coors, Protonic ceramic fuel cells for high-efficiency operation with methane. J Power Sources 118, 150–156 (2003)
70. K.-D. Kreuer, Proton-conducting oxides. Annu. Rev. Mater. Res. 33, 333–59 (2003)
71. F.W. Dynys, M.H. Berger, A. Sayir, Pulsed laser deposition of high temperature protonic films. Solid State Ionics 177, 2333–2337 (2006)
72. G. Meng, G. Ma, Q. Ma, R. Peng, X. Liu, Ceramic membrane fuel cells based on solid proton electrolytes. Solid State Ionics 178, 697–703 (2007)
73. W.A. Meulenberg, J.M. Serra, T. Schober, Preparation of proton conducting $BaCe_{0.8}Gd_{0.2}O_3$ thin films. Solid State Ionics 177, 2851–2856 (2006)
74. K.H. Ryu, S.M. Haile, Chemical stability and proton conductivity of doped $BaCeO_3$–$BaZrO_3$ solid solutions. Solid State Ionics 125, 355–367 (1999)
75. T. Shimada, C. Wen, N. Taniguchi, J. Otomo, H. Takahashi, The high temperature proton conductor $BaZr_{0.4}Ce_{0.4}In_{0.2}O_{3-\alpha}$. J. Power Sources 131, 289–292 (2004)

Chapter 12
Proton Conduction in Cerium- and Zirconium-Based Perovskite Oxides

Hiroshige Matsumoto

12.1 Introduction

Some perovskite-type oxides show protonic conduction and are useful for hydrogen-related electrochemical devices, including application to solid oxide fuel cells (SOFCs). Iwahara et al. reported protonic conductivity of strontium-cerate-based perovskite-type oxides in 1981 [1]. Since that time, various perovskite-type proton-conducting oxides have been found. For use of the proton-conducting perovskite oxides, we should understand not only their merits but also their weak points. This chapter concerns the proton-conducting properties of typical cerium- and zirconium-containing perovskite oxides from the points of view of conductivity, stability, electrode affinity, and dopant effect. Mixed conduction occurring in a special composition of the perovskite oxide is also introduced.

Alkaline earth cerates ($ACeO_3$) and zirconates ($AZrO_3$) are the typical host material of the proton-conducting perovskite oxides; A is an alkali earth metal, at least one of Ca, Sr, and Ba. Acceptor doping is essential for the protonic conduction to take place, i.e., tetravalent cerium or zirconium is substituted by a lower-valent cation, e.g., yttrium, ytterbium, or indium. Thus, $SrCe_{0.95}Yb_{0.05}O_{3-\alpha}$, $BaCe_{0.8}Y_{0.2}O_{3-\alpha}$, $BaZr_{0.9}Y_{0.1}O_{3-\alpha}$, and so forth are examples of proton-conducting perovskite oxides. They are designed to have oxide ion vacancies by acceptor doping; α in the above chemical formulae indicates the molar amount of oxygen vacancies. In moist environments, ambient water molecules are incorporated into the oxide to form protons that make hydrogen bonds to the lattice oxygen.

$$H_2O + V_O^{\bullet\bullet} + O_O^{\times} \rightleftarrows 2OH_O^{\bullet} \tag{12.1}$$

H. Matsumoto (✉)
INAMORI Frontier Research Center, Kyushu University, 744 Motooka, Nishiku, Fukuoka 819-0395, Japan
e-mail: matsumoto@ifrc.kyushu-u.ac.jp

T. Ishihara (ed.), *Perovskite Oxide for Solid Oxide Fuel Cells*,
Fuel Cells and Hydrogen Energy, DOI 10.1007/978-0-387-77708-5_12,

Kröger–Vink notation is used to express the defect equilibrium in Eq. (12.1); i.e., $V_O^{\bullet\bullet}$ and $O_O^{\times}$ express oxygen vacancy and lattice oxygen (oxide ion) that has a doubly positive charge and is electrically neutral, respectively, with regard to the charge of the doubly negative oxygen site. The species $OH_O^{\bullet}$ represents a lattice oxygen atom that folds one proton by the hydrogen bond; the electrical charge of the species is singly positive with respect to the electrical charge underlying at the oxygen site. The proton can move from that oxygen atom to another by breaking and forming the hydrogen bonds with the oxygen atoms. It should be noted that Eq. (12.1) expresses the defect equilibrium for the hydration reaction, so that $OH_O^{\bullet}$ in the equation is not the charge carrier but a species taking part in the defect equilibrium. It is a proton that acts as the ionic charge carrier. Equation (12.1) can alternatively be expressed in another form as Eq. (12.2) to emphasize that the mobile charge carrier is the interstitial proton:

$$H_2O + V_O^{\bullet\bullet} \rightleftarrows O_O^{\times} + 2H_i^{\bullet} \quad (12.2)$$

The hydration in Eq. (12.1) or (12.2) is an equilibrium reaction. For a material to be proton conducting, the equilibrium constant has to be sufficiently large; i.e., the equilibrium in either Eq. (12.1) or (12.2) has to go from the left-hand side to the right. In addition, for the protons to be mobile, the hydrogen bonds should not be so strong and the lattice oxygen should be packed and arranged well such that the protons can hop successively from oxygen to oxygen, resulting in macroscopic translational drift on applying an electric field to the proton-conducting oxides.

Proton-conducting perovskite materials are characterized by the occurrence of protonic conduction at high temperatures; $SrCeO_3$-based oxides were introduced to exhibit protonic conduction at 700°–1000°C in the Iwahara reports [1]. Accordingly, these materials are referred to as "high-temperature proton conductors" (HTPCs). Considering the fact that the working temperatures of most proton-conducting solids are restricted to below or near the boiling point of water, such high working temperature of the proton-conducting perovskite oxides is of both fundamental and practical interest. In particular, the SOFC is the most energy-effective type of fuel cell because it works at high temperatures. High operation temperature will generally bring smooth electrode kinetics, leading to an advantage of any electrochemical devices.

On the other hand, another characteristic of the proton-conducting perovskite oxide is the low temperature dependence of the proton conductivity. Activation energy of proton hopping between adjacent oxide ions is low; e.g., typical values of the activation energy of proton conductivity are around 0.6 eV in doped $SrCeO_3$ [2], 0.3–0.5 eV in doped $BaZrO_3$ [3], and 0.5–0.6 eV in $BaCeO_3$-based electrolytes [4]. Accordingly, the proton-conducting perovskite oxides are advantageous in intermediate to low temperature use; e.g., $BaCe_{0.9}Y_{0.1}O_{3-\alpha}$ has a conductivity a little less than 10^{-3} S/cm at 400°C [5]. These oxides can be candidate electrolyte materials for reducing the operation temperature of SOFCs. Ito et al. succeeded in observing quite high performance of a barium cerate-based fuel cell [6] that is described in another chapter.

Nevertheless, the commercial application of the proton-conducting perovskite oxides so far is limited: the only example is a hydrogen sensor for molten aluminum [7]. TYK Corp. (Tajimi, Japan) supplies the hydrogen sensor, using In-doped $CaZrO_3$, to analyze and control hydrogen activity in molten aluminum before casting to minimize generation of voids. There has been no commercial example for the proton-conducting oxide to be used in fuel cells, partly because the material is less than 30 years old, still a little too new. But at the same time, in the case of alkaline earth cerates and zirconates, their chemical stability should be well understood for practical use. In the case of the hydrogen sensor, TYK Corp. uses the $CaZrO_3$-based proton-conducting electrolyte, which in their opinion is stable. However, in the case of cerates, it is frequently stated that the material tends to react with carbon dioxide or water vapor to decompose into cerium oxide and the carbonate or hydrate of the alkaline earth metal, as discussed later [5, 8–10].

12.2 Conductivity

As just stated, cerium or zirconium is available as a B-site cation for the proton-conducting ABO_3 perovskite oxides. Figure 12.1 compares the temperature dependence of electrical conductivity of $BaCe_{0.9}Y_{0.1}O_{3-\alpha}$ and $SrZr_{0.9}Y_{0.1}O_{3-\alpha}$ in moist hydrogen and in moist oxygen. Both specimens show the typical conductivity behavior of the proton-conducting perovskite oxides. The conductivity in

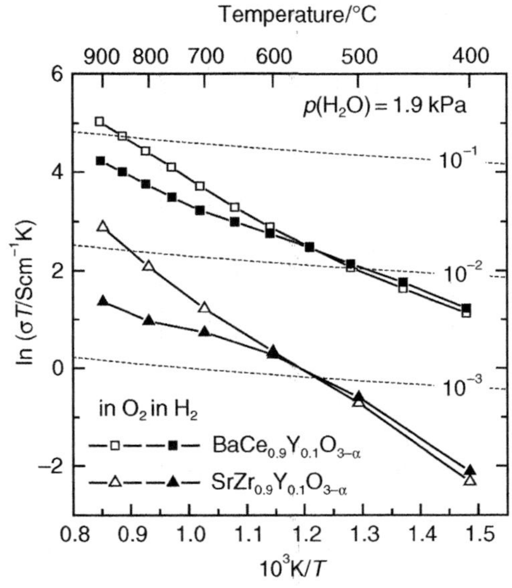

Fig. 12.1 Arrhenius plot of conductivity of $BaCe_{0.9}Y_{0.1}O_{3-\alpha}$ and $SrZr_{0.9}Y_{0.1}O_{3-\alpha}$ in moist H_2 (*closed symbols*) and in moist O_2 (*open symbols*) in the temperature range of 400°–900°C; $p(H_2O) = 1.9 \times 10^3$ Pa. *Dashed lines* indicate isoconductivity lines in S cm^{-1}

O_2 is higher than that in H_2 at 500–900°C due to the contribution of electron hole conduction in oxidative atmospheres [11–14]. An electron hole is generated in association with the following defect equilibrium:

$$V_O^{\bullet\bullet} + \frac{1}{2}O_2 \rightleftarrows O_O^{\times} + 2h^{\bullet} \tag{12.3}$$

Electron hole conduction, therefore, appears in oxidative atmospheres. The conductivity in H_2 is assignable to ionic conductivity, because the electron hole conductivity is much smaller in reducing atmospheres, as the equilibrium of Eq. (12.3) going from the right-hand side to the left. The Arrhenius plot of the conductivity in H_2 shows change in slope due to the change in the ionic charge carrier from proton to oxide ion with increasing temperature, i.e., the equilibrium constant of Eq. (12.1) decreases with increasing temperature [15, 16]. Such change in the major charge carriers can be determined by the measurements of gas concentration cells [17]. Figure 12.2 shows the protonic and oxide ionic transport number measured for $SrCe_{0.95}Yb_{0.05}O_{3-\alpha}$ by means of hydrogen and steam concentration cells. In this case, protonic conduction is dominant at 600°–700°C, but the oxide ionic transport number increases with temperature and is comparable to that of protons.

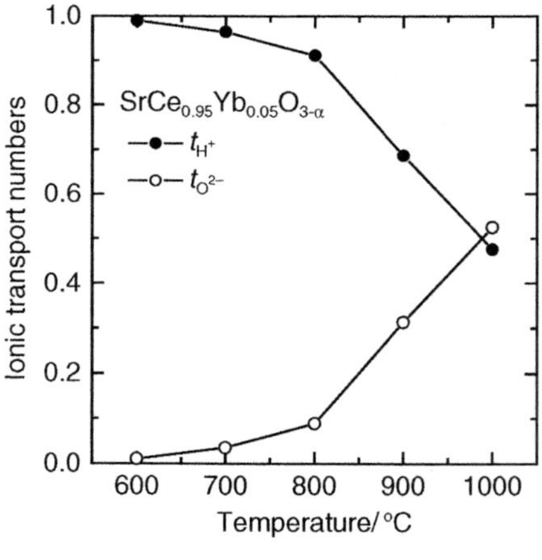

Fig. 12.2 The ionic (proton and oxide ionic) transport number of $SrCe_{0.95}Yb_{0.05}O_{3-\alpha}$ determined by hydrogen and steam concentration cells; $p(H_2)$ of 10^5 Pa and 10^3 Pa were used for the hydrogen concentration cells under $p(H_2O) = 2.1 \times 10^3$ Pa; $p(H_2O)$ of 6.1×10^2 Pa and 2.1×10^3 Pa were used for the steam concentration cells using 1% H_2–99% Ar mixture as a base gas

Usually, cerates have high protonic conductivity because of the high concentration of protonic charge carriers. In other words, cerates tend to have larger equilibrium constant than zirconates for the hydration reaction in Eq. (12.1) or (12.2). For the same kind and amount of doping, $BaCeO_3$ exceeds $SrZrO_3$ by roughly 1.5 orders of magnitude in terms of the conductivity in hydrogen, that is, ionic conductivity, partly because of the high basicity of cerium in comparison with zirconium. The equilibrium constant K for Eq. (12.1) or (12.2) is larger for cerates than for zirconates. Thus, cerates have higher proton concentration than zirconates when compared at the same temperature and water vapor pressure.

12.3 Activation/Deactivation of Electrodes

The B-site host cation, cerium or zirconium, affects not only the proton conductivity but also the activity of electrodes when they are used as the electrolyte for electrochemical devices. The following are the results of investigations on the hydrogen pumping properties of proton conductor cells using $SrCe_{0.95}Yb_{0.05}O_{3-\alpha}$ and $SrZr_{0.9}Y_{0.1}O_{3-\alpha}$ in comparison [18]. These results will be helpful to compare the affinity of electrode materials to cerates and zirconates.

A hydrogen pump is an electrochemical cell of a proton conductor in which hydrogen in the anode compartment can be electrochemically pumped to the cathode by sending a direct current. Experiments of hydrogen pumps were conducted with the following gas atmospheres.

$$\text{Moist } H_2,\ \text{Pt}|\text{electrolyte}|\text{Pt},\ \text{moist Ar} \tag{12.4}$$

The gases in both the anode and cathode compartments were moistened with water vapor. In general, slight moisture should be present in gases, both in the anode and cathode compartments in an electrochemical cell, for the proton-conducting oxides, because protons are formed by dissolution of water molecules into the oxide ion vacancy, as expressed in Eq. (12.1) or (12.2). In addition, water has a role to avoid an atmosphere that is too reducing when hydrogen is used; otherwise, irreversible reduction of the electrolyte will take place.

Figure 12.3 compares hydrogen pumps using $SrCe_{0.95}Yb_{0.05}O_{3-\alpha}$ and $SrZr_{0.9}Y_{0.1}O_{3-\alpha}$ as the proton-conducting electrolytes at 800°C [18]. Porous platinum was used as the electrodes. In the case of $SrCe_{0.95}Yb_{0.05}O_{3-\alpha}$, the evolution rate of hydrogen at the cathode agrees with Faraday's law up to around 600 mA/cm^2, as shown in Fig. 12.3(a). It is notable in the figure that electrode overpotentials are low for the cerate, as seen in Fig. 12.3(b). In contrast, a hydrogen pump using $SrZr_{0.9}Y_{0.1}O_{3-\alpha}$ has quite poor performance. As shown in Fig. 12.3(a), the hydrogen pumping rate deviates from Faraday's law at a current density as low as 20 mA/cm^2. Figure 12.3(b) indicates quite high overpotentials (>1 V) at both the anode and cathode for such a low current density. Therefore, the hydrogen pump using $SrCe_{0.95}Yb_{0.05}O_{3-\alpha}$ electrolyte shows much superior performance in both energy efficiency and hydrogen pumping rate.

We usually assume a three-phase boundary of gas–electrode–electrolyte for the electrode reactions to take place. Therefore, the foregoing experimental results suggest that the electrocatalytic activity of platinum for the cerate will be much higher than that for the zirconate. The electrode reaction zone will not be restricted to the three-phase boundary, which is mathematically one

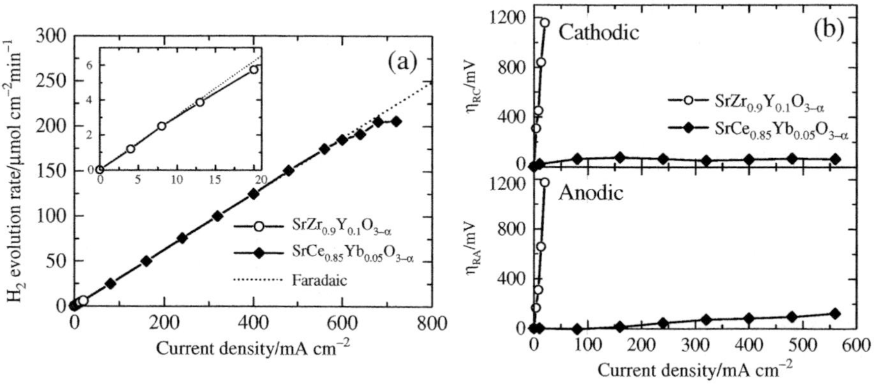

Fig. 12.3 (**a**) Evolution rate of hydrogen at the cathode and (**b**) electrode overpotentials in the ionic current region during hydrogen pumping at 800°C as functions of current density using $SrZr_{0.9}Y_{0.1}O_{3-\alpha}$ and $SrCe_{0.95}Yb_{0.05}O_{3-\alpha}$ electrolytes [18]; the *dotted line* in (**a**) indicates the theoretical evolution rate of hydrogen calculated from Faraday's law. The *left upper* graph in (**a**) is shown in smaller scale for $SrZr_{0.9}Y_{0.1}O_{3-\alpha}$, and concentration polarizations were excluded in (**b**). Porous platinum electrodes were used; anode gas = moist H_2 ($p(H_2O) = 2.3 \times 10^3$ Pa, fed at 30 mL min^{-1}); cathode gas = moist Ar ($p(H_2O) = 2.3 \times 10^3$ Pa, at 30 mL min^{-1}). (Reprinted from [18] with permission from Elsevier)

dimensional, but should spread to the gas–electrode interface by surface diffusion of ionic species on the electrode and/or to the gas–electrolyte interface by an electronic species locally present in the electrolyte. If one assumes similar microstructure of porous platinum electrodes for the two cases, the latter gas–electrolyte interface might be a reason for the large difference in the electrode overpotentials between the combinations of Pt/$SrCe_{0.95}Yb_{0.05}O_{3-\alpha}$ and Pt/$SrZr_{0.9}Y_{0.1}O_{3-\alpha}$.

Electrode performance will primarily be dependent on the choice of the electrode material, and specific properties should be determined and discussed for each material. However, the aforementioned result qualitatively predicts that cerates tend to have better performance when used for the electrode than zirconates.

12.4 Stability

Because alkaline earth metals are highly basic, the proton conductors tend to react with CO_2 [5, 8–10]. A carbonation reaction will take place for the perovskite oxides.

$$ABO_3 + CO_2 = ACO_3 + BO_2 \tag{12.5}$$

where A is Ba or Sr and B is Ce or Zr. As shown in Fig. 12.4, this reaction occurs with 1 bar CO_2 at temperatures less than a critical temperature (around

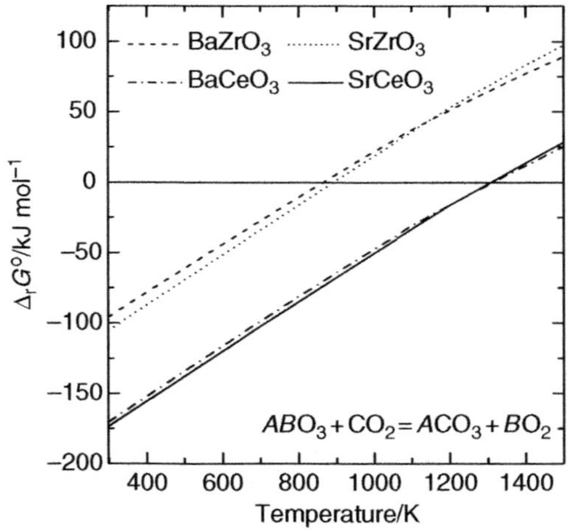

Fig. 12.4 Standard Gibbs free energy change of carbonation reactions of alkaline earth cerates and zirconates as a function of temperature calculated from a thermodynamic database [20]; standard pressure, $p^{\circ} = 1 \times 10^5$ Pa. (Reprinted from [24] with permission from Springer Science and Business Media)

1300 K for cerates and 850–900 K for zirconates) [19, 20]. Therefore, if hydrocarbon is used for a fuel cell with barium cerate-based electrolyte, the cell should be operated at a temperature higher than the critical temperature to avoid deterioration by the reaction with CO_2. Several studies suggest that cerate zirconate solid solutions are stable in CO_2 [21–23]. Figure 12.5 shows the X-ray diffraction (XRD) patterns of $SrCe_{0.95}Yb_{0.05}O_{3-\alpha}$, $SrZr_{0.9}Y_{0.1}O_{3-\alpha}$, and $BaCe_{0.6}Zr_{0.3}Nd_{0.1}dO_{3-\alpha}$ before and after the treatment with carbon dioxide; sintered perovskites were exposed at 800°C for 3 h in 100% CO_2 and the H_2–CO–CO_2 mixture. In the case of $SrCe_{0.95}Yb_{0.05}O_{3-\alpha}$, CeO_2 and $SrCO_3$ were formed on exposure to both CO_2 and H_2–CO–CO_2 mixture gases as revealed by Fig. 12.5(a). The electrolyte obviously reacted with CO_2, as shown in Eq. (12.6), to form $SrCO_3$ and CeO_2.

$$SrCeO_3 + CO_2 \rightarrow SrCO_3 + CeO_2. \tag{12.6}$$

Occurrence of the carbonation reaction is consistent with the thermodynamic data shown in Fig. 12.4. The XRD patterns of $SrZr_{0.9}Y_{0.1}O_{3-\alpha}$ before and after the CO_2 treatments are shown in Fig. 12.5(b); the intensity axis is expanded to confirm reflections from tiny components. The XRD patterns are entirely unchanged after the treatments in both atmospheres. In the case of $BaCe_{0.6}Zr_{0.3}Nd_{0.1}O_{3-\alpha}$ in Fig. 12.5(c), the XRD pattern after exposure in CO_2 shows almost no change. When the XRD patterns are seen minutely, faint reflection peaks at 33° and 47°, probably from CeO_2, are observed after the treatments with either CO_2 or a H_2–CO–CO_2 mixture. Decomposition of the perovskite structure (Eq. (12.6)) is suspected. Another possibility is evaporation of barium at the surface of the specimen, resulting in the formation of CeO_2–ZrO_2 fluorite. Long-term stability should be examined for the solid solution case.

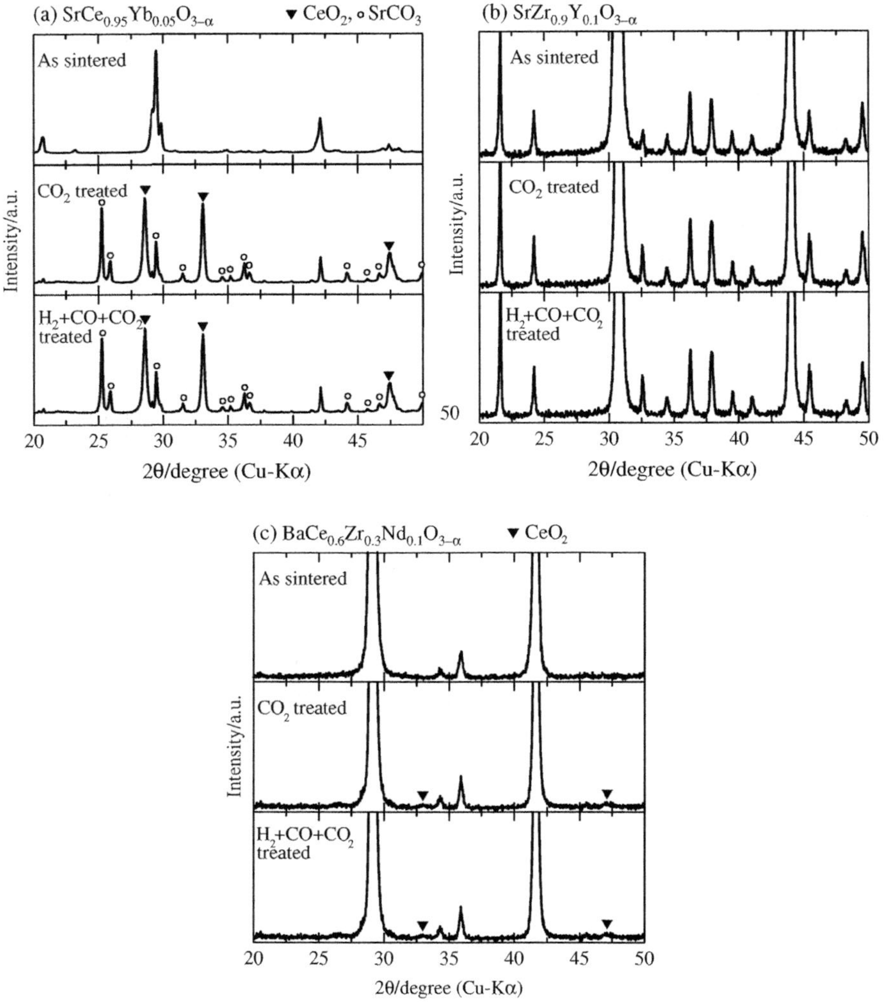

Fig. 12.5 XRD patterns of (**a**) $SrCe_{0.95}Yb_{0.05}O_{3-\alpha}$, (**b**) $SrZr_{0.9}Y_{0.1}O_{3-\alpha}$, and (**c**) $BaCe_{0.6}Zr_{0.3}Nd_{0.1}O_{3-\alpha}$ as prepared and of those treated in CO_2 and in H_2–CO–CO_2 mixture gases at 800°C for 3 h [24]. The mixture gas contains H_2 and CO with a molar ratio of 21 blended with 20% CO_2. The CO_2 and H_2–CO–CO_2 mixture gases were moistened with water vapor at p (H_2O) = 1.9×10^3 Pa (saturated at 17.0°C) Electrolyte specimens were exposed in the CO_2-containing atmospheres for 3 h and moderately quenched to room temperature (in about 10 min from 800° to 200°C), and the changes in crystal phase of the electrolyte surface were determined by XRD. (Reprinted from [24] with permission from Springer Science and Business Media)

12.5 Dopant

As mentioned, acceptor doping is essential for the protonic conduction to occur in the perovskite oxides, because hydration (Eq. (12.1)) should occur for the oxide to become proton conducting. For example, 10 mol% Y-doped $BaCeO_3$ can be written as $BaCe_{0.9}Y_{0.1}O_{2.95}$, which will be the case in a dry atmosphere and/or at too high temperature, and will become $BaCe_{0.9}Y_{0.1}H_{0.1}O_3$ when it is fully hydrated by moisture. Comparing the two formulae, it is clear that the maximum level of proton concentration is equal to the amount of introduced dopant, i.e., proton concentration is maximally 10 mol% because the Y dopant concentration is 10 mol%; when the valence of the dopant cation is doubly smaller, e.g., substitution of Ca (II) for Ce (IV), the maximum concentration of proton is also doubled. In this regard, the role of the dopant is simply the creation of the oxide ion vacancy, symbolized by α in the above written formulae, to receive hydration (entrance of water molecules into oxide ion vacancies) to generate protonic charge carriers. However, actually, the proton conductivity depends strongly on the type of dopant.

Figure 12.6 shows the temperature dependence of the conductivity of $BaCe_{0.9}M_{0.1}O_{3-\alpha}$ ($M = Y$, Tm, Yb, Lu, In, or Sc) measured in moist H_2 (closed symbols) and in moist O_2 (open symbols), respectively, in the temperature range of 400°–900°C [5]. All specimens show the typical conductivity behavior of proton conductor, as already explained in Fig. 12.1. The most significant effect of the dopant is on the magnitude of the conductivity. In cases whose the dopant cation is Y, Tm, Yb, or Lu, the conductivities are only slightly different. However, In or Sc doping causes a significant drop in conductivity. The

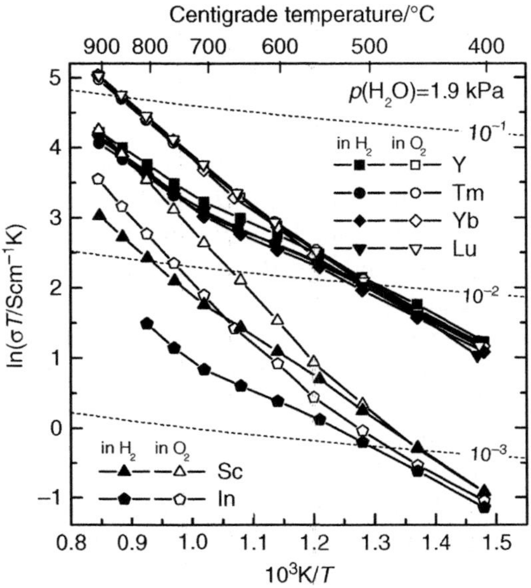

Fig. 12.6 Arrhenius plot of conductivity of $BaCe_{0.9}M_{0.1}O_{3-\alpha}$ ($M = Y$, Tm, Yb, Lu, In, or Sc) in moist H_2 (*closed symbols*) and O_2 (*open symbols*) in the temperature range of 400°–900°C; $p(H_2O) = 1.9 \times 10^3$ Pa. *Dashed lines* indicate isoconductivity lines in S cm^{-1}. (Reprinted from [5] with permission from Electrochemical Society)

conductivity in moist H_2 is plotted as a function of the ionic radius of the dopant cations in Fig. 12.7(a). showing a significant effect of the dopant cation size on the ionic conductivity. Y^{3+} (ionic radius, $r_i = 90$ pm), which is the largest dopant, was found to provide the highest conductivity. The decrease in the ionic radius up to Lu^{3+} ($r_i = 86.1$ pm) results in a gradual decrease in ionic conductivity. In case of Sc^{3+} ($r_i = 74.5$ pm) and In^{3+} ($r_i = 80$ pm), which are much smaller than the former four dopants, the ionic conductivity is lower by nearly one order of magnitude.

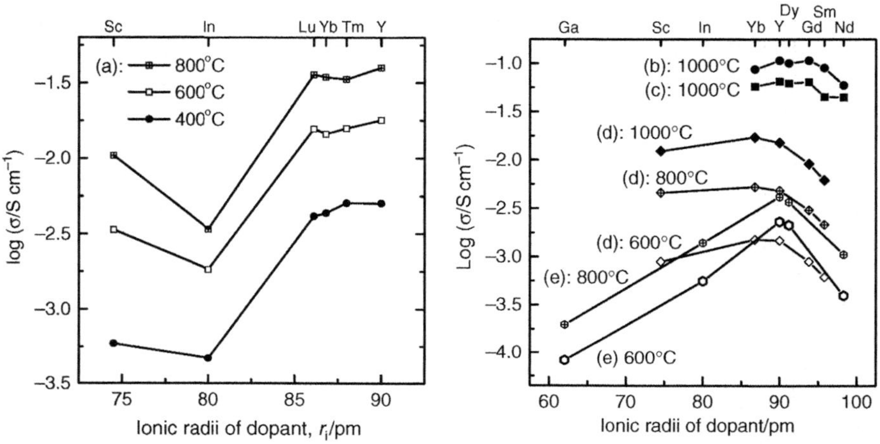

Fig. 12.7 Dependence of electrical conductivities of $AB_{1-x}M_xO_{3-\alpha}$ in moist atmospheres plotted on the ionic radii of the dopant M: **(a)** $BaCe_{0.9}M_{0.1}O_{3-\alpha}$ [5], **(b)** $BaCe_{0.8}M_{0.2}O_{3-\alpha}$ [25], **(c)** $BaCe_{0.9}M_{0.1}O_{3-\alpha}$ [25], **(d)** $SrCe_{0.95}M_{0.05}O_{3-\alpha}$ [26], **(e)** $BaZr_{0.95}M_{0.05}O_{3-\alpha}$ [27]. Conductivities were measured either in wet hydrogen or in wet argon. (The figure on the left reprinted from [5] with permission from Electrochemical Society)

Figure 12.7 shows the electrical conductivity of some proton-conducting perovskite oxides with different dopants in the same concentration collected from the literature [25–27]. We can see that there is a relationship between conductivity and the ionic size of the dopants, the same as the foregoing discussion, and that there is a suitable size of the dopant cation with respect to the host perovskite oxides. We should remember that the dopant is a minor component in the total chemical formulae of the proton-conducting perovskite oxide; as seen in the figure, even dopant at the level of 5 mol% can strongly affect the conductivity. Why is the conductivity so strongly dependent on the kind of dopants?

Optical absorption studies suggest that there will be at least two different kinds of sites for protons to reside in the oxide crystal lattice [28, 29]. As already stated, a proton can exist in the oxide by forming hydrogen bonds to the lattice oxygen. The O–H vibration modes can be observed in infrared absorption spectroscopy and are different according to the type of dopant. Yugami et al. and Omata et al. reported that the spectra in the ν_{OH} region can be decomposed into four bands [28, 29]. It is notable that the frequencies of the bands are not so different by the types of dopants, but the dopants affect the fraction of the bands [28], resulting in different shapes of the infrared absorption spectra with the type of the dopants. For example, doping of Sc, i.e., $SrZr_{0.95}Sc_{0.05}O_{3-\alpha}$ that has comparatively low conductivity, results in a high vibrational component of a high-frequency O–H band, whereas Y-doping that allows high conductivity for strontium zirconate leads to the growth of low-frequency bands. From these observations, Omata et al. concluded that the high-frequency O–H bands will be formed by less mobile protons and the low-frequency modes will be attributed to mobile protons [29].

It is the lattice oxygen that associates with a proton. Therefore, the difference in the mobility of protons will probably originate from the different character of the oxygen atoms with which protons are associated. In the perovskite structure, the B-site cation coordinates six oxygen atoms to form octahedra that connect with one another via oxygen atoms to form a crystal lattice. Accordingly, one oxygen atom has twofold coordination with B-site cations. Therefore, there are at least two kinds of oxygen atoms, i.e., an oxygen connecting to one dopant cation and one original B-site cation (O in *M*–O–*B*) and an oxygen connecting to two original B-site cations (O in *B*–O–*B*). Yugami et al. and Omata et al. suggested oxygen that has a bond to Sc will probably be trapping; a proton trapped to the oxygen atom at Sc–O–Zr has strong hydrogen bonds to be immobilized, resulting in the growth of high-frequency O–H vibration. The reverse will be true for the Y-doped case [28, 29].

Kreuer et al. proposed an idea of chemical matching; i.e., the dopant that provides the least effect on the basicity of adjacent oxygen best chemically matches, and both the ionic radius and electronegativity of the dopant will determine the chemical matching [30]. This idea is consistent with the foregoing discussion; i.e., the dopant that is indistinguishable from the original cations will add the least effect on the bonding oxygen, resulting in an environment for the protons similar to that around the original B-site cations.

With respect to the effect of dopants, there is an interesting observation about the relationship between the conductivity and chemical stability [5]. The dopant governs not only the conductivity but also the stability, and thus the two properties are likely to have a trade-off relationship. For evaluation of the chemical stability of the proton-conducting perovskite oxides, the reactivity with CO_2, i.e., the reaction shown in Eq. (12.5), can be used. As shown in Fig. 12.4, free energy change of the reaction is positive at high temperature, i.e., the perovskite is stable. The free energy change decreases with decreasing temperature and becomes negative at a temperature; that is, the carbonation reaction will actually occur below the temperature. The critical temperature decreases with increasing stability of the oxide so that the reaction temperature is a good measure for their chemical stability. The reaction temperature can be determined by thermogravimetry (TG) in the CO_2 stream. The result obtained for the specimens during cooling in CO_2 are shown in Fig. 12.8(b) together with the temperature program in Fig. 12.8(a). In the TG curves, an increase in the sample weight indicates the occurrence of carbonation in the reaction (Eq. (12.5)). The mass changes, $\Delta m/m_0$, were eventually around 15%, and this is a reasonable quantity to assume the reaction to have taken place. The TG curves are not so steep and thus the reaction temperature, evaluated by the inflection points, will include a significant deviation from equilibrium values to lower temperatures due to the contribution from the kinetics of the carbonation reaction. However, the TG curves apparently shift and reflect a change in the stability of the materials with variations in the dopants. Y-doping results in carbonation occurring at the highest temperature. The reaction temperature decreases with decreasing ionic size of the dopant cations. Figure 12.9 shows the CO_2 reaction temperature (open squares) as a function of the dopant cation radius; the temperature was taken upside down to show the upper plots having higher chemical stability. A monotonical relation can be seen between the chemical stability and the ionic size of the dopant.

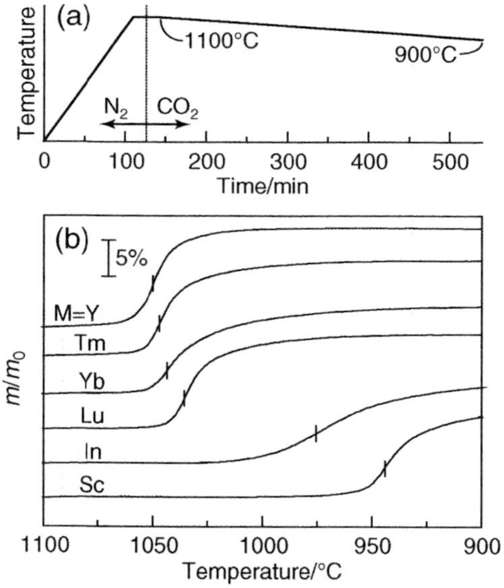

Fig. 12.8 CO_2-reactivity test of $BaCe_{0.9}M_{0.1}O_{3-\alpha}$ ($M = $Y, Tm, Yb, Lu, In, or Sc). (**a**) Temperature program and introduced gases during thermogravimetry (TG) and (**b**) mass change, $\Delta m/m_0$, during cooling the specimens at –0.5 K/min in CO_2 flow as a function of temperature; *bars* show inflection points that were assigned as reaction temperatures. A crushed sample of 20–30 mg in a Pt pan was heated in a N_2 stream up to 1100°C, and the TG analysis was then performed; an exposure to CO_2 during heating was avoided because the procedure will result in decomposition of the perovskite-type structure that cannot be recovered at 1100°C. (Reprinted from [5] with permission from Electrochemical Society)

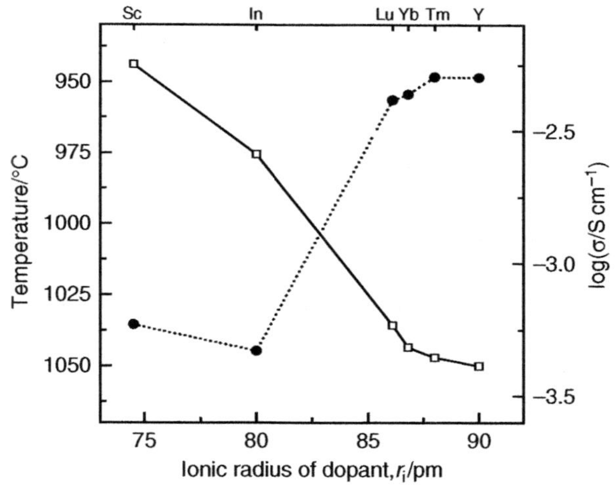

Fig. 12.9 CO_2 reaction temperature [*open squares*, assigned by the inflection points in Fig. 12.8(b)] and the conductivity isotherm at 400°C (the same as in Fig. 12.6) as functions of the dopant cation radius; the CO_2 reaction temperature was plotted upside down to ensure the upper plots having higher chemical stability. (Reprinted from [5] with permission from Electrochemical Society)

In Fig. 12.9, the conductivity isotherm at 400°C is also shown. There is a clear trade-off relationship between the ionic conductivity and chemical stability of the trivalent cation-doped barium cerate: Phenomenonologically, the conductivity and stability are connected and one will decrease in response to the other increasing. This kind of relationship has often been supposed for different base perovskites; e.g., cerates generally have high protonic conductivity and zirconates are chemically stable [31]. It has been demonstrated here that the relationship will also be true for the variation in the kind of dopant. Figure 12.9 suggests that not only the conductivity but also the chemical stability are governed by the ionic size of the dopant cation. From a practical perspective, the relationship between the electrical property and the chemical stability can be useful in optimizing the dopant cation in a proton-conducting oxide system.

12.6 Proton Hole Mixed Conduction

The foregoing discussion was made on the basis that the proton-conducting perovskite oxides generally have high ionic transport numbers in moist hydrogen. Hydrogen concentration cells generate electromotive forces in agreement with those predicted by the Nernst equation. For this reason, the oxide can be used for hydrogen sensors, hydrogen pumps, and fuel cells. Provided we can additionally introduce electronic conductivity in the proton-conducting solids, hydrogen permeation will occur through the materials. Mixed ionic-electronic conductors (MIECs) allow gas permeation by the codiffusion of ionic and electronic charge carriers, referred to in terms of ambipolar diffusion. There are many examples of mixed conduction in oxides for oxygen permeation membranes [32–34]. The MIEC having proton as the ionic charge carrier will serve as a hydrogen-separating membrane. No external electric power supply is necessary, so that the electrodes and the external electrical circuit are excluded to construct a simple hydrogen separator; the membrane uses the hydrogen potential gradient as the driving force for the hydrogen transport. In addition, the protonic mixed conductor works as electrodes in combination with the proton-conducting perovskite oxides to be used as the electrolyte.

It is likely that there have been no accepted materials so far as the protonic-electronic mixed conductors that allow high hydrogen transfer. Therefore, in contrast to the conventional high-temperature proton conductors (HTPCs), the protonic-electronic mixed conductors are not yet established. We recently found that the protonic-electronic mixed conduction occurs in ruthenium-doped perovskites, $BaCe_{0.9-x}Y_{0.1}Ru_xO_{3-\alpha}$ ($x = 0$–0.1) [35]; below is the summary of our findings.

Hydrogen permeation can be evaluated for a specimen using the following cell.

$$H_2 \text{ at } p(H_2, \text{supp}) \,|\text{specimen}|\, \text{Ar at } p(H_2, \text{ perm}) \tag{12.7}$$

If one assumes proton (H^+), oxide ion (O^{2-}), conduction electron (e), and electron hole (h) as ionic and electronic charge carriers, hydrogen evolution at the permeating side will obey the ambipolar diffusion mechanism, expressed by Eq. (12.17) [36]:

$$J_{H_2} \approx \frac{RT}{4F^2L} \frac{(\sigma_{H^+} + \sigma_{O^{2-}})(\sigma_e + \sigma_h)}{\sigma_T} \ln \frac{p(H_2, \text{supp})}{p(H_2, \text{perm})} \tag{12.8}$$

in which σ_j and σ_T are the partial conductivity of j species and the total conductivity, respectively, and L is the thickness of the mixed conductor; the equation assumes equal water vapor pressure in the two compartments and constant partial conductivities. The ambipolar hydrogen flux is proportional to $\ln(p(H_2,\text{supp})/p(H_2,\text{perm}))$.

Figure 12.10 shows the hydrogen flux through $BaCe_{0.9-x}Y_{0.1}Ru_xO_{3-\alpha}$ ($x = 0.075$, 0.1) at 800°C measured with the cell in Eq. (12.7); hydrogen permeation was not experimentally observed for the Ru-undoped sample ($x = 0$). It is notable that the hydrogen flux is proportional to $\ln(p(H_2,\text{supp})/p(H_2, \text{perm}))$, suggesting that the hydrogen transport occurred in accordance with the ambipolar diffusion mechanism in Eq. (12.8). It is evident that the hydrogen permeation occurs via the ionic-electronic mixed conduction in the Ru-doped perovskites. The maximum hydrogen flux obtained is 6.5×0^{-8} mol/s cm^2 ($x = 0.1$), equivalent to the internally short-circuited current density of 12.5 mA/cm^2.

As explained in previous sections, protonic and oxide-ionic change carriers are assumable for the proton-conducting perovskite oxides; the $SrCe_{0.95}Yb_{0.05}O_{3-\alpha}$ case has been shown in Fig. 12.2. By observing the change in water vapor pressure in the permeating compartment, 0.5 of the ionic

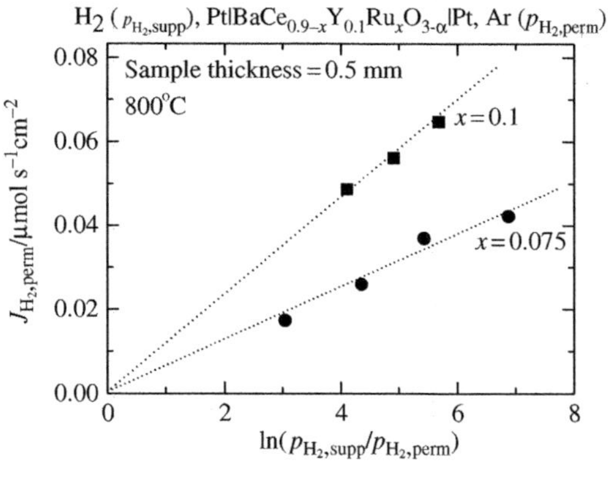

Fig. 12.10 Hydrogen permeation rate through $BaCe_{0.825}Y_{0.1}Ru_{0.075}O_{3-\alpha}$ and $BaCe_{0.8}Y_{0.1}Ru_{0.1}O_{3-\alpha}$ at 800°C as a function of the logarithm of the hydrogen pressure ratio (hydrogen supply side to permeate side) [35]. Sample thickness = 0.5 mm. H_2/Ar mixtures were fed to the left compartment of cell (12.7), and argon gas was passed to sweep the hydrogen evolved in the right compartment. The gases contained water vapor at 1.9×10^3 Pa and were fed at 30 mL/min. The partial pressures of hydrogen in the Ar sweep gas, $p(H_2$, perm) was determined by gas chromatography. (Reprinted from [35] with permission from Electrochemical Society)

conduction was estimated to be due to the oxide ions and the other half could be attributed to protons in the case of $BaCe_{0.8}Y_{0.1}Ru_{0.1}O_{3-\alpha}$ at 800°C. This fraction is quite similar to that of the base HTPC of the acceptor-doped barium cerate. Therefore, the doping of Ru will provide an electronic conduction without changing the protonic/oxide-ionic fraction of conductivity in the acceptor-doped $BaCeO_3$.

To investigate if electron holes or conduction electrons are responsible for the electronic conduction, soft X-ray absorption spectroscopy (XAS) was employed, expecting that the electron hole can be observed by this technique [37, 38]. This technique elucidates unoccupied density of states, and thus electron holes, unoccupied states in the valence band, can be detected. Figure 12.11 shows the XAS spectra of $BaCe_{0.9-x}Y_{0.1}Ru_xO_{3-\alpha}$ ($x = 0$ and 0.075) as sintered in air and after annealed in moist 1% H_2 at 800°C for 10 h. In the case that the specimens were untreated after sintering in ambient air containing humidity, the spectra of both undoped and Ru-doped specimen have peaks just below the top of the valence band. The peaks at this position are attributable to electron holes present by the defect equilibrium shown in Eq. (12.3). The undoped specimen annealed in moist 1% H_2 has no peak at the position, due to the equilibrium leaning to the left as $p(O_2)$ decreases. However, in the case of the Ru-doped sample, the hole peak remains even after the treatment in the reducing atmosphere. This fact suggests that the electronic species for the mixed conduction will be attributed to the hole formed in the Ru-doped materials.

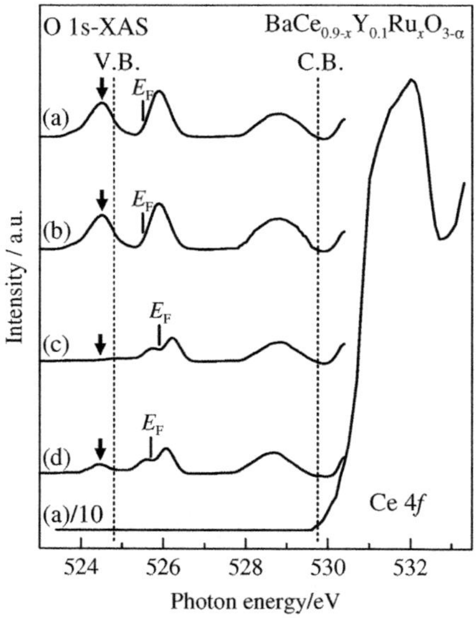

Fig. 12.11 O 1s X-ray absorption spectroscopy (XAS) spectra of $BaCe_{0.9}Y_{0.1}O_{3-\alpha}$ and $BaCe_{0.825}Y_{0.1}Ru_{0.075}O_{3-\alpha}$ measured at room temperature. The spectra (**a**) and (**b**) are of $BaCe_{0.9}Y_{0.1}O_{3-\alpha}$ and $BaCe_{0.825}Y_{0.1}Ru_{0.075}O_{3-\alpha}$, respectively, untreated after sintering in moist air and thus reflecting the nature in an oxidative atmosphere; the spectra (**c**) and (**d**) are of those annealed in moist 1%H_2 diluted by Ar at 800°C for 10 h exhibiting the electronic structure in the reducing atmosphere. *V.B.* and *C.B.*, the positions of the top of the valence band and the bottom of the conduction band, respectively [35]. The measurements were conducted at the revolver undulator beamline BL-19B at the Photon Factory of the High Energy Accelerator Research Organization, Tsukuba, in Japan. (Reprinted from [35] with permission from Electrochemical Society)

It is not clear how the Ru doping results in the formation of an electron hole in reducing atmosphere. However, it is possible to assume a kind of energy level produced by the Ru-4*d* orbitals, which locate just above the valence band and accept electrons from the valence band to generate electron holes. Additional experiments, e.g., determination of the valence of Ru and the conductivity dependences on $p(O_2)$ and $p(H_2O)$, will provide a definite model for the protonic-electronic mixed conduction in Ru-doped perovskites.

References

1. H. Iwahara, T. Esaka, H. Uchida, N. Maeda, Solid State Ionics 3–4, 359 (1981)
2. T. Scherban, A.S. Nowick, Solid State Ionics 35(1–2), 189 (1989)
3. H.G. Bohn, T. Schober, J. Am. Ceramic Soc. 83(4), 768 (2000)
4. J.F. Liu, A.S. Nowick, Solid State Ionics 50(1–2), 131 (1992)
5. H. Matsumoto, Y. Kawasaki, N. Ito, M. Enoki, T. Ishihara, Electrochem. Solid-State Lett. 10(4), B77 (2007)

6. N. Ito, M. Iijima, K. Kimura, S. Iguchi, J. Power Sources 152, 200 (2005)
7. T. Yajima, K. Koide, H. Takai, N. Fukatsu, H. Iwahara, Solid State Ionics 79, 333 (1995)
8. M.J. Scholten, J. Schoonman, J.C. van Miltenburg, H.A.J. Oonk, Solid State Ionics 61(1–3), 83 (1993)
9. S. Gopalan, A.V. Virkar, J. Electrochem. Soc. 140(4), 1060 (1993)
10. N. Bonanos, K.S. Knight, B. Ellis, Solid State Ionics 79, 161 (1995)
11. H. Iwahara, N. Maeda, H. Uchida, Solid State Ionics 11, 117 (1983)
12. N. Fukatsu, N. Kurita, T. Yajima, K. Koide, T. Ohashi, J. Alloys Compd. 231(1–2), 706 (1995)
13. N. Bonanos, B. Ellis, K.S. Knight, M.N. Mahmood, Solid State Ionics 35(1–2), 179 (1989)
14. T. He, K.D. Kreuer, Y.M. Baikov, J. Maier, Solid State Ionics 95(3–4), 301 (1997)
15. H. Iwahara, T. Yajima, H. Ushida, J. Electrochem. Soc. 140(6), 1687 (1993)
16. H. Iwahara, T. Yajima, H. Ushida, Solid State Ionics 70–71(pt 1), 267 (1994)
17. M. Wider?e, N. Kochetova, T. Norby, Dalton Transactions (19), 3147 (2004)
18. H. Matsumoto, T. Shimura, H. Iwahara, T. Higuchi, K. Yashiro, A. Kaimai, T. Kawada, J. Mizusaki, J. Alloys Compd. 408–412, 456 (2006)
19. H. Yokokawa, S. Yamauchi, T. Matsumoto, Calphad 26(2), 155 (2002)
20. Thermodynamic-database-MALT-group, Thermodynamic database for Windows. Kagaku Gijyutu-sha (2005)
21. K. Katahira, Y. Kohchi, T. Shimura, H. Iwahara, Solid State Ionics 138 (1–2), 91 (2000)
22. S. Wienstroer, H.D. Wiemhofer, Solid State Ionics 101–103 (PART 1), 1113 (1997)
23. K.H. Ryu, S.M. Haile, Solid State Ionics 125(1), 355 (1999)
24. H. Matsumoto, S. Okada, S. Hashimoto, K. Sasaki, R. Yamamoto, M. Enoki, T. Ishihara, Ionics 13(2), 93 (2007)
25. H. Iwahara, T. Hibino, M. Yamada, Proc. 3rd Int. Symp. Solid Oxide Fuel Cells 93–94, 137 (1993)
26. A. Mineshige, S. Okada, M. Kobune, T. Yazawa, Solid State Ionics 177 (26–32), 2443 (2006)
27. H. Iwahara, T. Yajima, T. Hibino, K. Ozaki, H. Suzuki, Solid State Ionics 61 (1–3), 65 (1993)
28. H. Yugami, Y. Shibayama, S. Matsuo, M. Ishigame, S. Shin, Solid State Ionics 85(1–4), 319 (1996)
29. T. Omata, Y. Noguchi, S. Otsuka-Yao-Matsuo, Solid State Ionics 176(39–40), 2941 (2005)
30. W. Munch, K.D. Kreuer, S. Adams, G. Seifert, J. Maier, Phase Transitions 68(3), 567 (1999)
31. K.D. Kreuer, Solid State Ionics 97(1–4), 1 (1997)
32. T. Ishihara, T. Yamada, H. Arikawa, H. Nishiguchi, Y. Takita, Solid State Ionics 135 (1–4), 631 (2000)
33. Y. Teraoka, H.M. Zhang, K. Okamoto, N. Yamazoe, Materials Res. Bull. 23 (1), 51 (1988)
34. H. Takamura, K. Enomoto, Y. Aizumi, A. Kamegawa, M. Okada, Solid State Ionics 175(1–4), 379 (2004)
35. H. Matsumoto, T. Shimura, T. Higuchi, H. Tanaka, K. Katahira, T. Otake, T. Kudo, K. Yashiro, A. Kaimai, T. Kawada, J. Mizusaki, J. Electrochem. Soc. 152(3), (2005)
36. T. Norby, Solid State Ionics 125(1), 1 (1999)
37. T. Higuchi, H. Matsumoto, T. Shimura, K. Yashiro, T. Kawada, J. Mizusaki, S. Shin, T. Tsukamoto, Jpn. J. Appl. Phys. Part 1: Regular Papers and Short Notes and Review Papers 43(8 A), 5419 (2004)
38. T. Higuchi, T. Tsukamoto, N. Sata, T. Hattori, S. Yamaguchi, S. Shin, Solid State Ionics 175(1–4), 549 (2004)

Chapter 13
Mechanisms of Proton Conduction in Perovskite-Type Oxides

K.D. Kreuer

13.1 Introduction

The proton is the only ion of chemical significance that has no electron shell on its own. It, therefore, strongly polarizes its environment, which leads to high binding energies. Speaking in terms of transport, this corresponds to a high degree of self-localization and implies a strong coupling of long-range proton motion to the dynamics of the proton environment. This is actually one common feature of all proton conduction mechanisms in quite diverse families of compounds [1].

In the most trivial case, the proton is just diffusing as part of a protonated species (e.g., H_3O^+), which allows for an overall proton transport provided the unprotonated species (e.g., H_2O) are also mobile (vehicle mechanism [2, 3]). In most proton conductors, however, the proton is actually transferred between different binding sites. As just pointed out, this has to involve a strong "proton–phonon coupling," which was first recognized for proton conduction in ice by Fischer et al. in the late 1960s [4]. Their considerations were surely inspired by the mechanisms of electron transfer in solution, i.e., Marcus theory [5], and by the theory of incoherent phonon-assisted hydrogen tunneling in metals that was developed at the same time [6]. All these mechanisms have in common that the mobile species (proton, electron, hydrogen) are following a moving structural pattern (e.g., proton solvation shell), and it is the activation of this "structural diffusion" that is assumed to control the total activation enthalpy of, e.g., proton diffusion. In the case of proton mobility within hydrogen-bonded liquids, this is usually the rearrangement of the hydrogen bond pattern, which requires thermally activated hydrogen bond-breaking and bond-forming processes. This effect has particularly been studied for the mobility of excess protons in water. Today's understanding of proton conduction in water is

K.D. Kreuer (✉)
Max-Planck-Institut für Festkörperforschung, Heisenbergstr. 1, D-70569, Stuttgart, Germany
e-mail: Kreuer@fkf.mpg.de

T. Ishihara (ed.), *Perovskite Oxide for Solid Oxide Fuel Cells*,
Fuel Cells and Hydrogen Energy, DOI 10.1007/978-0-387-77708-5_13,

essentially based on simulation and experimental work, which has been performed independently in 1995 by Tuckerman et al. and Agmon [7, 8]. Essential features of the suggested mechanism are that (i) intermolecular proton transfer occurs through rather short hydrogen bonds; (ii) this transfer is highly coupled to hydrogen bond-breaking and bond-forming processes in the weaker bonded second hydration sphere of the protonated species (H_3O^+ or $H_5O_2{}^+$); and (iii) the position of the protonic charge follows the center of symmetry of the hydrogen bond pattern [9]. In other words, proton transfer and structural reorganization, which together form an uninterrupted trajectory for proton translocation, take place in different parts of the hydrogen-bonded structure. In this way, both proton transfer and structural reorganization, the two essential ingredients of "structure diffusion" may take place at very high rates [9]. Later, a similar mechanism has been found for proton conduction in imidazole [10], an amphoteric molecular compound, which shows high intrinsic proton conductivity in the liquid state. As in the case of phosphoric and phosphonic acid, this is also the result of the high degree of self-dissociation, which is low in the case of water.

Albeit the proton conduction mechanisms in hydrogen-bonded networks comprise the interactions of many species, proton conduction mechanisms in oxides are usually less complex; this has to do with the fact that protons are usually present at low concentrations as part of noninteracting [11], positively charged point defects (generally hydroxide ions residing on oxide ion sites, $OH_O^{\bullet}$). These defects show significant hydrogen bond interaction with neighboring oxide ions, and it is the local dynamics of this hydrogen bonding that turned out to be the clue to the understanding of proton conduction mechanisms in perovskite-type oxides. Only in very few perovskite-type oxides, such as $Ba_2YSnO_{5.5}$ [12], very high proton content may add some complexity to the proton conduction mechanism.

In the first part of this short chapter, it is discussed where protons find energetically favored sites in oxides with the perovskite structure, before the mechanisms of proton diffusion via these sites are described in detail. While this discussion is restricted to structure and dynamics of protonic defects in ideal (cubic) perovskites, the following section addresses complications such as symmetry reduction and the effect of the presence of dopants, before, finally, a few implications for the development of proton-conducting electrolytes for fuel cell applications are discussed.

13.2 Proton Sites

The possible positions for protons in perovskite-type oxides are strongly constrained through the strong binding interaction with the oxygen, i.e., the most electronegative species in this type of compound, with which protons form hydroxide ions. In other words, protons are localized within the valence electron density of the oxygen, which attains some H1s character. The orientation of the hydroxide ion is then mainly determined by the hydrogen bonding with other, neighboring oxide ions, which is an attractive interaction, and the interaction with the cations of the structure, which, of course, is repulsive. Although clear signatures of hydrogen bonding were found in the OH-stretching vibration [13–18], i.e., typical red-shifted continua, there have been different suggestions for the orientation of these bonds. Within the simple cubic perovskite structure, i.e., a framework of corner-sharing undistorted oxygen octahedra, each oxygen has eight nearest oxygen neighbors with separations corresponding to the edge length of the octahedra, and four next nearest neighbors, which reside on the vertices of the four neighboring octahedra in the same plane.

A proton jump width of 170 pm obtained from quasi-elastic neutron scattering (QNS) spectra of Y-doped $SrZrO_3$ was interpreted as the separation between proton sites on the edges of neighboring octahedra corresponding to linear hydrogen bonding of the hydroxide ions to their nearest oxygen neighbors [19] Similar positions have been favored by first-principles molecular dynamics simulations of protons in Sc-doped $SrTiO_3$ [20, 21]. From a neutron and X-ray powder diffraction study on deuterated $Ba_3Ca_{1+y}Nb_{2-y}O_{9-\delta}$, however, the deuterons were refined to point toward the four next nearest oxygens [22]. These positions actually do not violate the empirical Badger–Bauer rule [23], which is excluding hydrogen bonding within the same coordination polyhedron. The uncertainty of the proton sites is actually reflecting the locally flat free energy surfaces, especially in the case of perovskites with small lattice constants such as $SrTiO_3$. In the very first quantum molecular dynamics study of protons in $BaCeO_3$ [24], the trace of the proton at $T = 900$ K was found to be confined to a kind of "donut" with the hydroxide oxygen defining the center. This "donut" actually contains all the sites just discussed, and from a careful analysis of the proton probability density function for different perovskite-type oxides, proton sites have been obtained [25, 26]. For perovskites with large lattice constants such as $BaCeO_3$ and $BaZrO_3$, sites close to the octahedra edges were identified (Fig. 13.1a), whereas in perovskites with small lattice constants such as $SrTiO_3$ and $CaTiO_3$, the proton sites are closer to the plane formed by the hydroxide ion and its four next nearest oxygen neighbors (Fig. 13.1b). As illustrated in Fig. 13.1, there are always 8 sites per hydroxide ion (24 per unit cell), but for large lattice constants the only hydrogen bond interaction is within octahedra, while for small lattice constants, also hydrogen bonding to oxygen of neighboring octahedra appears to be possible. Hydrogen bonding is actually indicated by a small contraction (a few pm) of the average separation between the hydroxide ion and its eight nearest and four next nearest oxygen neighbors (e.g., [27]); it is most likely the repulsion with the highly charged B-site cation (Ti^{4+}) and the attraction from the next nearest oxygen, which is pushing the proton out of the octahedron edge. The resulting degree of hydrogen bond bending (for the average configuration) surely has severe implications on the proton conduction mechanisms, which are discussed in the next section.

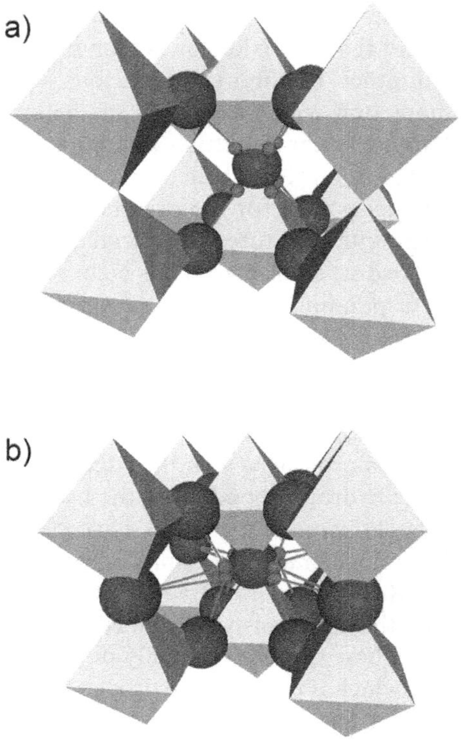

Fig. 13.1 Illustration of proton sites (*small spheres*) as obtained from molecular dynamics simulations for cubic perovskite-type oxides with large lattice constants (**a**) and small lattice constants (**b**) [24, 25]. Only the oxygens (*dark*) interacting with the proton are shown. Possible hydrogen bond interactions are indicated by *grey lines*

13.3 Mechanisms of Proton Conduction (Undoped, Cubic Perovskites)

The elementary reactions underlying the mobility of protonic defects have been investigated in great detail, experimentally as well as by numerical simulations from quite different points of view. It is near at hand to even consider proton tunneling, but as in any fast proton conductors tunneling effects are anticipated to be negligible [1]. Even without taking lattice relaxation (polarization) effects around the proton into consideration, proton tunneling is estimated to become significant at extremely low temperature ($T < 14$ K) only [28].

Apart from the strong localization of the proton within the hydroxide ion, the proton self-localization also involves a slight contraction of the average OH/O separation (see above) and an expansion of the separation between the hydroxide ion and the two neighboring B-site cations (e.g., Ce^{4+}; Fig. 13.1) [24–26]. Any proton conduction mechanism must at least partially compensate for these relaxation effects; i.e., it must involve some host lattice adjustment in the transition state configuration. In a first attempt to relate lattice dynamics to proton transfer between adjacent oxide ions, the coupling of proton transfer to the O–B–O bending mode was investigated numerically [29]. The rate of proton transfer was found to increase, and the H/D isotope effect was decreasing with increasing anharmonicity of this mode. The influence of the hydrogen bonding on the local dynamics had been neglected, and only proton transfer had been treated in this early study, but the importance of the oxide ion dynamics had already been clearly recognized. Generally speaking, the local lattice relaxation may be treated as a diffusing small polaron, which the proton is following; this is nothing but structure diffusion (see Introduction), which was treated analytically by several authors [30, 31, 32]. However, these approaches are either phenomenological or they make severe assumptions on the relevant modes (e.g., O–B–O bending). In particular, they do not give any specific information on chemical interactions and how they determine the evolution of configurations involved in the proton conduction mechanism.

Our current more detailed understanding of proton conduction mechanisms in perovskite-type oxides actually started to emerge from independent quantum molecular dynamics simulations [23], quasi-elastic neutron scattering [33], and μ-SR experiments [34, 35]. They all clearly show rapid rotational diffusion of the proton within a "donut" around the oxygen with which the proton is making a covalent bond (Fig. 13.2). In this local dynamics the strong covalent bond remains intact, and it is only the orientation of the hydroxide ion that is changing. Because of the free energy barriers separating the different orientations, this dynamics is stochastic, i.e., diffusional in nature. Of course, it involves forming and breaking of hydrogen bonds (see red lines in Fig. 13.1), which obviously costs very little excitation, which is explained later. These early

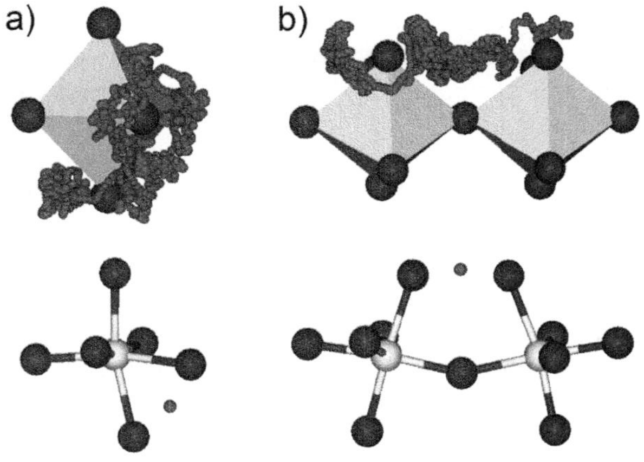

Fig. 13.2 Proton traces sampled by quantum molecular dynamics simulations involving intra-octahedron transfer (**a**) and inter-octahedra proton transfer (**b**) [24, 25]. The transition state configurations for proton transfer are shown for both cases: In the case of intra-octahedron transfer, this is characterized by B–O bond elongations and strong contraction of the OH–O separation; in the case of inter-octahedra transfer, severe tilting of the participating octahedra is involved (see text)

results also indicate that it is the proton transfer from a hydroxide ion to a neighboring oxide ion that is rate-limiting, long-range proton transport.

Considering the large average oxygen separations (e.g., compared to hydrogen bonding in water) of typically 290–320 pm, this appears to be reasonable, but it seemed to be at odds with the strongly red-shifted OH-stretching absorptions in the IR spectra ([12] and references therein), indicating strong hydrogen bond interactions, which favor fast proton transfer reactions rather than fast reorientation processes, the latter requiring the breaking of such bonds. As the structural oxygen separation is larger than 290 pm in most perovskite-type oxides and strong hydrogen bonds may only be formed for significantly lower separations, the free energy the system gains by hydrogen bond formation is competing with the free energy required for the lattice distortion involved in the OH/O contraction. A reanalysis of a quantum-MD simulation of a protonic defect in cubic $BaCeO_3$ [25, 36, 37] demonstrates that these two free energy contributions are almost balanced for a wide range of oxygen separations (approximately 250–300 pm). In this way, short oxygen separations, which favor proton transfer, and large oxygen separations, which allow rapid bond breaking involved in rotational diffusion, correspond to similar free energies of the entire system and, therefore, have similar probabilities of occurring. Indeed, the simulation finds the protonic defect to form short but transient hydrogen bonds with all eight nearest oxygen neighbors. In the time-averaged picture seen in diffraction experiments, this leads only to a slight reduction of the structural OH/O separations (see Fig. 13.1), but in most instant configurations one of the eight OH/O separations is reduced to about 280 pm as a result of hydrogen bonding [24]. Although the hydrogen bond interaction has a stabilizing effect of about 0.5 eV on this configuration [38], the bond is a soft high-energy hydrogen bond with extended bond length variations; this also leads to configurations where the protonic defect behaves almost like a free OH with small OH stretching amplitudes compared to the extended stretching vibrations in the hydrogen-bonded state.

From the thermodynamics of such "dynamical hydrogen bonds" one may actually expect an activation enthalpy of long-range proton diffusion not more than 0.15 eV provided that the configuration O–H. . .O is linear, for which the proton transfer barrier vanishes at O/O separations less than about 250 pm. However, the mobility of protonic defects in cubic perovskite-type oxides has activation enthalpies of the order of 0.4–0.6 eV [35], which raises the question of which interactions are controlling the activation enthalpy of proton transfer.

As already pointed out, in the average configuration the hydrogen bonds are bent to some extent (see Fig. 13.1). This change suppresses proton transfer for two reasons: there is a remaining proton transfer barrier even at low oxygen separations, and any proton transfer requires both energy and momentum transfer in this bent configuration. An analysis of a few transition state configurations obtained from molecular dynamics simulations showed that the B–O bonds are elongated, which reduces the repulsion between the B-site cation and the proton and allows for the formation of an almost linear, short hydrogen bond in the transition state complex [Fig. 13.2(a)]. The proton transfer in this configuration probably occurs over some remaining barrier, as indicated by the experimentally observed H/D isotope effects [28, 39]. Although the H/B repulsion is reduced in this configuration, major contributions to the activation enthalpy arise from the B–O bond elongation and the proton transfer barrier.

The importance of the H/B repulsion is also evidenced by the finding that the activation enthalpies of proton mobility in cubic perovskites with pentavalent B-site cations (I–V perovskites) are significantly higher than for perovskites with tetravalent B-site cations (II–IV perovskites) [36]. On this background, proton mobility in III–III perovskite may be even higher, provided the oxide shows cubic symmetry.

So far, we have just considered the mechanism of proton conduction in perovskite-type oxides with large lattice constants, and these are commonly the ones with potentially high concentrations of protonic defects and therefore also high proton conductivity [35], but perovskite-type oxides with small lattice constant (e.g., $SrTiO_3$, $CaTiO_3$) may also exhibit high proton mobilities. In these cases, hydrogen bonding even to the next nearest oxygen, between the vertices of the octahedra, becomes possible [Fig. 13.1(b)], opening another proton transfer path as observed in MD simulations [20, 24]. The transition state complexes then involve a tilting of neighboring octahedra between which the transfer takes place (Fig. 13.2). While only 25% of the observed proton transfers are occurring between octahedra (inter-octahedra transfer) in the case of $SrTiO_3$, proton transfer in $CaTiO_3$ is observed to be dominated by inter-octahedra transfers (70%). It should be noted that for the latter the highest proton mobility has been found in the simulations [25], but small concentration of protonic defects under common conditions prevents titanates from showing high proton conductivity [37].

The mechanisms just described provide a qualitative explanation for the empirical finding that the highest proton conductivities are observed in oxides with the cubic perovskite structure [40]. The framework of corner-sharing BO_6 octahedra shows high coordination numbers for both cation sites (12 for the A site and 6 for the B site). There is only 1 oxygen site in the ideal perovskite structure, with each oxygen surrounded by 8 nearest and 4 next nearest oxygens. Generally speaking, the high coordination numbers lead to low bond strengths and low angles between the bonds, which is in favor of the above-described dynamics. For example, the rotational diffusion of the protonic defect corresponds to a dynamical hydrogen bonding of the OH with the 8 or 12 oxygens, forming the "reaction cages" (see Fig. 13.1). The angles between the possible orientations are small enough that the effective barriers for bond-breaking and bond-forming processes are usually very low, and the equivalence of all oxygens guarantees the absence of any extra contribution to the activation energy that may originate from differences in site energies. The latter is especially important for the proton transfer reaction, which may be biased to a particular oxygen site, which then may act as a kind of proton trap. Also, the high number of equivalent transfer paths directly enters into the pre-exponential factor of proton mobility.

13.4 Complications (Symmetry Reduction, Doping, Mixed Site Occupancy)

From these considerations, it is anticipated that any perturbation of the highly symmetric perovskite structure may lead to a decrease of the rate of long-range proton transport. This may not only be brought about by the reduction of the symmetry of the crystallographic (time/space-averaged) structure, but also locally by the presence of point defects (e.g., acceptor dopants, oxide ion vacancies), mixed site occupations, or higher dimensional defects, such as dislocations or grain boundaries.

The effects of the tilting and twisting of the BO_6 octahedra, i.e., the reduction of the crystallographic symmetry, has been investigated in detail by comparing structural and dynamical features of protonic defects in Y-doped $BaCeO_3$ and $SrCeO_3$ [41]. The large orthorhombic distortion of $SrCeO_3$ has tremendous effects on the arrangement of the lattice oxygen, which leads to the appearance of shorter and longer O/O separations and also changes the chemical character of the oxygen. The cubic oxygen site splits into two sites with probabilities of 1/3 (O1) and 2/3 (O2). As a result of different chemical interactions with the cations, especially the strontium on the A site, the oxygens on these sites show distinctly different electron densities (basicities) and therefore different binding energies for the proton. Although in $SrCeO_3$ the most basic oxygen is O1, it is O2 in $BaCeO_3$. Assuming that protons are associated with these sites for most of the time, they may show long-range proton transport via the more frequent O2 sites in $BaCeO_3$, whereas isotropic long-range proton transport in $SrCeO_3$ must involve transfer between chemically different O1 and O2 sites. Much more rapid transport is anticipated to occur between neighboring O1 oxygens. However, they just form one-dimensional arrays as a result of octahedra tilting [Fig. 13.3(a)], and transport along these paths is expected to be very sensitive to any perturbation. Even for the fast rotational diffusion step, the orthorhombic distortion is clearly inducing a biasing of the hydroxide orientation toward the more basic oxygen neighbors, O1 [Fig. 13.3(b)]. All these observations are thought to be the reason for the higher activation enthalpy and lower conductivity in $SrCeO_3$ compared to $BaCeO_3$. Similar simulations were later carried out on orthorhombic $CaZrO_3$ [42], and also in this case octahedra tilting is opening the proton transfer between the vertices of neighboring octahedra, whereas no intra-octahedron proton transfer was detected in the simulation.

The mobility of protons is not only very sensitive toward reduction of the crystallographic symmetry but also toward local structural and chemical perturbations induced by the acceptor dopant or by mixed occupancy on the B site. To allow for the formation of protonic defects by dissociative absorption of water, perovskite-type oxides are commonly doped with aliovalent ions (acceptor dopants) matching the ionic radius of the B-site cation (e.g., In^{3+} for Zr^{4+}). Indeed, this simple concept has been proven successful, e.g., for oxide ion conductors (Sc-doped zirconia shows higher oxide ion conductivity than

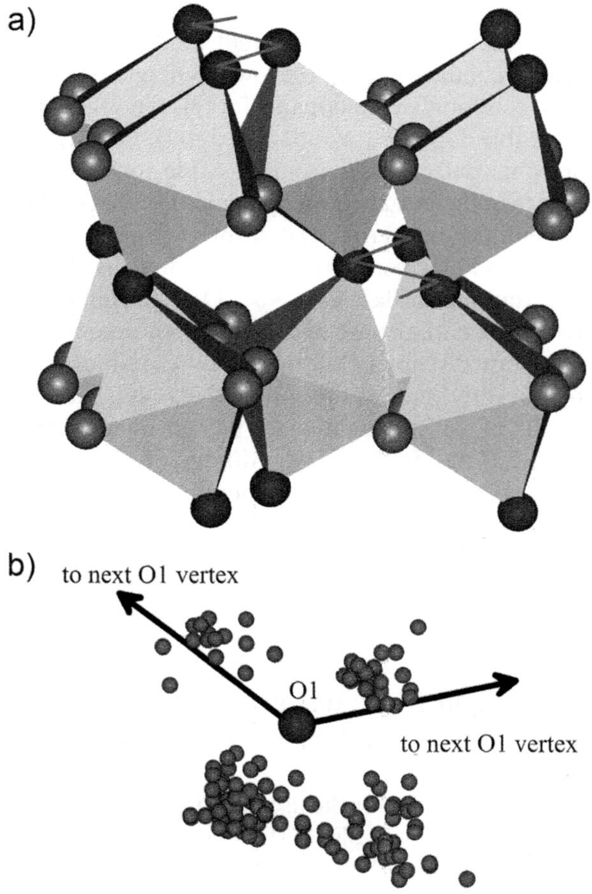

Fig. 13.3 Proton diffusion path in orthorhombic $SrCeO_3$ as obtained from a quantum molecular dynamics simulations [40]. The proton rotational diffusion around O1 is distinctly biased toward the neighboring O1 (**b**), which is part of a one-dimensional proton diffusion path formed by O1 oxygen only (**a**). Proton transfer between O1 (*dark*) and the less basic and more frequent O2 (*grey*) is anticipated to control long-range, isotropic proton diffusion (see text)

Y-doped zirconia). But when it comes to proton conductivity in oxides, this approach clearly fails. Although Sc^{3+} and In^{3+} are matching Zr^{4+} with respect to their ionic radii, $BaZrO_3$ shows much lower proton mobility when doped with Sc or In compared to Y as an acceptor dopant with a significantly higher ionic radius [39]. Only for the latter, the proton mobility and its activation enthalpy are virtually independent on the dopant concentration. Although Y on the Zr site expands the lattice locally and on the average and even leads to tetragonal distortions at concentrations above 5 mol%, it actually leaves the acid–base properties of the coordinating oxygen almost unchanged [24]. Obviously, this chemical match of the dopant makes it so to say "invisible" to the diffusing proton. The most common observation, however, is a decreasing proton mobility and an increasing activation enthalpy with increasing dopant concentration, as observed in Y-doped $BaCeO_3$, for example [11, 37], for which the local distortions around the acceptor dopant has recently been studied experimentally [43].

In an attempt to calculate association energies for the formation of clusters of protonic defects and different kinds of acceptor dopants in $SrCeO_3$-based proton conductors [44], the lowest energies have been found for Y and Yb, the most commonly used dopants for this type of compound. For $CaZrO_3$, the most favorable dopants are anticipated to be Ga, Sc, and In [45] Even interaction of protonic defects with residual oxide ion vacancies has been anticipated to influence the local proton dynamics in $SrTiO_3$ [20].

With this background, it is not surprising that also cation ordering on the B site of complex perovskites may become disadvantageous for proton mobility. For example, in $Ba_3Ca_{1+x}Nb_{2-x}O_{3-\delta}$, proton conductivity is higher with lower activation enthalpy when chemically and geometrically different Ca and Nb are randomly distributed on the B site. Ca/Nb ordering occurring after annealing significantly reduces the conductivity [46]. This observation is in line with the results of an investigation of proton conductivity in the system $SrTiO_3$–$BaTiO_3$–$SrZrO_3$–$BaZrO_3$ [39]. Although Sr/Ba mixing on the A site actually left some space for a materials optimization, Zr/Ti mixing on the B site led to a significant suppression of the proton conductivity. The highest proton mobilities in this system were actually found for the end members $SrTiO_3$ and $BaZrO_3$, both showing the simple cubic perovskite structure.

13.5 Implications for the Development of Proton-Conducting Electrolytes for Fuel Cell Applications

The requirements for the use of proton-conducting oxides as separator material in solid oxide fuel cells (SOFC) not only comprise a high mobility of protonic defects but also a high concentration of such defects and stability under fuel cell operating conditions. At the moment, only electrolytes based on highly Y-doped $BaZrO_3$ seem to combine these properties in a unique way [35]. Surprisingly, the development of this material required very little compromising. $BaZrO_3$ is actually the cubic perovskite with the highest lattice constant. The high symmetry is essential for the high solubility limit of protonic defects and for the high isotropic proton mobility. The high lattice constant ($a > 420$ pm for Y-doped samples [39]) goes along with a high stability of protonic defects, and the covalency of the Zr–O bond reduces the Zr/H-repulsive interaction and, therefore, the activation enthalpy of the mobility of protonic defects. Only $Ba_3CaNb_2O_9$ has a similarly favorable lattice constant, but Ca/Nb ordering on the B site leads to a symmetry reduction. Less densely packed $BaCeO_3$ already shows a significant orthorhombic lattice distortion.

Another key feature is the availability of a nearly perfect acceptor dopant (i.e., a dopant that leaves the oxygen basicity almost unchanged). Although in all other reported cases the increase of the acceptor dopant concentration leads to a reduction of the proton mobility and an entropic destabilization of protonic defects, both the proton mobility and the thermodynamics of hydration are practically unchanged for dopant levels up to 20% Y in $BaZrO_3$ [37]. High proton mobility and entropically stabilized protonic defects even at high dopant concentrations and the high solubility limit lead to the high proton conductivity of this material. For temperatures below about 700°C and a water partial pressure of 23 hPa, this exceeds the oxide ion conductivity of the best oxide ion conductors. Although the bulk conductivity of Y-doped $BaZrO_3$ is even slightly higher than the proton conductivity of $BaCeO_3$-based oxides, the chemical stability is by far more advantageous, as expected from the higher electronegativity of Zr compared to Ce and the higher covalency of the Zr–O bond. For the CO_2 partial pressure of air (38 Pa corresponding to 380 ppm), pure $BaZrO_3$ is stable above 300°C, which is only slightly higher than for $BaTiO_3$ and $SrTiO_3$, which are known for their superior stabilities [12].

High bulk proton conductivity, high stability, and a wide ionic domain [47] therefore make Y-doped $BaZrO_3$ an interesting parent compound for the development of proton-conducting electrolytes for SOFC applications. Unfortunately, the unfavorable brittleness, the grain boundary impedance, and the increasing phase instability with increasing Y-dopant level remain problems to be solved. The addition of small amounts of $BaCeO_3$ or a compromise in the choice of the kind of dopant may help to reduce these problems.

In any case, the potentially very high proton mobility in perovskite-type oxides, which was discussed in detail in this chapter, may become a clue to the reduction of the unfavorably high operation temperature of conventional SOFCs.

Acknowledgments I thank R. Merkle for reading the proofs and U. Traub for preparing the figures.

References

1. K.D. Kreuer, Chem. Mater. 8, 610–641 (1996)
2. K.D. Kreuer, W. Weppner, A. Rabenau, Solid State Ionics, 3/4, 353–358 (1981)
3. K.D. Kreuer, A. Rabenau, W. Weppner, Angew. Chem. Int. Ed. Engl. 21, 208 (1982)
4. S.F. Fischer, G.G. Hofacker, M.M. Ratner, J. Chem. Phys. 52, 1934 (1970)
5. R.A. Marcus, J. Chem. Phys. 24, 966 (1956)
6. C.P. Flynn, A.A. Stoneham, Phys. Rev. 1, 3966 (1970)
7. M.E. Tuckerman, D. Marx, M.M. Klein, M. Parrinello, J. Chem. Phys. 103, 45 (1995)
8. N. Agmon, Chem. Phys. Lett. 244, 456 (1995)
9. K.D. Kreuer, Solid State Ionics 136–137, 149–160 (2000)
10. W. Münch, K.K. Kreuer, W. Silvestri, J. Maier, G. Seifert, Solid State Ionics, 145, 437 (2001)
11. K.D. Kreuer, W. Münch, M. Ise, A. Fuchs, U. Traub, J. Maier, Ber. Bunsenges. Phys. Chem. 101, 1344 (1997)
12. P. Murugaraj, K.K. Kreuer, T. He, T. Schober, J. Maier, Solid State Ionics 98, 1 (1997)
13. K.D. Kreuer, Solid State Ionics 97, 1 (1997)
14. H. Yugami, Y. Shibayama, T. Mattori, M. Ishigame, Solid State Ionics 79, 171 (1995)
15. T. Scherban, Y.Y. Baikov, E.E. Sharkova, Solid State Ionics 66, 159 (1993)
16. S. Shin, H.H. Huang, M. Ishigame, Solid State Ionics 40/41, 910 (1990)

17. H. Engstrom, J.J. Bates, L.L. Boatner, J. Chem. Phys. 73, 1073 (1980)
18. N. Sata, K. Hiramoto, M. Ishigame, S. Hosoya, S. Shin, N. Niimura, Phys. Rev. B54, 15795 (1996)
19. N. Sata, S. Shin, K. Shibata, M. Ishigame, J. Phys. Soc. Jpn. 68, 3600 (1999)
20. F. Shimojo, K. Hoshino, H. Okazaki, Solid State Ionics 113–115, 319 (1998)
21. F. Shimojo, K. Hoshino, Solid State Ionics 145, 421 (2001)
22. I. Sosnowska, R. Przenioslo, W. Schäfer, W. Kockelmann, K. Wysocki, J. Alloys Compd. 328, 226 (2001)
23. R.M. Badger, S.S. Bauer, J. Chem. Phys. 5, 839 (1937)
24. W. Münch, G. Seifert, K.K Kreuer, J. Maier, Solid State Ionics 86–88, 647 (1996)
25. W. Münch, K.K.Kreuer, G. Seifert, J. Maier, Solid State Ionics 125, 39 (1999)
26. W. Münch, K.K. Kreuer, G. Seifert, J. Maier, Solid State Ionics 136–137, 183 (2000)
27. K.D. Kreuer, W. Münch, U. Traub, J. Maier, Ber. Bunsenges. Phys. Chem. 102, 552 (1998)
28. E. Matsushita, Solid State Ionics 145, 445 (2001)
29. K. D. Kreuer, A. Fuchs, J. Maier, Solid State Ionics 77, 157 (1995)
30. A. L. Samgin, Russ. J. Electrochem. 35, 287 (1999)
31. T. Ishii, A. Ichigozaki, J. Phys. Soc. Jpn. 73, 130 (2004)
32. T. Tomoyose, N. Shimoji, K. Wakamura, J. Phys. Soc. Jpn. 74, 3011 (2005)
33. M. Pionke, T. Mono, W. Schweika, T. Springer, H. Schober, Solid State Ionics 97, 497 (1997)
34. T. Matzke, U. Stimming, C. Karmonik, M. Soetramo, R. Hempelmann, F. Güthoff, Solid State Ionics 86–88, 621 (1996)
35. R. Hempelmann, M. Soetramo, O. Hartmann, R. Wäppling, Solid State Ionics 107, 269 (1998)
36. K.D. Kreuer, Solid State Ionics 125, 285 (1999)
37. K.D. Kreuer, Solid State Ionics 136–137, 149 (2000)
38. K.D. Kreuer, Annu.Rev. Mater. Res. 33, 333 (2003)
39. A.S. Nowick, Solid State Ionics 97, 17 (1997)
40. K.D. Kreuer, S. Adams, A. Fuchs, U. Klock, W. Münch, J. Maier, Solid State Ionics 145, 295 (2001)
41. W. Münch, K.K. Kreuer, S. Adams, G. Seifert, J. Maier, Phase Transitions 68, 567 (1999)
42. M.S. Islam, Solid State Ionics, 154–155, 75 (2002)
43. A. Longo, F. Giannici, A. Balerna, C. Ingrao, F. Deganello, A. Martorana, Chem. Mater. 18, 5782 (2006)
44. G.C. Mather, M.M. Islam, Chem. Mater. 17, 1736 (2005)
45. M.S. Islam, R.R. Davies, J.J. Gale, Chem. Mater. 13, 2049 (2001)
46. A.S. Nowick, D. Yang, K.K. Liang, Solid State Ionics 125, 303 (1999)
47. T. He, K.K. Kreuer, P. Murugaraj, Y.Y. Baikov, J. Maier, Proc. Eurosolid 4, 60 (1997)

Chapter 14
Intermediate-Temperature SOFCs Using Proton-Conducting Perovskite

Naoki Ito

14.1 Introduction

Today, most vehicles are driven by internal combustion engines with liquid fuel. The efficiency of internal combustion engines is only 15%–25%, and a power source with higher efficiency is needed from the aspects of energy conservation and CO_2 emissions. A fuel cell is one leading candidate for this alternative power source. There are several types of fuel cells, but the polymer electrolyte membrane fuel cell (PEFC) is the top runner now, and auto makers are putting most of their research and development resources into a PEFC + on-board hydrogen storage system [1]. Regarding the PEFC, it is well known that several problems remain, such as high cost due to the use of platinum, liquid water management, and the durability of the polymer electrolyte. However, the two major issues of this system lie on the fuel side, hydrogen infrastructure and hydrogen storage. Although hydrogen itself has very high power density, it is very difficult to store hydrogen with high physical density. Many researchers are working on the search for better hydrogen storage technology, such as a high-pressure tank, hydrogen storage metal, carbon nano-tubes, and liquid hydrogen. However, so far the driving range of fuel cell (FC) vehicles is much less than that of vehicles with internal combustion engines. The issue regarding hydrogen infrastructure is more a political and economic one. "The hydrogen economy" needs a huge expenditure to build up the network of hydrogen distribution and refueling stations, but no one has a vision of this huge cost yet.

As discussed above, there are two major issues on the hydrogen fuel side. If a FC vehicle propelled with liquid fuel is realized, that can be the main trend of a future vehicle because liquid fuel does not have such issues (Table 14.1). The solid oxide fuel cell (SOFC) is the only fuel cell that can be directly driven by liquid fuel; other fuel cells need fuel processors that produce hydrogen from

N. Ito (✉)
Fuel Cell System Development Division, Toyota Motor Corporation, 1200, Mishuku, Susono, Shizuoka 410-1193, Japan
e-mail: naoki@ito.tec.toyota.co.jp

T. Ishihara (ed.), *Perovskite Oxide for Solid Oxide Fuel Cells*,
Fuel Cells and Hydrogen Energy, DOI 10.1007/978-0-387-77708-5_14,

Table 14.1 Fuel options for FC vehicle

	Hydrogen	Liquid fuels
Fuel	*Requires new infrastructure and storage technology*	Easy handling and infrastructure
Vehicle	Simple system	*Complex system and low efficiency*

liquid fuel. So, in the case of liquid fuel, an on-board fuel processor is required, and that is a major challenge. "Reforming" is a build-up technology in the area of petroleum industry, but the challenge here is to make the system so compact as to be on board. Toyota made a prototype vehicle with a system that has a conventional fuel processor and PEFC in 2002, named the FCHV-5. The system configuration is shown in Fig. 14.1(a). The high-temperature shift reactor, low-temperature shift reactor, and a PReferential OXidation reactor should be placed after a reformer to minimize CO content in fuel gas because PEFC can be easily poisoned by less than 100 ppm of CO. These components and the heat exchangers to connect them made the system very complicated and huge. As a result, it was proved that this complicated system is not practical.

As shown in Fig. 14.1(b), if the operating temperature of the fuel cell is close to the reforming temperature of liquid fuel, a simple and attractive system is possible. Another advantage of this alternative system is high efficiency because the exhaust heat from the fuel cell can be used in the endothermic reforming reaction. On the other hand, a very high power density fuel cell is needed for the main power source of a vehicle because the maximum power required for a passenger vehicle is as high as 50–100 kW. These requirements are summarized

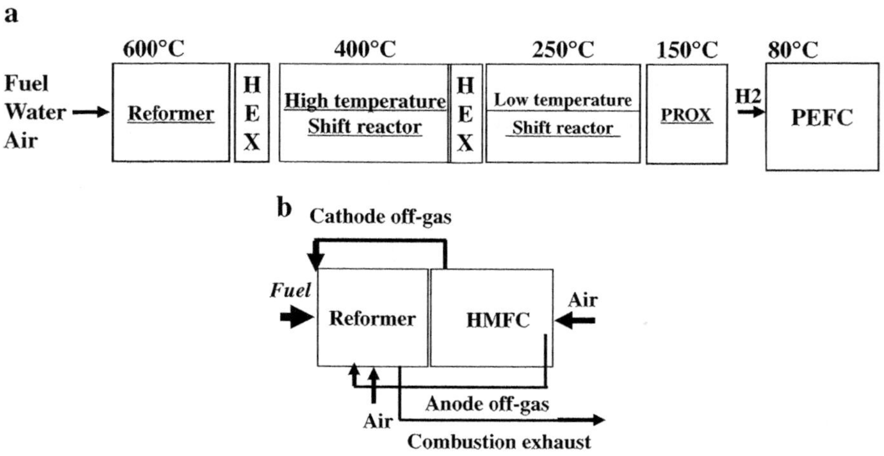

Fig. 14.1 **(a)** PEMFC + conventional fuel processor system; **(b)** Alternative system with intermediate Temperature FC

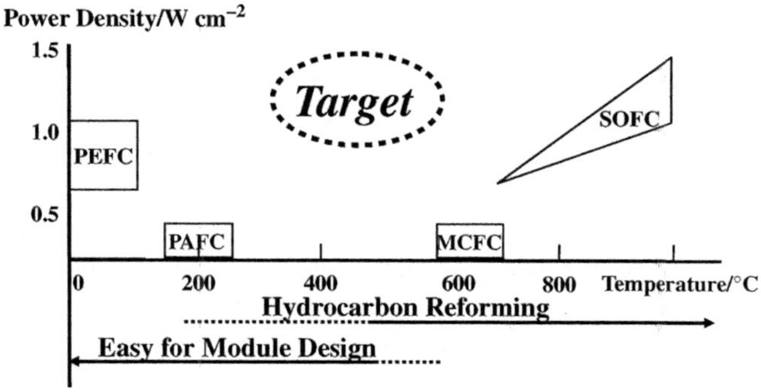

Fig. 14.2 Target of hydrogen membrance fuel cell

in Fig. 14.2. An on-board reforming FC vehicle can be realized only with a fuel cell in this target range. PEFCs have high power density, but they cannot be operated above 100°C because the polymer electrolyte needs liquid water. SOFCs also have high power density above 800°C. However, below 800°C power density decreases rapidly, even with a thin-film electrolyte system.

Intermediate-temperature fuel cells with high power density have not been realized because there is not a “good” electrolyte with high ion conductivity in this temperature region. So, research has been focused on decreasing electrolyte resistance because a major part of fuel cell resistance is derived from the electrolyte resistance. There have been two main research approaches for intermediate-temperature fuel cells: (1) the search for a new electrolyte with high ionic conductivity, and (2) search for a thin-film electrolyte.

One approach is the search for a new electrolyte material with high ion conductivity. Ishihara et al. found that $LaGaO_3$-based perovskite oxides are good oxygen ion conductors at intermediate temperatures, and they are working on SOFCs based on these electrolyte materials [2, 3]. The major challenge of this approach is a trade-off between conductivity and chemical stability. Haile et al. proposed proton-conducting solid acids, such as $CsHSO_4$ and CsH_2PO_4, for intermediate-temperature fuel cells [4, 5]. Poor physical properties, chemical stability, and temperature limit (around 250°C) are three major issues for solid acids.

The other approach is an attempt for a thinner electrolyte supported on a porous electrode. Recently, many researchers have been working on a thin-film-type SOFC to realize an intermediate-temperature fuel cell [6–9]. In this approach, the electrolyte must be supported on either the anode or cathode. The main difficulty regarding this approach lies in making a solid thin film on a porous electrode. To get a dense electrolyte layer, the electrolyte thickness should be several times larger than the maximum pore size of the support electrode. Thus, the preparation method of the support electrode with good pore-size distribution is a key technology of this approach; not only it is difficult, but also it needs expensive techniques to commercialize. If the pore size of the support electrode is too small, concentration polarization can be a problem.

There is a group of metals that is known to be used as a hydrogen permeation membrane. Their mechanism is as follows. First, hydrogen dissolves into metal as hydrogen atoms at the surface of one side of the film. Second, hydrogen atoms move in the metal film by diffusion. Finally, hydrogen atoms recombine into hydrogen at the surface of the other side of the film. The driving force of the hydrogen permeation is the hydrogen partial pressure gradient, and these metals are a "short-circuited proton conductor" in reality. The hydrogen permeation ability of this group of metals is shown in Fig. 14.3 [10]. Their hydrogen permeability is several orders of magnitude higher than perovskite proton conductors. So, these metals can be used as a dense anode.

The essence of our new concept is an ultrathin proton conductor electrolyte supported on solid hydrogen membrane. We named this new type of fuel cell a hydrogen membrane fuel cell (HMFC). The schematic diagram of the HMFC structure is shown in Fig. 14.4. A much thinner electrolyte can be easily realized because it is formed on a solid, nonporous membrane. Another advantage of this approach is an ease of high-density stacking because the physical base of this fuel cell is a metal film, not ceramics as in the case of SOFCs. The HMFC can use only proton conductors as an electrolyte because the hydrogen membrane layer only allows hydrogen to permeate.

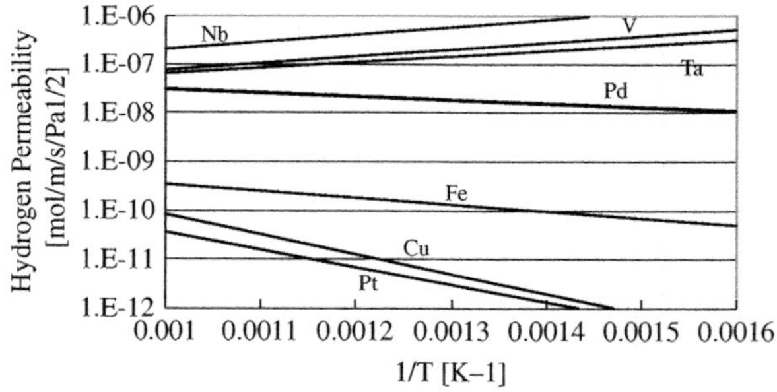

Fig. 14.3 Hydrogen permeability of metals

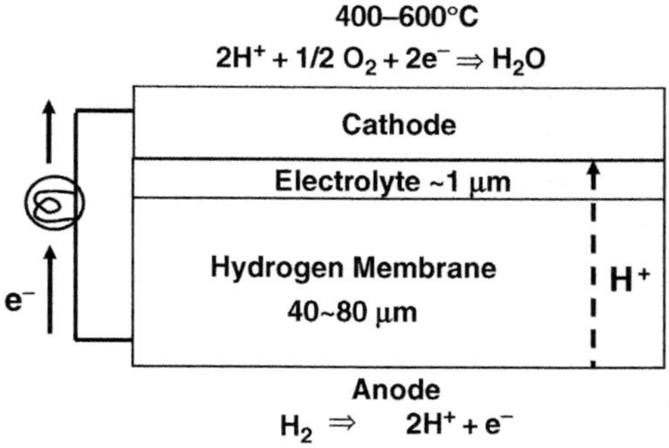

Fig. 14.4 Schematic diagram of hydrogen membrane fuel cell structure

14.2 Preparation of Fuel Cells

$BaCe_{0.8}Y_{0.2}O_3$ was chosen as the electrolyte material because it has the highest proton conductivity in the intermediate-temperature region. $BaCe_{0.8}Y_{0.2}O_3$ is a pure proton conductor under 600°C, so it is a good electrolyte material to prove the HMFC concept [11]. Palladium was chosen as the hydrogen membrane material because it has enough hydrogen permeation capability and chemical stability. Some other metals have higher hydrogen permeation capability, but they have problems regarding hydrogen embrittlement or oxidation. Pd is a very expensive metal, and we should consider switching to cheaper metals before the stage of commercialization. The pulse laser deposition method is chosen to make the electrolyte film. This method has an advantage regarding crystal quality and film composition and is a good technique to prove the HMFC concept.

The $BaCe_{0.8}Y_{0.2}O_3$ electrolyte was deposited on a Pd film by pulse laser deposition. Then, the perovskite cathode paste was screen printed on the coated Pd structure and dried by a heat gun. This procedure was repeated three times to make a complete cathode layer that is 30 μm thick. The size of the Pd film was 15 × 15 mm. The electrolyte and electrode were coated on the Pd film as a circle 6 mm in diameter to make an effective fuel cell area of 0.283 cm^2. The pulse laser deposition apparatus is shown in Fig. 14.5. Two types of test cells, cell A and cell B, were prepared for comparison. Their characteristics are summarized in Table 14.2.

14.3 Characterization of Fuel Cells

Figure 14.6a shows the cross-section scanning electron microscopy (SEM) image of the electrolyte and hydrogen membrane. Before preparation of the cross-section sample, the tungsten protection layer was coated on the electrolyte layer to avoid damage. As seen, a solid and uniform electrolyte layer without

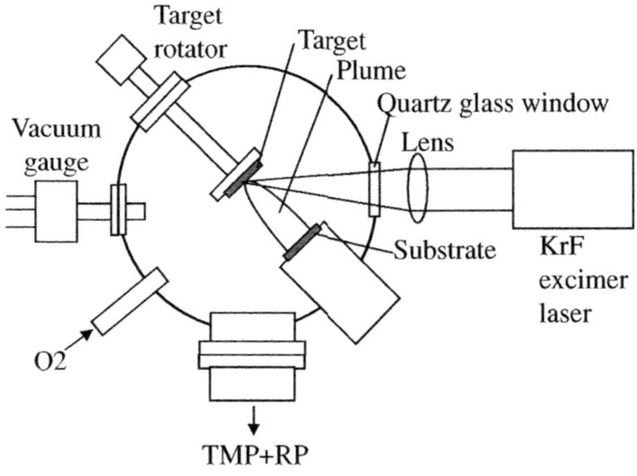

Fig. 14.5 Schematic diagram of pulse laser deposition apparatus

Table 14.2 Characteristics of the HMFC test cells

	Test cell A	Test cell B
Electrolyte	$BaCe_{0.8}Y_{0.2}O_3$ 2 μ m	$BaCe_{0.8}Y_{0.2}O_3$ 0.7 μ m
Hydrogen Membrane	Pd 80 μ m	Pd 40 μ m
Cathode	$La_{0.5}Sr_{0.5}MnO_3$ + Pt(0.5 wt%) Screen printing	$La_{0.6}Sr_{0.4}CoO_3$ Ceramic coating

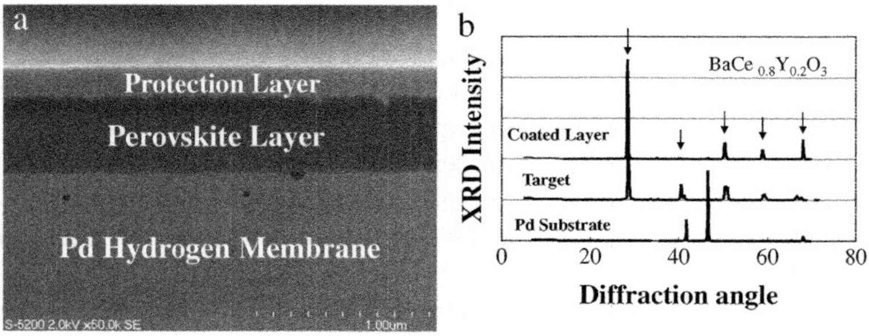

Fig. 14.6 (a) SEM cross section image of the perovskite thin film on Pd substrage; **(b)** X-ray diffraction pattern of the perovskite thin film on Pd substrate

any pores, peeling, or flaking was formed on the Pd substrate. Figure 14.6(b) shows the X-ray diffraction pattern of the electrolyte film. All the peaks of the electrolyte sample match the $BaCe_{0.8}Y_{0.2}O_3$ target peaks when the Pd peaks are excluded. From these data, it is confirmed that a good perovskite layer can be formed on Pd film by pulse laser deposition. Figure 14.7 shows the cross-section transmission electron microscopy (TEM) image around the electrolyte and hydrogen interface. It is shown that long columnar crystals are made over the microcrystal around 100 nm from the interface. This columnar crystal structure is typical for the vapor deposition methods.

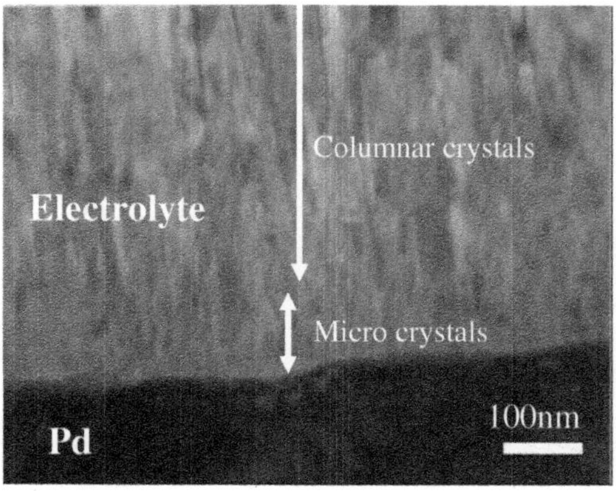

Fig. 14.7 TEM cross section image of the Pd and electrolyte interface

14.4 Operation and Evaluation of Fuel Cells

The V–I characteristic of a single cell A is shown in Fig. 14.8(a). Although the power density has not reached the target, it is proved that the HMFC concept works without any serious problems. The cell resistance was divided into an IR resistance and a polarization resistance by the AC impedance method. The composition of cell resistance is shown in Fig. 14.8(b). The polarization resistance is further analyzed by an AC impedance method because it is difficult to introduce a reference electrode to a fuel cell with a thin-film electrolyte [12]. AC impedance spectra were measured for various P_{H2} of anode side and P_{O2} of the cathode side to determine the polarization (Fig. 14.9). In all conditions, two polarization semicircles were seen. One semicircle has a peak around 3 kHz (Ra), and the other has a peak around 30 kHz (Rb). Both of these are affected by P_{O2} of cathode gas, and none of these is affected by P_{H2} of anode gas. So, both Ra and Rb correspond to the cathode polarization, and the anode polarization is very small compared to the cathode polarization. As mentioned, Pd has very high hydrogen permeation capability, and it is reasonable that Pd has high activity as a fuel cell anode. The P_{O2} dependence of Ra and Rb is analyzed

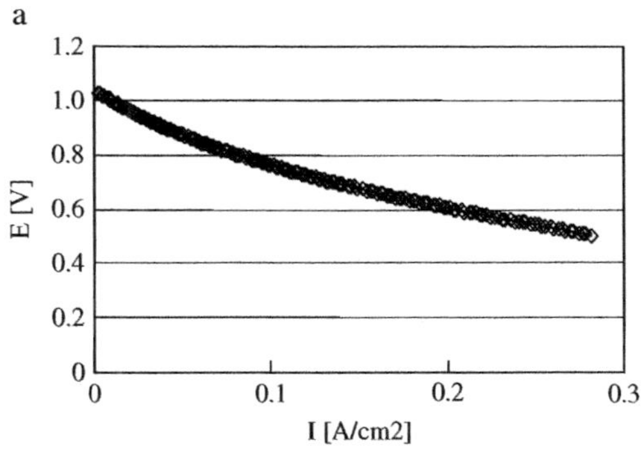

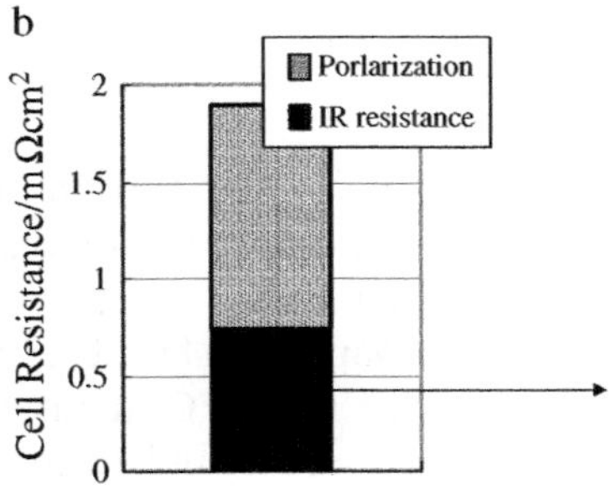

	HMFC (Estimated from IR)	Literature
Proton Conductivity [S/cm]	2.7×10^{-4}	1.5×10^{-2}

Fig. 14.8 **(a)** I–V characteristics of test cell A at 500°C. Anode gas was moist H_2, and Cathode gas was moist air (Both 40°C humidified.); **(b)** Composition of cell resistance of test cell A

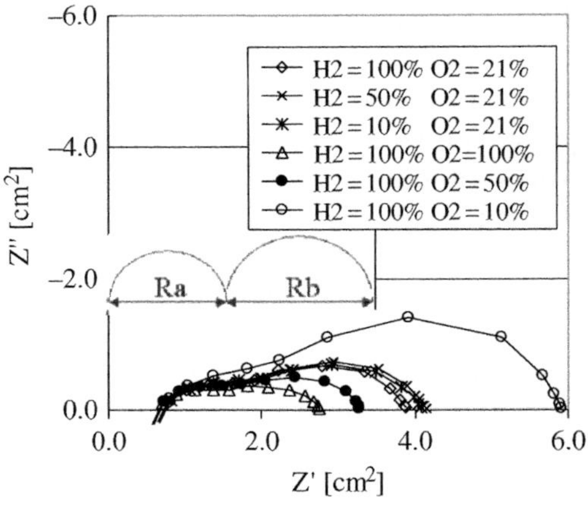

Fig. 14.9 AC Impedance spectra of test cell A

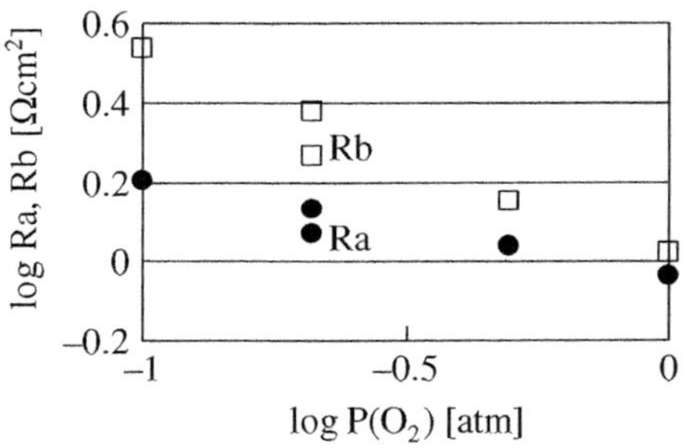

Fig. 14.10 P(O2) dependance of cathode polarization

in Fig. 14.10. Log Ra showed 1/4 slope to log P_{O2} and log Rb showed 1/2 slope to P_{O2}. Below is the cathode reaction mechanism proposed for an LSM-based electrode.

Step 1 $1/2\ O_2 \longleftrightarrow O_{ad}$ (equilibrium)
Step 2 $O_{ad} + 2e^- \rightarrow O^{2-}{}_{ad}$
Step 3 $O^{2-}{}_{ad} \rightarrow O^{2-}{}_{TPB}$
Step 4 $O^{2-}{}_{TPB} + 2H^+ \rightarrow H_2O$

Either step 2 or 3 is most likely to be the rate-determining step [13]. In the case of HMFC, the polarization Ra can correspond to step 3 and the polarization Rb can correspond to step 2 from P_{O2} dependence analysis.

Although the cathode material used in our fuel cell is popular in SOFCs, the cathode polarization is much smaller than that reported in SOFC literature using bulk electrolyte [14]. Hibino et al. also reported anode and cathode polarization of fuel cells using a perovskite proton conductor as a bulk electrolyte, and the polarization was much larger than our result [15]. These facts suggest that electrolyte thickness may affect the polarization.

Table 14.3 Open circuit voltage of test cell B in various conditions

Temperature °C	Anode gas	Cathode gas	Measured OCV mV	Theoretical voltage mV	Transport number
440	H2 = 100%	Air	1103	1140	0.968
440	H2 = 50%	Air	1082	1120	0.966
440	H2 = 10%	Air	1036	1073	0.966
530	H2 = 100%	Air	1081	1120	0.965
610	H2 = 100%	Air	1051	1100	0.955

Performance of cell B was measured in the next step. Cell B has a thinner electrolyte, hydrogen membrane, and more active cathode than cell A for better performance. The open circuit voltage (OCV) of a single cell was measured at various temperatures and hydrogen concentrations. The results of OCV measurement are shown in Table 14.3. Proton transport numbers were calculated as measured OCVs divided by theoretical voltages. Measured OCVs are close to theoretical voltages and much higher than typical PEFCs. The difference between the measured OCVs and theoretical values is probably the result of hydrogen leak through cracks of the electrolyte because measured transport numbers are almost independent from operating conditions. The high OCV of the HMFC is a substantial advantage over PEFCs, especially in the low current density region.

The V-I characteristic of the cell B is shown in Fig. 14.11. The power densities were 0.9 W/cm^2 and 1.4 W/cm^2 at the operating temperatures of 400° and 600°C, respectively. The temperature dependence of the HMFC performance is smaller than that of the SOFCs [16], which can be explained by the fact that the proton conductivity has smaller temperature dependence than oxygen ion conductivity. It is shown that the HMFC has a relatively wide operation temperature window and that further decrease of the operating temperature is possible if system design or fuel choice requires. The composition of the cell resistance of cell B is shown in Fig. 14.12. In Fig. 14.13, the conductivity of the film electrolytes is calculated from measured IR resistances and compared with the bulk

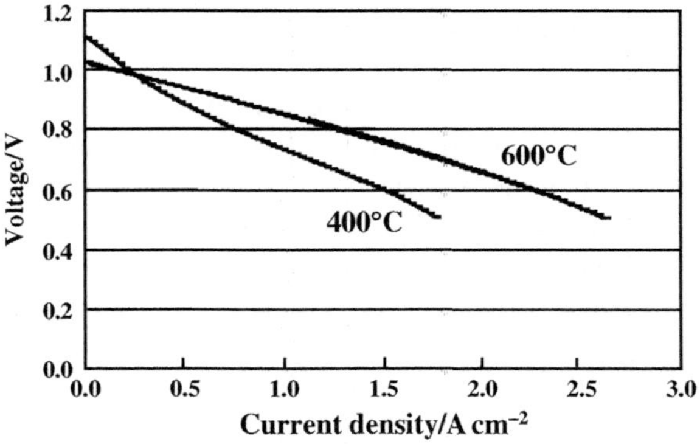

Fig. 14.11 I–V characteristics of test cell B. Anode gas was moist H_2, and Cathode gas was moist air (Both 40°C humidified.)

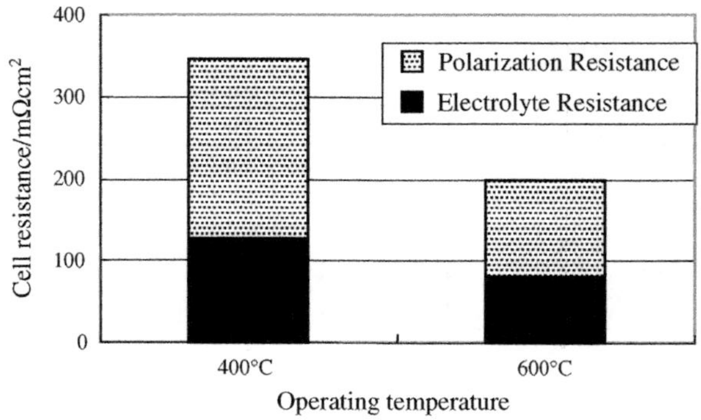

Fig. 14.12 Composition of cell resistance of test cell B

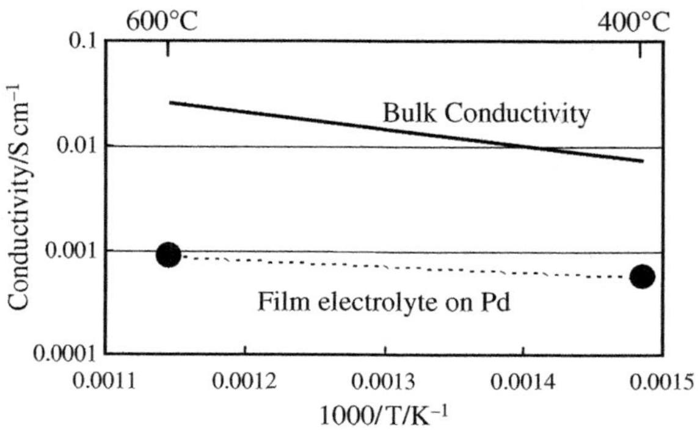

Fig. 14.13 Temperature dependence of electrolyte conductivity of test cell B

conductivity reported in the literature [17]. Even though the HMFC structure can decrease the total electrolyte resistance, the conductivity of the film electrolyte is about one order of magnitude lower than that of bulk electrolyte, and the temperature dependence is more moderate than bulk electrolyte. These results suggest that measured IR resistance may include an interface resistance.

14.5 Conclusion

A new concept, the hydrogen membrane fuel cell (HMFC), is proposed for intermediate-temperature fuel cells. The essence of the HMFC is an ultrathin proton conductor electrolyte supported on a metal hydrogen membrane. The performance of the HMFC is as high as that of high-temperature SOFCs, even though it is operated at intermediate temperatures. The conductivity of the film electrolyte is smaller than that of the bulk electrolyte, and it suggests the existence of some interface resistance. From the polarization analysis, it was shown that most of the polarization derives from the cathode. It is shown that HMFC is an attractive new technology not only for vehicle applications but also for stationary applications. Improving the conductivity of the film electrolyte and the search for a cathode with higher activity are two major challenges for higher performance of the HMFC.

References

1. C. Bernay, M. Marchand, M. Cassir, J. Power Sources, 108, 139–152 (2002)
2. T. Ishihara, M. Honda, T. Shibayama, H. Minami, H. Nishiguchi, Y. Takita, J. Electrochem. Soc. 145, 3177–3183 (1998)
3. T. Ishihara, T. Shibayama, M. Honda, H. Nishiguchi, Y. Takita, J. Electrochem. Soc. 147, 1332–1337 (2000)
4. S.M. Haile, D.A. Boysen, C.R.I. Chisholm, R.B. Merle, Nature 410, 910–913 (2001)
5. D.A. Boysen, T. Uda, C.R.I. Chisholm, S.M. Haile, Sci. Exp. 303, 68–70 (2003)
6. S. de Souza, S.J. Visco, L.C. de Jonghe, Solid State Ionics, 98, 57–61 (1997)
7. R. Doshi, Von L. Richards, J.D. Carter, X. Wang, M. Krumpelt, J. Electrochem. Soc., 1273–1278 (1999)
8. R. Peng, C. Xia, X. Liu, D. Peng, G. Meng, Solid State Ionics, 152–153, 561–565 (2002)
9. J. Will, A. Mitterdorfer, C. Kleinlogel, D. Perednis, L.J. Gauckler, Solid State Ionics, 131, 79–96 (2000)
10. R.E. Buxbaum et al., Ind. Eng. Chem. Res. 35, 530–537 (1996)
11. H. Iwahara, T. Yajima, H. Ushida, Solid state Ionics, 70/71, 267–271 (1994)
12. S.B. Adler, J. Electrochem. Soc. 149(5) E166–E172 (2002)
13. S.H. Chan et al., J. Electrochem. Soc. 151(1) A164–A172 (2004)
14. J.M. Ralph, J.T. Vaughey, M. Krumpelt, Proc. of the VIIth Symposium on SOFC, Vol. 2001–16, 466–467 (2001)
15. T. Hibino, A. Hashimoto, M. Suzuki, M. Sano, J. Electrochem. Soc., 149, A1503–A1508 (2002)
16. D. Ghosh, G. Wang, R. Brule, E. Tang, P. Huang, Proc. of the VIth Symposium on SOFC, Vol. 99–19, 822–823 (1999)
17. H. Iwahara, T. Shimura, H. Matsumoto, Electrochemistry, 68, 154–161 (2000)

Chapter 15
$LaCrO_3$-Based Perovskite for SOFC Interconnects

Teruhisa Horita

15.1 Introduction

Lanthanum chromite-based perovskite oxides ($LaCrO_3$) have been widely recognized as promising interconnect materials for solid oxide fuel cells (SOFCs). The interconnects must separate fuel and oxidant gases and also have high electronic conductivity at high temperature (773–1273 K). Therefore, interconnects should meet the following requirements:

1. Density and gas tightness
2. High electronic conductivity without oxygen electrochemical leak
3. Chemical stability in both oxidant and fuel atmospheres
4. Thermochemical compatibility with the other cell components

To meet the foregoing requirements, the composition of $LaCrO_3$ was modified by doping of lower valence alkaline ions, such as Ca^{2+}, Mg^{2+}, and Sr^{2+}, at the La^{3+} or Cr^{3+} sites. The substitution of La site and Cr site with the other elements can decrease the sintering temperature and increase the electronic conductivity. So far, a number of papers have been published regarding the physical and chemical properties of doped $LaCrO_3$. The present chapter describes the overview of $LaCrO_3$-based ceramics for SOFC interconnects. The following topics are introduced and discussed:

1. Sintering properties and chemical compatibility with the other components
2. Electronic conductivity
3. Defect chemistry and oxygen electrochemical leak
4. Lattice expansion in reduction and temperature change
5. Mechanical strength

T. Horita (✉)
National Institute of Advanced Industrial Science and Technology (AIST), AIST Central 5, 1-1-1 Higashi, Tsukuba, Ibaraki, 305-8565, Japan
e-mail: t.horita@aist.go.jp

T. Ishihara (ed.), *Perovskite Oxide for Solid Oxide Fuel Cells*,
Fuel Cells and Hydrogen Energy, DOI 10.1007/978-0-387-77708-5_15,

15.2 Sintering Properties and Chemical Compatibility with the Other Components

Because interconnects must separate fuel and oxidant gases, the $LaCrO_3$-based perovskite should be a dense body. However, alkaline earth-doped or nondoped $LaCrO_3$ with stoichiometric compositions (La/Cr = 1) are known to be difficult to sinter in an air atmosphere [1–6] because of the vaporization of chromium vapor-phase transports, which causes grain growth without densification. Doping of alkaline earth with liquid-phase formation can reduce the sintering temperature and improve the densification. Sintering with liquid-phase formation was reported by several authors. Doping of Sr into $LaCrO_3$ ($La_{0.84}Sr_{0.16}CrO_3$) markedly increases the sintered density when 3–5 mol% excess $SrCO_3$ is added as a sintering aid before firing in a reducing atmosphere around 2200 K [7]. Sakai et al. first found that Ca-excess or Cr-deficient $(La_{1-x}Ca_x)Cr_{1-y}O_3$ remarkably increases the sintering properties in air atmosphere, as shown in Fig. 15.1 (densification more than 96% of theoretical density at lower than 1773 K) [8–10]. When the Cr-deficient (y) was about 2%–3%, a significant increase of densification was observed because of the depression of CrO_3 vapor pressures, which improves the diffusion of cations (La and Cr) and densification of $(La,Ca)CrO_3$ [11]. Also, liquid-phase $CaCrO_4$ or $Ca_m(CrO_4)_n$ can play an important role in the densification of $(La,Ca)CrO_3$ in air atmosphere (as shown in Fig. 15.1). Some authors observed small amounts of $CaCrO_4$ in samples of $La_{0.7}Ca_{0.32}CrO_3$ that had been calcined at around 1273 K [12–14]. Liquid-phase formation was also observed in $La_{0.76}Sr_{0.24}CrO_3$ and $Y_{0.6}Ca_{0.4}CrO_3$ [12, 13]. Melting chromates, $SrCrO_4$ and $CaCrO_4$, were precipitated from the perovskite structure at intermediate temperatures because of limited alkaline earth solubility. These liquid-phase chromates enhance the densification of doped $LaCrO_3$ at high temperatures.

Although the second phase-forming $LaCrO_3$ can be dense after sintering, the long-term stability should be considered at operation temperature because of the reaction of the second phase with the other cell components. For sintered $(La,Ca)CrO_3$ dense ceramics, $Ca_m(CrO_4)_n$ melt is observed in the air side and CaO or $CaCr_2O_4$ is observed in the fuel side. These second phases can react with the YSZ electrolyte and $LaMnO_3$ cathode. Some authors have reported interdiffusion of elements between the second phase and the other cell components [15, 16]. Reaction of $LaCrO_3$ (LCC) with the YSZ electrolyte is not so significant. However, interdiffusion of elements is observed between $(La,Ca)CrO_3$ and $(La,Sr)MnO_3$ (LSM) cathode. Especially, Ca diffusion into $(La, Sr)MnO_3$ and Sr diffusion into $(La, Ca)CrO_3$ were observed around the LCC–LSM interfaces. This phenomenon is related to the cation diffusion in $LaCrO_3$ perovskite structures. A-site diffusion of cation is reported to be faster than that of B-site diffusion in the doped $LaCrO_3$ (ABO_3 structures) [17–20].

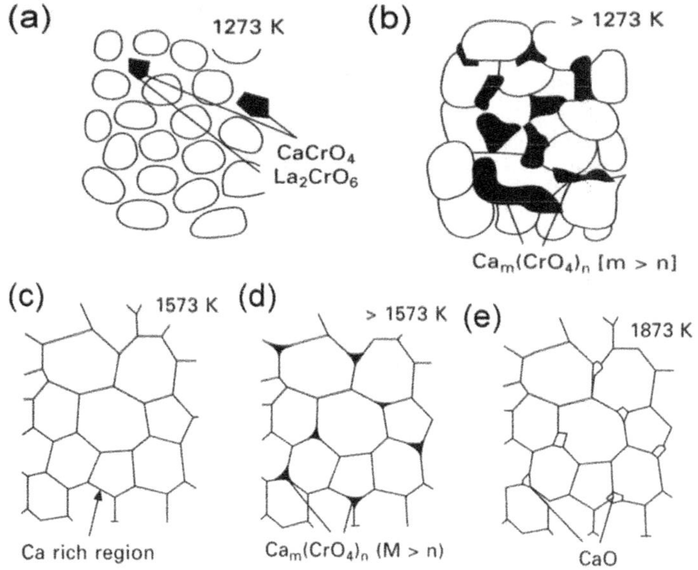

Fig. 15.1 Schematic expression of speculation about the effect of second phase on sintering of $La_{1-x}Ca_xCrO_3$. (**a**) Second phase exists as a form of $CaCrO_4$ or La_2CrO_6 before sintering. (**b**) At around 1373 K, the second phase changes to $Ca_m(CrO_4)_n$, whose Ca:Cr ratio is slightly grater than unity as a result of $CaCrO_4$ and La_2CrO_6. $Ca(CrO_4)_n$ reacts with $La^{1-x}Ca_xCrO_3$ grains and enhances their grain growth and particle joining. Ca:Cr ratio increases with temperature. (**c**) After the densification at 1573 K, most second phases migrate to the surface as a form of $Ca_5(CrO_4)_3$. Calcium-rich substance remains in the vicinity of the grain boundary of $La_{1-x}Ca_xCrO_3$. (**d**) Above 1573 K, $Ca_m(CrO_4)_n$ begins to be segregated at triple junctions, and finally it remains as a form of CaO at 1873 K (**e**). (Reproduced by permission of The American Ceramic Society [10])

15.3 Electronic Conductivity

The interconnects should posses high electronic conductivity greater than 10 S/cm at operating temperatures (873–1273 K) under reducing and oxidant atmospheres $P_{O2} = 10^{-16}$ Pa-10^5 Pa. Doped $LaCrO_3$ is a p-type conductor, and the electronic conductivity increases with a concentration of low-valence cations, such as Sr^{2+} or Ca^{2+} in La^{3+} sites in the following reaction:

$$LaCrO_3 + xAEO + 0.25xO_2 \rightarrow La_{1-x}AE_xCr^{3+}{}_{1-x}Cr^{4+}{}_xO_3 + 0.5xLa_2O_3 \quad (15.1)$$

An electronic hole will be formed on Cr^{4+} sites, and the conduction mechanism is a small polaron hopping process via the Cr^{4+} sites. The electronic conductivity is about 10–100 Scm^{-1} at 1273 K in air. Figure 15.2 shows electronic conductivity of doped $LaCrO_3$ as a function of temperature [21, 22]. The electronic conductivity

Fig. 15.2 Electrical conductivity of $La_{1-x}Ca_xCrO_{3-\delta}$ (x = 0.1–0.3) as a function of inverse temperature. (Reproduced by permission of The Electrochemical Society [22])

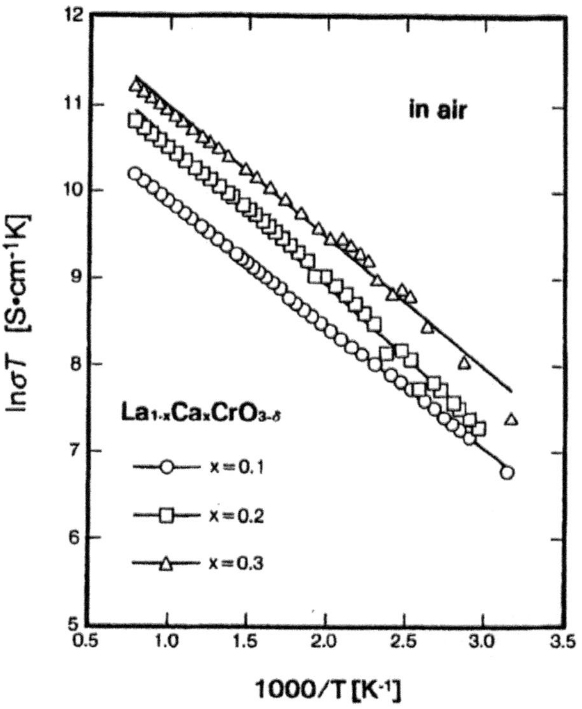

increases with increasing temperature, suggesting semiconductor temperature dependence. Increasing the Ca concentration in $La_{1-x}Ca_xCrO_{3-\delta}$ enhanced the electronic conductivity because of increased Cr^{4+} concentration. There are some deviations of electrical conductivity among the examined alkaline earth elements: Ca-doped $LaCrO_3$ shows a larger electrical conductivity than Sr-doped $LaCrO_3$. This difference was reported to be due to the difference of lattice distortion and phase stability. The activation energy for conductivity was 0.12–0.14 eV and mobility was 0.066–0.075 $cm^2/V/s$ at 1173–1323 K.

Electronic conductivity decreases with a reduction of oxygen partial pressure because of the decrease of Cr^{4+} concentration in a reducing atmosphere as follows:

$$La_{1-x}AE_xCr^{3+}_{1-x}Cr^{4+}_{x}O_3 \rightarrow La_{1-x}AE_xCr^{3+}O_{3-\frac{x}{2}} + \frac{x}{4}O_2 \qquad (15.2)$$

Figure 15.3 shows electrical conductivity as a function of oxygen partial pressure at 1273 K (1000°C). As predicted above, the electrical conductivity decreases with a reduction of oxygen partial pressure. The electrical conductivity is proportional to ${P_{O2}}^{1/4}$, which is consistent with the defect chemistry of $La_{1-x}Ca_xCrO_{3-\delta}$ [22]. A doping of B site has been also considered by several authors [1–5]. The typical dopant cation is Mg^{2+} replacement into Cr^{3+} sites.

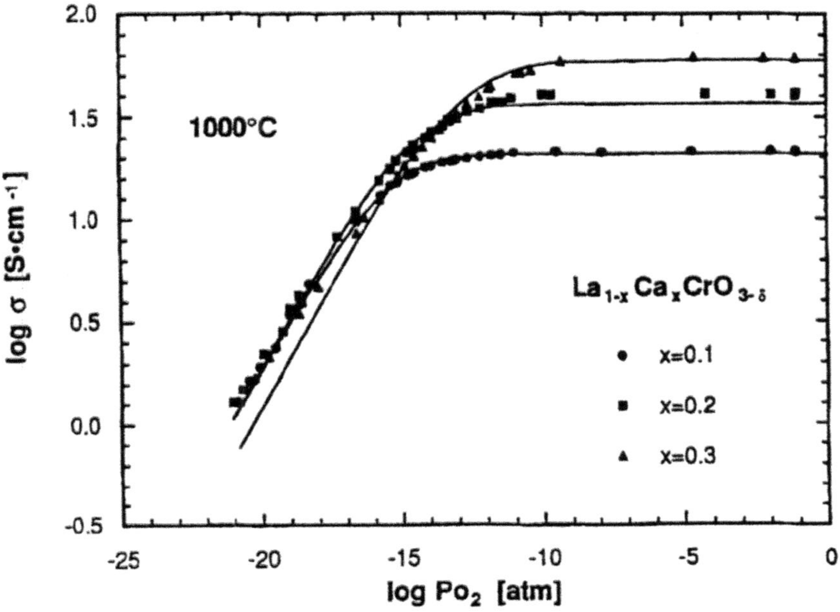

Fig. 15.3 Electrical conductivity of $La_{1-x}Ca_xCrO_3$ ($x = 0.1, 0.2, 0.3$) as a function of oxygen partial pressures at 1273 K (*solid lines* are calculated lines from the defect chemistry). (Reproduced by permission of The Electrochemical Society [22])

This substitution also increases the concentration of Cr^{4+}, which eventually increases the electrical conduction.

15.4 Defect Chemistry and Oxygen Electrochemical Leak

Doped $LaCrO_3$ generates Cr^{4+} at high oxygen partial pressures and oxygen vacancies in reducing atmospheres. The important function of interconnects is electronic conduction without electrochemical oxygen leak. Thus, the defect chemistry associated with Ce^{3+}/Ce^{4+} transition and oxygen vacancy formation should be clarified. Several authors have treated the defect chemistry of doped $LaCrO_3$: Weber et al. and Mizusaki et al. treated Sr-doped $LaCrO_3$ [21, 23], Yasuda et al. and Onuma et al. treated Ca-doped $LaCrO_3$ [22, 24], and Oishi et al. treated $LaCr(M)O_3$ [25]. Using Kröger–Vink notation, the reaction for oxygen vacancy formation can be expressed by the following:

$$2Cr_{Cr}^{\bullet} + O_O^X = 2Cr_{Cr}^X + V_O^{\bullet\bullet} + \frac{1}{2}O_2(g) \tag{15.3}$$

The equilibrium constant for the above reaction can be written as

$$K = \frac{[Cr_{Cr}^{X}]^2 [V_O^{\bullet\bullet}]}{[Cr_{Cr}^{\bullet}][O_O^X]} \cdot (p(O_2))^{1/2} \quad (15.4)$$

The charge neutrality condition is expressed as follows:

$$[M_{La}'] = [Cr_{Cr}'] + 2[V_O^{\bullet\bullet}] \quad (15.5)$$

Using the chemical formula of $La_{1-x}M_xCrO_{3-d}$, the following equations can be assumed:

$$[Cr_{Cr}^{X}] + [Cr_{Cr}^{\bullet}] = 1 \quad (15.6)$$

$$[M_{La}'] = x \quad (15.7)$$

$$[V_O^{\bullet\bullet}] = d \quad (15.8)$$

$$[O_O^X] = 3 - d \quad (15.9)$$

The equilibrium constant for the reaction is expressed by using the relationships (15.6)–(15.9):

$$K = \frac{(2d - x + 1)d}{(3 - d)(y - 2d)^2} \cdot (p(O_2))^{1/2} \quad (15.10)$$

The value of K can be obtained by fitting to the measured data. For $La_{0.7}Ca_{0.35}CrO_{3-\delta}$, the best fitted value of K was 9×10^8 atm$^{0.5}$. Figure 15.4 shows measured oxygen numbers as a function of oxygen partial pressure. Oxygen numbers are decreased from 3 with decreasing oxygen partial pressure. Therefore, the oxygen vacancies can be formed in the $LaCrO_3$ lattice, which are the diffusion paths of oxide ions. Since the interconnect material is placed in a large oxygen potential gradient, oxygen can permeate through the $LaCrO_3$-based materials via oxygen vacancies ($V_O^{\bullet\bullet}$) [26–30]. When oxide ions can migrate from high to low oxygen partial pressures, electrons can move in the opposite direction. In accordance with the convention, the oxygen permeation current density is negative when oxygen moves from the left (high oxygen partial pressure) to the right (low oxygen partial pressure) side, as illustrated in Fig. 15.5. The oxygen permeation current density can be calculated from the following equation:

$$J_{O2-}(Am^{-2}) = -\frac{1}{4FL} \int_{p_{O2(x=0)}}^{p_{O2(x=L)}} \frac{(\sigma_{O^{2-}} \sigma_{e-})}{(\sigma_{O^{2-}} + \sigma_{e-})} d\mu_{O_2} \quad (15.11)$$

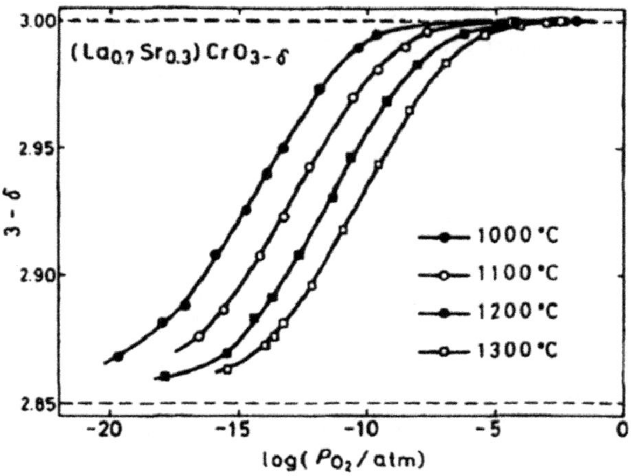

Fig. 15.4 Measured oxygen numbers of $La_{0.7}Ca_{0.3}CrO_{3-\delta}$ as a function of oxygen partial pressures at 1273–1573 K. (Reproduced by permission of Elsevier Science [23])

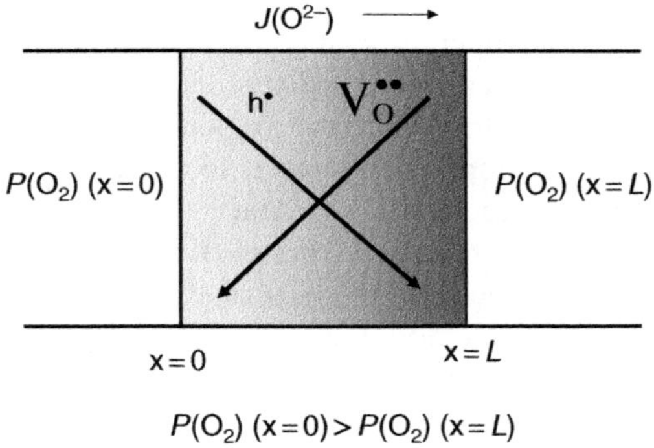

Fig. 15.5 Schematic drawing of oxygen permeation through LaCrO$_3$ interconnects

where F is the Faraday constant, μ_{O2} is the chemical potential of oxygen, σ_{O2-} is the oxide ionic conductivity, σ_e is the electronic conductivity and L is the thickness of LaCrO$_3$. If the electronic conductivity is high enough, the above equation can be simplified as follows:

$$J_{O2-}(Am^{-2}) = -\frac{1}{L}\int_{p_{O2(x=0)}}^{p_{O2(x=L)}} \sigma_{O^{2-}} \frac{RT}{4F} d\ln P_{O_2} \qquad (15.12)$$

where R is the gas constant and T is the temperature. The above equation indicates that oxygen permeation current density can be calculated from the conductivity of oxide ion through the dense LaCrO$_3$. To evaluate the conductivity of oxide ion, the following equation can be applied:

$$\sigma_{O2-}(\Omega^{-1}m^{-1}) = \frac{4F[V_O^{\bullet\bullet}]D_V}{RTV_m} \qquad (15.13)$$

where R is the molar gas constant (J mol^{-1} K^{-1}), T is temperature (K), $[V_O^{\bullet\bullet}]$ is the oxygen vacancy mole fraction, D_v is the oxygen vacancy diffusion coefficient (m^2 s^{-1}), and V_m is the molar volume of $LaCrO_3$. The vacancy concentration can be determined from the experimental and calculated oxygen nonstoichiometry data. A precise analysis was made of the oxygen chemical potential distribution in the $LaCrO_3$-based oxides and oxygen electrochemical permeation (leak). The calculated oxygen permeation current ($J(O^{2-})$/mA cm^{-2}) is shown in Fig. 15.6. At 900°C (1173 K), the permeation current density is expected to be less than 80 mA/cm^2 in any total current densities examined (in the case of $La_{0.7}Ca_{0.3}CrO_{3-\delta}$). However, at high temperatures, above 950°C (1223 K), the leakage current densities are more than 100 mA/cm^2, which are above 10% of total current densities [28, 29]. This value is significantly large compared with the operating current densities. For the experimental determination of oxygen permeation through the $LaCrO_3$-based oxide ceramics, the electronic blocking electrochemical method and isotope oxygen exchange method ($^{16}O/^{18}O$ exchange) were applied [30, 31]. The measured oxygen permeation current density was about 3–10 mA/cm^2 at 10^{-13} Pa (about 10^{-18} atm) at the temperature of 1273 K (the thickness is assumed to be 3 mm). The measured current density is considerably small compared with the calculated value due to the surface oxygen reactivity of $LaCrO_3$. The oxygen permeation needs an ionization process to oxide ion (O^{2-}) from oxygen molecules. A low surface reactivity can reduce the permeation flux through the $LaCrO_3$ and eventually reduce the permeation current density. The oxygen vacancy diffusion coefficients in $La_{0.7}Ca_{0.3}CrO_3$ were determined by the isotope oxygen exchange method: the oxygen vacancy diffusion coefficient (D_v) was measured to be around 10^{-5} cm^2 s^{-1} at 1273 K [31]. The D_v values with different Ca concentration are almost the same level (10^{-5} cm^2 s^{-1}), and the activation energy for D_v is around 77–142 kJ/mol [28].

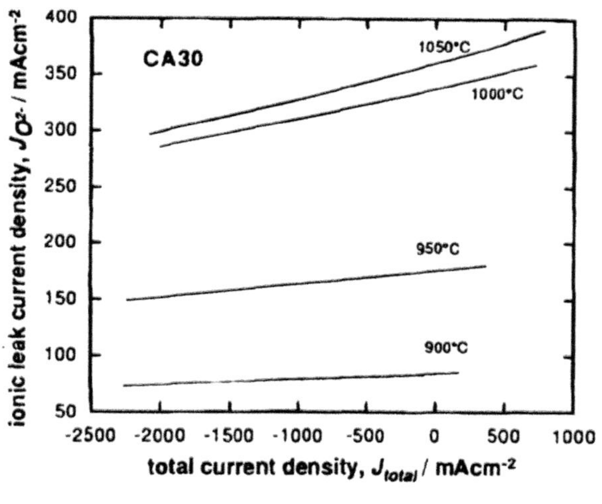

Fig. 15.6 Oxygen permeation current density through $LaCrO_3$ interconnects. Relationship between ionic leak current density and total current density in $La_{0.7}Ca_{0.3}CrO_3$ under oxygen potential gradient. The thickness of the $LaCrO_3$ plate is assumed to be 3 mm. (Reproduced by permission of The Electrochemical Society [28])

15.5 Lattice Expansion During Reduction and Temperature Change

Doped lanthanum chromites show lattice expansion under high-temperature reducing atmospheres [32–34] because of the formation of oxygen vacancies in the lattice under reducing atmosphere. In planar-type solid oxide fuel cells, the $LaCrO_3$-based interconnect plate is placed in a large oxygen potential gradient (air and fuel conditions). The lattice expansion and deformation of the $LaCrO_3$ plate are significant under operating conditions, and this phenomenon has been reported in real stacks and modules [1–3]. A numerical model was proposed, and deformation was calculated from the profiles of oxygen vacancy concentration. Static and transient stress calculation was carried out for plate and rectangular specimens. When the plate size is $100 \times 100 \times 3$ mm, the warp of the plate is calculated by a numerical model. The maximum displacement of the center part normal to the originally flat surface was calculated to be about 0.77 mm. The calculated tensile stress coming from the transient deformation was as high as 50 MPa or more [32].

Because interconnects contact with the other cell components, the thermal expansion behaviors should be matched within the acceptable level. Doping of Ca increases the thermal expansion coefficient (TEC) values from 8.5×10^{-6}/K to 10.0×10^{-6}/K (at $x = 0.3$ in $La_{1-x}Ca_xCrO_3$) [1–5]. The TEC value of Ca-doped $LaCrO_3$ is close to that of YSZ (about 10×10^{-6}/K), which can reduce the thermal stress during heating and fabrication. Doping of Sr is also effective to increase the TEC values to match the TEC values of YSZ. The thermal expansion behavior is dependent on oxygen partial pressure as predicted by the defect chemistry. For example, Mori [35] measured the TEC values for several kinds of $La_{0.8}Sr_{0.2}Cr_{0.8}M_{0.2}O_3$ (M = dopants) in air and in hydrogen. Figure 15.7 shows thermal expansion behavior of Sr-, Ti-, and V-doped $LaCrO_3$ [35]. Doping of V decreased the TEC values to be matched with YSZ. In reducing atmosphere, the TEC values are a little bit larger than those in air atmospheres. The lattice expansion in the reducing atmosphere is attributed to the expansion of chromium ion because of the difference of ionic radius between Cr^{3+}(VI) (0.0615 nm) and Cr^{4+}(VI) (0.055 nm).

15.6 Mechanical Strength

Mechanical strength is one of the critical issues for using $LaCrO_3$-based oxide ceramics. During operation, temperature change of the SOFC system is inevitable because of the starting and stopping of the system. Also, under operating conditions, an oxygen potential gradient is generated between air and fuel atmospheres. Thus, the mechanical strength of $LaCrO_3$ should be high enough under thermal stress and reducing conditions. The reported mechanical strengths are 20–130 MPa for $La_{1-x}Ca_xCrO_3$ ($x = 0.1$–0.3), 50–80 MPa for $La_{1-x}Sr_xCrO_3$ ($x = 0.1$–0.5), and 80–170 MPa for $LaCr_{0.9}Mg_{0.1}O_3$ at 1273 K in air. Generally, the mechanical strength of Sr-doped $LaCrO_3$ showed a higher

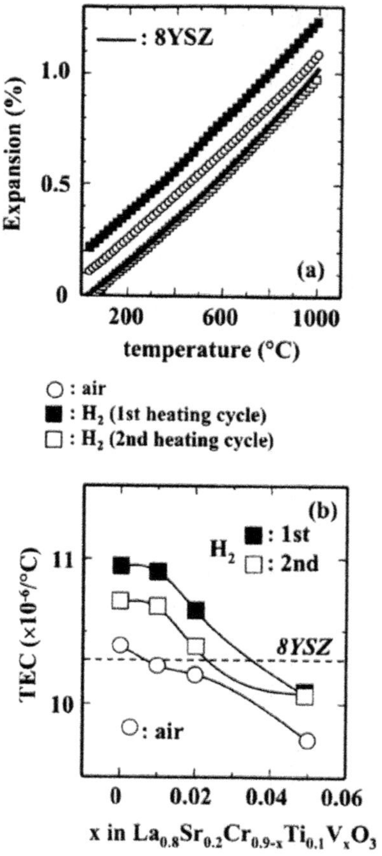

Fig. 15.7 (**a**) Thermal expansion behaviors of some doped $LaCrO_3$ in air and in the H_2 atmosphere in comparison with 8YSZ electrolyte. (**b**) Thermal expansion coefficients of $La_{0.8}Sr_{0.2}Cr_{0.9-x}Ti_{0.1}V_xO_3$ in the temperature range 50°–1000°C in air or in the H_2 atmosphere. (Reproduced by permission of The Electrochemical Society [35])

value than that of Ca-doped $LaCrO_3$ in air. However, for reasons of processing of $LaCrO_3$ powders and sintered bodies, Ca-doped $LaCrO_3$ or Mg-doped $LaCrO_3$ is adopted in the developing cells and stacks.

15.7 Summary

$LaCrO_3$ based perovskite is one of the promising materials for interconnects of solid oxide fuel cells (SOFCs). Some cations doping into the A site or B site of $LaCrO_3$ increased the sintering properties, electrical conductivity, and mechanical strength to the applicable level. Gas tightness can be achieved by the sintering of $LaCrO_3$. However, electrochemical oxygen leak should be considered under a large oxygen potential gradient. Although the measured oxygen leak value is negligible in a real operation condition, some oxygen can permeate through the $LaCrO_3$ plate during operation via oxygen vacancies. Defect chemistry of $LaCrO_3$-based perovskite is introduced because it correlates with electrical conductivity, lattice expansion in reducing atmosphere, thermal expansion, and oxygen permeation.

References

1. N.Q. Minh, Ceramic fuel cells. J. Am. Ceram. Soc. 76 (3) (1993)
2. N.Q. Minh and T. Takahashi, Science and Technology of Ceramic Fuel Cells. Elsevier, Amsterdam, pp. 165–198 (1995)
3. S.C. Singhal, Advances in solid oxide fuel cell technology. Solid State Ionics 135, 305–313 (2000)
4. H.U. Anderson and F. Tietz, Chapter 7: Interconnects. High Temperature Solid Oxide Fuel Cells, S.C. Singhal, K. Kendall (eds.), pp. 173–195. Elsevier Advanced Technology, UK (2003)
5. J.W. Fergus, Lanthanum chromite-based materials for solid oxide fuel cell interconnects. Solid State Ionics 171, 1–4 (2004)
6. N. Sakai, H. Yokokawa, T. Horita, K. Yamaji, Lanthanum chromite-based interconnects as key materials for SOFC stack development. Int. J. Appl. Ceramic Technol. 1 (1), 23–30 (2004)
7. D.B. Meadowcroft, Properties of strontium doped lanthanum chromite. Br. J. Appl. Phys. Ser. 2, 2, 1225–1233 (1969)
8. N. Sakai, T. Kawada, H. Yokokawa, and M. Dokiya, Sinterability and electrical conductivity of calcium-doped lanthanum chromites. J. Material Sci. 25, 4531 (1990)
9. N. Sakai, T. Kawada, H. Yokokawa, M. Dokiya, T. Iwata, Thermal expansion of some chromium deficient lanthanum chromites. Solid State Ionics 40/41, 394–397 (1990)
10. N. Sakai, T. Kawada, H. Yokokawa, M. Dokiya, I. Kojima, Liquid-phase assisted sintering of calcium-doped lanthanum chromites. J. Am. Ceram. Soc. 76(3), 609–616 (1993)
11. H. Yokokawa, N. Sakai, T. Kawada, and M. Dokiya, Chemical thermodynamic considerations in sintering of $LaCrO_3$-based perovskites. J. Electrochem. Soc. 138 (4), 1018–1027 (1991)
12. J.L. Bates, L.A. Chick, and W.J. Weber, Synthesis, air sintering and properties of lanthanum and yttrium chromites and manganites. Solid State Ionics 52, 235–242 (1992)
13. L.A. Chick, J. Liu, J.W. Stevenson, T.R. Armstrong, D.E. McCready, G.D. Maupin, G. W. Coffery, C.A. Coyle, Phase transitions and transient liquid-phase sintering in calcium-substituted lanthanum chromite. J. Am. Ceram. Soc. 80 (8), 2109–2120 (1997)
14. M. Mori, T. Yamamoto, T. Ichikawa, Y. Takeda, Dense sintered conditions and sintering mechanisms for alkaline earth metal (Mg, Ca, and Sr)-doped $LaCrO_3$ perovskites under reducing atmosphere. Solid State Ionics 148, 93–101 (2002)
15. H. Nishiyama, M. Aizawa, H. Yokokawa, T. Horita, N. Sakai, M. Dokiya, T. Kawada, Stability of lanthanum chromite-lanthanum strontium manganite interface in solid oxide fuel cells. J. Electrochem. Soc. 143(7), 2332–2341 (1996)
16. J.D. Carter, C.C. Appel, M. Mogensen, Reaction at calcium doped lanthanum chromite-yttria stabilized zirconia interface. J. Solid State Chem. 122, 407–415 (1996)
17. T. Horita, K. Yamaji, N. Sakai, M. Ishikawa, H. Yokokawa, M. Dokiya, Cation diffusion in $(La,Ca)CrO_3$ perovskite by SIMS. Solid State Ionics 108, 383–390 (1998)
18. T. Horita, M. Ishikawa, K. Yamaji, N. Sakai, H. Yokokawa, M. Dokiya, Calcium tracer diffusion in $(La,Ca)CrO_3$ by SIMS. Solid State Ionics 124, 301–307 (1999)
19. T. Akashi, T. Maruyama, T. Goto, Transport of lanthanum ion and hole in $LaCrO_3$ determined by electrical conductivity measurements. Solid State Ionics 164, 177–183 (2003)
20. N. Sakai, K. Yamaji, T. Horita, H. Negishi, H. Yokokawa, Chromium diffusion in lanthanum chromites . Solid State Ionics 135, 469–474 (2000)
21. W.J. Weber, C.W. Griffin, J.L. Bates, Effects of cation substitution on electrical and thermal transport properties of $YCrO_3$ and $LaCrO_3$. J. Am. Ceram. Soc. 70 (4), 265–270 (1987)
22. I. Yasuda, T. Hikita, Electrical conductivity and defect structure of calcium-doped lanthanum chromites. J. Electrochem. Soc. 140 (6), 1699–1704 (1993)

23. J. Mizusaki, S. Yamauchi, K. Fueki, A. Ishikawa, Nonstoichiometry of the perovskite-type oxide $La_{1-x}Sr_xCrO_{3-\delta}$. Solid State Ionics 12, 119–124 (1984)
24. S. Onuma, K. Yashiro, S. Miyoshi, A. Kaimai, H. Matsumoto, Y. Nigara, T. Kawada, J. Mizusaki, K. Kawamura, N. Sakai, H. Yokokawa, Oxygen nonstoichiometry of perovskite-type oxide $La_{1-x}Ca_xCrO_{3-\delta}$ ($x = 0.1, 0.2, 0.3$). Solid State Ionics 174, 287–293 (2004)
25. M. Oishi, K. Yashiro, J.-O. Hong, Y. Nigara, T. Kawada, J. Mizusaki, Oxygen nonstoichiometry of B-site doped $LaCrO_3$. Solid State Ionics 178 (3–4), 307–312 (2007)
26. B.A. van Hassel, T. Kawada, N. Sakai, H. Yokokawa, M. Dokiya, Oxygen permeation modeling of $La_{1-y}Ca_yCrO_{3-\delta}$. Solid State Ionics 66, 41–47 (1993)
27. B.A. van Hassel, T. Kawada, N. Sakai, H. Yokokawa, M. Dokiya, H.J.M. Bouwmeester, Oxygen permeation modeling of perovskite. Solid State Ionics 66, 295–305 (1993)
28. I. Yasuda, M. Hishinuma, Electrochemical properties of doped lanthanum chromites as interconnectors for solid oxide fuel cells. J. Electrochem. Soc. 143 (5), 1583–1590 (1996)
29. I. Yasuda, M. Hishinuma, Precise determination of the chemical diffusion coefficient of calcium-doped lanthanum chromites by means of electrical conductivity relaxation. J. Electrochem. Soc. 141 (5), 1268–1273 (1994)
30. N. Sakai, K. Yamaji, T. Horita, H. Yokokawa, T. Kawada, M. Dokiya, K. Hiwatashi, A. Ueno, M. Aizawa, Determination of the oxygen permeation flux through $La_{0.75}Ca_{0.25}CrO_{3-\delta}$ by an electrochemical method. J. Electrochem. Soc. 146 (4), 1341–1345 (1999)
31. T. Kawada, T. Horita, N. Sakai, H. Yokokawa, M. Dokiya, Experimental determination of oxygen permeation flux through bulk and grain boundary of $La_{0.7}Ca_{0.3}CrO_3$. Solid State Ionics 79, 201–207 (1995)
32. H. Yakabe, M. Hishinuma, I. Yasuda, Static and transfer model analysis on expansion behavior of $LaCrO_3$ under an oxygen potential gradient. J. Electrochem. Soc. 147 (11), 4071–4077 (2000)
33. H. Yakabe, I. Yasuda, Model analysis of the expansion behavior of $LaCrO_3$ interconnector under solid oxide fuel cell operation. J. Electrochem. 150 (1), A35–A45 (2003)
34. F. Boroomand, E. Wessel, H. Bausinger, K. Hilpert, Correlation between defect chemistry and expansion during reduction of doped $LaCrO_3$ interconnects for SOFCs. Solid State Ionics 129, 251–258 (2000)
35. M. Mori, Enhancing effect on densification and thermal expansion compatibility for $La_{0.8}Sr_{0.2}Cr_{0.9}Ti_{0.1}O_{3-\delta}$ based SOFC interconnect with B-site doping. J. Electrochem. Soc. 149 (7), A797–A803 (2002)

Index

GPSR Compliance

The European Union's (EU) General Product Safety Regulation (GPSR) is a set of rules that requires consumer products to be safe and our obligations to ensure this.

If you have any concerns about our products, you can contact us on ProductSafety@springernature.com

In case Publisher is established outside the EU, the EU authorized representative is:

Springer Nature Customer Service Center GmbH
Europaplatz 3
69115 Heidelberg, Germany

Batch number: 10755284

Printed by Printforce, the Netherlands